EXPERIMENTAL DESIGN AND ANALYSIS

A Series of Books in Psychology

WAYNE LEE

EXPERIMENTAL DESIGN AND ANALYSIS

W. H. Freeman and Company
San Francisco

Library of Congress Cataloging in Publication Data

Lee, Wayne, 1935–
Experimental design and analysis.

1. Experimental design. 2. Analysis of variance.
I. Title.
QA279.L4 519.5′35 74-12241
ISBN 0-7167-0772-1

Printed in the United States of America

10 9 8 7 6 5 4 3 2 1

CONTENTS

Chapter 3 Factors and Designs

Chapter 4 The Score Model

Chapter 5 Sums of Squares

Chapter 6 Expected Mean Squares and Significance Tests

Chapter 7 Latin Square Designs

Appendix A Statistical Tables

Appendix B Power Charts

PREFACE

This book is an introductory treatment of multifactor experimental design and analysis of variance. It is aimed at students and researchers in the behavioral sciences, particularly psychology. The only prerequisite to this book is a one-semester course in introductory statistics or its equivalent, which should have exposed the student to probability distributions and hypothesis testing. There are, of course, a variety of other textbooks aimed at the same market as this one. I felt justified in writing another one, however, because experimental design remains a difficult topic to master, and I intended to incorporate some unique features to assist the student and user of experimental designs.

My dissatisfaction with textbooks on experimental design began when I was a graduate student in psychology at The Johns Hopkins University in the early 1960's, and was just beginning to study and use experimental designs. Although the textbooks I studied were very useful and much appreciated, I was bothered by two things: (1) It was frequently difficult to understand why various computations were made, and (2) I felt incompetent to alter or enlarge the designs presented—I could use only those designs that were spelled out. Although more recent textbooks have shown some improvements in these respects, since that time I have come to see other possibilities that have not been utilized.

One distinguishing feature of this book is its organization. The standard method of organization is to begin with a comprehensive discussion of a simple one-factor design, including the analysis, and then to proceed to a similar discussion of more complicated designs. In this book, after introductory and review chapters, Chapter 3 discusses the kinds of factors that enter into designs, the relations among them, a method for symbolizing designs, a classification of designs, and a comparison of the purposes and relative merits of the different classes. There is no discussion of data analysis in this chapter. Such an approach has the advantage of helping the student to view the various kinds of factors and designs as a whole. In other books, this overall view tends to get lost among the formulas, tables, and calculations for specific designs as they are presented chapter by chapter.

Chapter 4 is devoted to score models. Typically, the score models are presented, if at all, in the separate chapters that are devoted to the different designs. The concentration of models in one chapter allows comparison and contrast and hence should facilitate understanding of the models. The various stages in the analysis of variance can be readily interpreted in terms of the score model but are difficult to comprehend otherwise. I believe, therefore, that the especial attention given to score models is justified. Chapter 5 is devoted to sums of squares and Chapter 6 focuses on expected mean squares and F-tests. Again, these topics are concentrated within specific chapters rather than being strewn about in chapters devoted to different designs.

The structure of this book is meant to further the book's aim of instilling general methods that allow the construction of a suitable design, if necessary; other books simply present a collection of designs from which an experimenter must choose. Chapter 3 tells how to symbolize a design, given a collection of factors that are related by crossing and nesting. Chapter 4 explains how to derive the score model for any design so symbolized, no matter how many factors are used. Chapter 5 gives rules for deriving sum of squares formulas for any score model developed in Chapter 4, and Chapter 6 gives general rules for specifying the F-ratios. Although the general rules given in Chapters 5 and 6 also appear in other books, these rules assume that the score model is available. The methods for design symbolization and model derivation given in Chapters 3 and 4 are vital, then, to a flexible and general method for experimental design and analysis of variance; these methods, to my knowledge, were first developed by me[1] and have not previously appeared in any textbook.

The general procedures of Chapters 3 to 6 apply only to designs in which all pairs of factors are related either by crossing or nesting. Chapter 7 presents a number of specific designs based on the Latin square arrangement. In all, the book covers the large majority of multifactorial experimental designs most often used by psychologists. Although general methods are emphasized, layouts and computational formulas for many of the simpler, frequently used designs are provided, together with numerical examples. A disadvantage of the organization used is that, except for the Latin square designs, the layout, score model, and formulas pertaining to a particular design are not grouped together in one chapter. The student or researcher can easily find the relevant material for a particular design, however, by consulting the Design Index.

[1] W. Lee, "Experimental Design Symbolization and Model Derivation," *Psychometrika* 31:397–412 (1966).

Some other features of the present book include a special symbolism for model terms that facilitates comprehension of which factors are used and "sum of squares elements" tables that systematize the computation of sums of squares. In addition, an accompanying *Workbook for Experimental Design and Analysis* is available that includes questions and numerical exercises. The exercises involve only a small number of scores, so that students require only an inexpensive electronic calculator (costing less than $100) for a course based on this book. A more elaborate and expensive calculator permitting automatic squaring and accumulation of squares is naturally more suitable for analyzing the scores from actual experiments. Some instructors will undoubtedly wish to teach students how to use the local computer facility for analysis of variance. If suitable computational devices are available, the instructor may very well wish to supplement the exercises given in the workbook with exercises that utilize more data.

The main reason I wrote this book was to present the comprehensive and general methods of Chapters 3 to 6 for creating and analyzing the designs having factors related by crossing and nesting. Certainly the book is more comprehensive than any other in its treatment of such designs—not necessarily in terms of the specific designs and formulas presented, but in terms of the designs that the student will be able to create and use. I never intended that this book should be a comprehensive handbook of statistical formulas, designs, and methods. For example, analysis of covariance, trend analysis, and multiple comparisons are given only introductory treatments. Nonparametric methods, quasi-experimental designs, and multivariate analysis of variance are not treated at all. Although the book attempts to provide an understanding of the reasons for various calculations and conclusions, it does not aim to present very much statistical theory or proof. Nonetheless, sufficient material is presented here to allow the student who masters it to use and analyze the same kinds of multifactor designs he normally finds in research reports. The book is brief enough to be covered in one semester; this is not true of most competing books.

Although I have some familiarity with the various schools of probability,[2] I found that I could accomplish my aims in this book by adhering to the traditional frequentist position. However vulnerable this position may be to criticism, it serves as a convenient vehicle for explaining the inferential techniques currently used for interpreting multifactor experiments.

The basic ideas for this book were developed during the middle 1960's while I was a member of the faculty of the Department of Psychology at the University of California, Berkeley. I drafted several chapters of the book at Lund University in Sweden when I was a guest professor there in 1972. I am indebted, therefore, to both institutions. The manuscript was critically read by Eleanor W. Willemsen, Karen Stark, and Leonard A. Marascuilo, much to its benefit. I cannot claim, however, that the final result is entirely to their satisfaction. I am indebted to the Literary Executor of the late Sir Ronald A.

[2] W. Lee, *Decision Theory and Human Behavior* (New York: Wiley, 1971).

Fisher, F.R.S., to Dr. Frank Yates, F.R.S., and to Longman Group Ltd., London, for permission to reprint Appendix Table A.4 from their book *Statistical Tables for Biological, Agricultural and Medical Research.* Finally, I would like to express my appreciation to Wendell R. Garner, whom I was fortunate to have guide my first steps as an experimentalist when I was in graduate school.

Albany, California
April 1974

Wayne Lee

EXPERIMENTAL DESIGN AND ANALYSIS

CHAPTER ONE

Introduction

1.1 ABOUT THIS BOOK

This book was written primarily for students and researchers in psychology and related behavioral sciences. The main topics covered are experimental design and analysis of variance—topics with which every behavioral researcher and consumer of such research should be familiar. The student is assumed to have a background in statistics equivalent to that obtained from a one- or two-semester course. Some review of the relevant background is included in Chapter 2.

Experimental design concerns the arrangement of various *conditions* or situations to which experimental *subjects* (for example, people or rats) will be exposed. Observations are made of the responses of the subjects and an *analysis of variance* is performed on the data gathered. The purpose of these procedures is to allow valid scientific inferences to be made about the effects on behavior of the conditions investigated.

The statistical methodology used in behavioral research can be categorized into two broad classes, the *experimental* and *correlational* methods, both of which are discussed in Section 1.2. The experimental methodology includes the topics of experimental design and analysis of variance, to which this book is addressed. There are many investigations in psychology, however, that utilize, instead, "correlational" techniques. Particularly because any investigation is likely to be called an experiment, it is well for the student to recognize that the "experimental" techniques of "experimental design" and "analysis of variance" are not used in all psychological "experiments."

Experimental design and analysis of variance are part of a larger project, the *experimental investigation*. Section 1.3 considers the stages of experimental investigation and shows how experimental design and analysis of variance contribute to the project as a whole.

The student has already familiarized himself with the simplest experimental designs during his studies of elementary statistics. For example, he learned to perform a t-test to determine whether persons exposed to two different conditions react differently. (This test is reviewed in Section 2.4.) The random assignment of persons to two such groups is the simplest example of an experimental design. A somewhat more complex design is obtained by establishing more than two conditions and assigning a different group of persons to each. Many students reading this book have had some exposure to such designs as well.

This book is primarily concerned with still more complex designs, called *multifactor designs* to contrast them with the one-factor designs just mentioned. The basic nature of multifactor designs and their advantages are discussed in Section 1.4. Section 1.5 provides an overview of the main body of the book, which begins with Chapter 2.

1.2 EXPERIMENTAL VERSUS CORRELATIONAL INVESTIGATIONS

As previously explained, statistical methodology in the behavioral sciences can be divided into the *experimental* and *correlational* methods. It would be well for the student to understand these two areas and how they differ.

When the terms "experimental design" and/or "analysis of variance" appear in a textbook title, the book concentrates on experimental methods and plays down correlational methods. Obviously, the present book falls in this class. The titles of methodological books for correlationists characteristically include such terms as "correlational methods," "factor analysis," "psychological measurement," and "theory of mental tests." "Survey research" can also be considered part of correlational methodology. Many introductory methodological books include discussion of both the experimental and the correlational approaches. Their titles usually include the terms "statistics" or "statistical," and exclude the other terms that are often included in higher-level discussions of the experimental or correlational approaches.

The distinction between the experimental and correlational methods is of considerable importance, but it becomes blurred, partly because the term "experimental" is often interpreted rather loosely to apply to scientific investigations of both types. Either kind of investigation is likely to be called an experiment. The "experimental" approach, in the more narrow sense that applies in this book, is used by investigators who wish to discover how various experimental conditions affect the behavior of some class of subjects. They are primarily concerned with the effect of a condition, the estimation of which includes averaging the scores across all subjects exposed to that condition. The "correlational" approach, on the other hand, is used by investigators who are interested in differences among persons and the relationships among different kinds of measurements made on people.

Different subjects can react quite differently to the same condition. To the "experimental" investigator, these differences are a nuisance; they make his job of finding differences among conditions more difficult. He ignores the individual differences as much as possible and concentrates on the mean scores across subjects. The experimentalist must be careful that naturally occurring differences among subjects are not confused with the effects of conditions. One protection he takes to guard against this is to assign subjects to conditions on a random basis, when this is feasible. Then the naturally high-scoring or low-scoring subjects will tend to be equally distributed among the conditions.

Whereas individual differences constitute a hazard and an interference for the experimentalist, the correlational investigator concentrates on these differences, rather than on the effects of various conditions. One of his major contributions is the development of psychological tests that give accurate measures of individual differences. One sign of a good measure is reliability; if a measure is reliable, individual scores on the test are stable from occasion to occasion. Measures are not going to be reliable if an individual's score is easily affected by various situations or experiences to which he may be exposed between testing. What most interests the experimentalist, then—the effects of conditions on scores—is a hazard and an interference for the correlationist, who is attempting to create reliable tests. Just as the experimentalist normally overlooks individual differences in spite of their effects on his study, the correlationist typically overlooks the effects of conditions on scores. If the test is highly reliable, the effects of conditions must be reasonably small.

It is standard working procedure for the correlationist to obtain more than one score—oftentimes many scores—on the same subject, under identical conditions (insofar as this is possible). The scores may be obtained from the same scale; that is, they may be obtained from the same or a very similar test, with some time interval between successive observations. The purpose of such an investigation might be to determine the reliability of the test or to observe systematic changes in individual scoring over time. The scores may be on different scales, and if so, are collected over a relatively brief period. The correlationist would usually subject such scores to some kind of *multivariate analysis*, such as *factor analysis*.

We must clearly differentiate between the *factors of factor analysis* and the *factors of experimental design*. The former are basic attributes underlying a large set of test scores; they are discovered through calculations on the scores obtained from tests. The latter are variables (often called *independent variables*) chosen by the investigator before execution of the experiment; *levels* (various possible values) of these factors (independent variables) are combined to form different *conditions* to which the subjects are exposed.

Whereas it is standard working procedure for the correlationist to obtain several scores on the same subject, designs used by the experimentalist frequently require only one score per subject. But the experimentalist, too, often obtains a number of scores on each subject—in so-called repeated measurements designs. In these designs, however, the scores are normally obtained under differing conditions, in contrast to the standard procedure for the correlationist. In a way, learning experiments might be thought to be an exception to this statement, since in such experiments the experimenter may expose subjects to the same stimuli over successive trials. However, we conceive that the differences in a subject over trials resulting from learning constitute, by definition, changes in "condition." The experimenter may obtain several scores on different scales on the same subject under one condition. His investigation, however, would include different conditions, whereas that of the correlationist would usually have only one condition.

There is an important difference in the outlooks of experimentalists and correlationists. Generally speaking, experimentalists tend to view behavior and personality as environmentally determined, whereas correlationists are more impressed with the hereditary determinants. The nature-nurture controversy sometimes becomes heated, but all psychologists recognize that both nature and nurture are important determinants—the question is one of relative importance.

The distinction that has been described between experimental and correlational approaches to psychological inquiry is very real and important, but not absolute. For example, correlations between different test scores are required in *multivariate analysis of variance* and the *analysis of covariance*, techniques of the experimentalist that will be considered later. In addition, it is possible to combine multifactor experimentation with factor analysis. Such an approach relates equally to experimental and correlational psychology, but such analyses are very rare. The designs discussed in this book often include a blocking factor (Section 3.2.3), the levels of which include subjects grouped together according to some pre-experimental basis, such as sex or age. When a design has a blocking factor, the experimentalist may draw conclusions about differences among different kinds of subjects, and hence shows an interest in individual differences that, as previously explained, characterizes the correlationist. A blocking factor, however, would normally be of secondary interest in an experimental design, and is frequently included only to increase the efficiency of the design for investigating other factors (Section 8.3).

All psychologists who wish to perform or consume research should understand experimental methodology—even psychologists whose primary interests are correlational. Experimentation can, for example, demonstrate what effects different instructions or other conditions can have on average IQ scores, or whether subjects under instruction to "fake good" on a personality inventory can do so, relative to a control group.

To summarize, many investigations rely on correlational methodology which is not treated in this book. Books on "experimental design and analysis of variance," such as the present one, do not cover those statistical methods that primarily concern correlational investigations. Instead, such books are concerned with "experimental" investigations in which subjects are assigned to different conditions in order to assess how different conditions affect the mean scores for some population of subjects.

1.3 STAGES OF AN INVESTIGATION

An investigation in the behavioral sciences usually proceeds through the following six stages:

1. Conception
2. Design
3. Preparation
4. Execution
5. Analysis
6. Dissemination

These six stages are common to both experimental and correlational investigations, though the detailed steps taken in a stage will often differ depending on the class of investigation being undertaken. Each of these stages will be briefly considered, with emphasis placed on the steps that are appropriate for the experimentalists. This book mainly concerns only two of these stages, design and analysis.

1.3.1 Conception

An experimental investigation commences when the experimenter decides that he wishes to learn more about the effects that experimental variables (factors) have on some measure of behavior. Initially, he may not be sure exactly what experimental factors will be included in the experiment or exactly what measure he will use. In stage (1), conception, he collects and considers various possibilities.

The investigator might instigate the investigation with some practical application in mind. For example, an investigation might show that one condition produces superior learning, or superior clinical results, or an increase in work accomplished. Alternatively, differences found between experimental conditions might be of interest for strictly scientific reasons—because the results bear on some theory or improve methodology, or simply because the results represent an advance of empirical knowledge.

1.3.2 Design

Stage (1) merges into stage (2), design. An experimental design is a plan for applying different experimental *conditions* to experimental *units* to determine how such conditions affect some measure of behavior, usually known as the *criterion* or *dependent variable*. In this book this measure for a particular unit is referred to as the *score*. In the fields of psychology and education, a unit is usually a person or an animal and is called a *subject*; but the unit could be something else, such as a class of students or a school. For the sake of concreteness and simplicity, however, the experimental unit will henceforth be referred to as the "subject." Although "subject" is often abbreviated S elsewhere, the term will always be spelled out in this book, and S will be reserved for another meaning that will be explained later.

The different conditions are the distinct situations or treatments to which a subject may be exposed. These conditions arise, as previously mentioned, because the factors (independent variables) take different values, called *levels*. In some experimental designs each subject is exposed to only one experimental condition. In other designs, known as *repeated measurements designs*, each subject is exposed to more than one condition—sometimes to all conditions in the design.

By the time that stage (2) is completed, the experimenter will have specified exactly what experimental factors are to be used and what the levels are of each, how many subjects will be required, from what population they will be obtained (for example, undergraduate students, secretaries, military officers), and to what condition or conditions each subject will be exposed. If each subject is to be exposed to more than one condition, the order of presentation of different conditions must be specified for each subject. The criterion to be employed will also be specified, as well as the method for obtaining the criterion measures. There are other specifications that should be established in the design stage, and they will be discussed in subsequent sections.

1.3.3 Preparation

In stage (3), the investigator must prepare the stimuli required by the conditions and the instructions to be presented to the subjects. He must ensure that whatever room and equipment needed for the experiment are available, and, if necessary, reserve them for the required times. Subjects must be contacted, and arrangements must be made for their appearance. If an assistant is to run the subjects and collect the data, his availability must be ascertained and he should be given whatever information, instructions, and practice he requires.

1.3.4 Execution

Stage (4) can also be referred to as "collecting the data" or "running the experiment." Unless the procedures are quite similar to those used in previous experiments the investigator has run, it is a good idea to try out the experiment on one or a few subjects of the type specified for the experiment before collecting "real" data. Flaws or difficulties in the procedures, stimuli, or equipment thus can be corrected before proceeding. Sometimes a small-scale "pilot experiment" is run. Data are collected and analyzed to guide the design of a more extensive but similar experiment.

1.3.5 Analysis

Stage (5), analysis, can be divided into two substages: score derivation and analysis of variance. A *score* is a single number, a value of the criterion, which indicates, if imperfectly, the effect of the condition on the subject. *Analysis of variance* is a widely used term that nonetheless lacks precisely defined boundaries; in this book it means the various statistical methods and calculations applied to scores to determine what effects the conditions have on the subjects.

Analysis of data is broken down into two substages because the data (observations) for each subject-condition combination often fail to meet the requirement for analysis of variance that the result for each subject-condition combination be a single-valued number (a "score") along some dimension (the "criterion"). For example, the data for a subject-condition might be fifty "true" and fifty "false" responses to a series of questions about material presented for learning. In and of itself, the data would not be suitable for analysis of variance. Instead of a single score along a dimension, we have fifty dichotomous responses. However, if the responses are compared to the acceptable ones, we can obtain a score, "the number of correct responses." All the individual responses are now summarized in a single measure along a dimension. (The score could also have been "the percentage of correct responses" without having any basic effect on the succeeding analysis.)

Neither this book nor others like it that concentrate on experimental design and analysis of variance have very much advice to offer either on what data to collect or how they should be combined to obtain scores. These decisions are too highly dependent on the particular experiment being conducted. The main concern of this book, then, is the second substage, analysis of variance.

Each subject-condition combination yields exactly one score, although a particular condition often has a number of scores associated with it, derived from separate subjects. In repeated measurements designs, each subject has a number of scores associated with him, derived from different conditions. The score for each subject might be some objective measure, such as the time required to complete a task, the number of test questions answered correctly in a given period, a galvanic skin response as recorded on a meter, or the number of trials required to reach some standard of performance, such as ten consecutive correct responses in performing a concept-learning task. The score might be a subjective reaction by the subject; for example, he might rate how interesting he found some passage he read, or how fatigued or depressed he feels. The score might be a subjective reaction toward the subject by another person—a so-called "judge"; for example, the judge might rate the originality of some product created by the subject.

The scores should exist along some dimension because the technique for analyzing the experimental data we wish to use assumes they do. This is not to say that an experimenter must never perform experiments that fail to meet this requirement. If he does, however, he will forego the use of the standard, powerful quantitative techniques for dealing with his scores that are provided by analysis of variance. This does not mean that other valid statistical techniques cannot or should not be used. The so-called *nonparametric* techniques can often serve the experimenter very nicely when scores do not conform to the requirements for analysis of variance. They do not, however, provide the power and flexibility that analysis of variance does. Nonparametric statistical methods are beyond the scope of the present book.

Strictly speaking, the analysis of variance assumes that the scores exist on a continuum, that is, that scores may differ by infinitesimal amounts; but this assumption, as well as certain other ones that will be considered later, is relaxed in practice. For example, the score, number of test questions answered correctly, is not from a continuum, since the scores can only be integral numbers. Likewise, rating scales typically allow only a relatively few possible responses along some gradation. Nonetheless, we accept such scoring instruments as long as they allow some reasonable spread of scores along some gradation of response.

Even allowing such responses, the requirement that scores exist along some dimension is limiting for behavioral research. For example, the subjects responses may instead form a dichotomy, that is, they may fall only into two possible categories. After exposing a subject to speeches by two "political candidates," we may wish to know which candidate he prefers. His response, then, would not be suitable for analysis of variance. Oftentimes, however, a response vehicle that appears to be, by nature, dichotomous, can be converted into a dimensional response vehicle. In the preceding case, for example, the subject might be required to rate his strength of preference for one candidate over the other. Conversion of the basically dichotomous responses of a true-false test to a single score has already been demonstrated. The technique used requires as a minimum that a series of dichotomous responses be available for each condition.

Another way of dealing with dichotomous data is to use a group of subjects, rather than the individual subject, as the experimental unit. Then the percentage of subjects in the group responding one way or the other constitutes a score on a dimension for the unit (group). This is not an ideal method, however, since it is extravagant in the use of subjects. Actually, the best approach to use with dichotomous data is often to use a nonparametric technique rather than to translate them into scores along a continuum.

Only the criterion (dependent variable) is constrained to form a continuum in analysis of variance. The design factors (independent variables) have no such constraint. The levels of a factor may simply be categories, such as different stimulus colors or shapes. Nothing forbids the levels of a factor to lie along a continuum, however.

A condition can affect a subject in many different ways, so to observe responses along only one dimension can be extremely limiting. Actually, it is not uncommon for experimenters to score subjects on several dependent variables. The requirement already stated for a single score resulted from a limitation in the scope of this book, not from a limitation in the statistical methodology for handling such situations. The statistical methodology for dealing with such multiple scores is known as the *multivariate analysis of variance*. In contrast, this book concerns only the *univariate analysis of variance*. Multivariate analysis of variance requires considerably more computation than univariate analysis, as well as additional assumptions to which real scores often fail to conform. The vast majority of experimental analyses of variance reported in the literature are univariate, so the student who uses this book should not feel particularly confined by such a limitation. When several different scores are obtained on each subject-condition, the experimenter lacking the knowledge or facilities for a multivariate analysis can perform separate, independent univariate analyses on each kind of score. Whereas this method may not be ideal, an experimenter certainly should not forego observing auxiliary dependent variables he feels might be informative, simply because he lacks the facilities for a multivariate analysis.

1.3.6 Dissemination

By the end of stage (5), the experimenter will have reached conclusions about how the different factors affect the criterion measures. He may decide at this point that the results do not bear further consideration, and may stop the project there. It is to be hoped, however, that the results are worth disseminating to his colleagues. Indeed, the suspicion is widespread that dissemination is not restricted to worthy results. In any case, the investigator may now proceed to the final stage, dissemination, if he so desires.

Dissemination can assume differing degrees of formality. For example, it may consist only of an oral presentation, in a university seminar, to an industrial committee meeting, or at a meeting of a scientific organization. With or without the oral presentation as a preliminary, the results may be written up in a form suitable for distribution.

All written reports of experimental investigations tend to be cast in similar form. The Introduction gives the background and purpose of the investigation. Next is the Method section, which describes the experimental design, subjects, procedures, and measures. There follows a Results section describing the way the data were analyzed and giving the results of the analysis. Next comes a Discussion section, which describes how the results relate to the purpose of the investigation, as given in the Introduction, and the implications of the results for the scientific field as a whole. The report may contain a final, usually brief, section labeled Conclusions or Summary. In addition, a report often will have an Abstract section preceding the Introduction that gives a very brief summary of purpose, method, results, and conclusions.

The report may be sent only to the investigator's colleagues, or it may be part of a series of reports representing the work of some organization such as a university institute, a government agency, or a corporate laboratory. Such reports may or may not be available for public consumption. The results of an investigation will be available to a wider audience for a longer period if they are published in a scientific journal.

After he prepares and promulgates whatever oral and written versions of his experiment he cares to, the investigator's role in dissemination recedes. His findings may become a part of the main body of knowledge in his field, but even publication in a journal does not assure an experimenter that his work will receive any attention. A large percentage of investigations published in journals might as well be buried in the ground, for all the influence they will have on psychology, as shown in Figure 1.3-1. The effect of a publication is gauged by its citation by other investigators and by inclusion of its findings in survey articles and textbooks.

1.4 MULTIFACTOR EXPERIMENTATION

Let A and B each represent factors. The levels of factor A are symbolized $a_1, a_2, \ldots, a_a, \ldots, a_A$. The levels of factor B are $b_1, b_2, \ldots, b_b, \ldots, b_B$. Lower-case letter a, when subscripted, stands for a level of factor A. Lower-case letter a, when used as a subscript, is a variable going from 1 to A in integral steps. Thus a_a is used to represent any arbitrary level of factor A.

Figure 1.3-1. The equivalent fate of much published research.

Capital letter A also has a dual use. It represents the factor itself, that is, a collection of levels, and it also represents the number of levels of the factor. Capital B and lower-case b have comparable meanings for factor B. The different meanings of the capital and lower-case letters will be clear from the context in which they are used.

A condition to which a subject is exposed in a one-factor design with factor A would be symbolized simply as either a_1, a_2, or some other a_a. Experimental investigations can be, and often are, conducted with one-factor experimental designs.

Now consider a two-factor design with factors A and B. A condition to which a subject is exposed would now be symbolized ab_{11}, ab_{21}, or some other ab_{ab}. The first subscript in ab_{ab} gives the level of factor A, and the second subscript gives the level of factor B. In a two-factor design with factors A and B, any condition to which a subject is exposed consists of one level of factor A and one level of factor B. It is never true, for example, that one subject in such a design receives simply, say, a_4, with no b_b applicable, whereas another receives simply b_3, with no a_a applicable. Some level of each design factor applies to each condition of the experiment.

In a three-factor design with factors A, B, and C, a condition would be described as abc_{234}, abc_{124}, or some other abc_{abc}.

When an experimenter wishes to investigate the effects of two or more factors on some measure of behavior, he can perform either a series of one-factor experiments or a single multifactor experiment. (If many factors are involved, a series of multifactor experiments is a third possibility.) For example, suppose the experimenter wishes to assess the ability of some group of subjects to absorb the information in a prose passage by reading (a_1) or by listening (a_2). Furthermore, he wishes to compare learning when the information is presented in the form of short, simple sentences (b_1) with learning when information is presented in the form of longer, complicated sentences (b_2). First a one-factor experiment could be conducted with A, and then a one-factor experiment could be conducted with B. Alternatively, a two-factor experiment could be conducted, including A and B.

The multifactor experiment is preferable for several reasons:

(1). It would probably be simpler to study both factors in one experiment than to set up two separate experiments.

(2). The relative effectiveness of a_1 versus a_2 may depend on the complexity of the sentences in the passage used. (Even if B is not included as a factor in the experiment, the passage does have some fixed level of complexity.)

If B is not a factor (that is, if it is not varied with the experiment), then we cannot determine whether the relative effectiveness found for a_1 versus a_2 holds for various sentence complexities, or whether it holds just for the one used. If B is a factor, we could make such a determination. Conversely, if we were to investigate B in a one-factor experiment, we would use either reading (a_1) or listening (a_2), so we could not tell if our results applied to both modes. If the level of a second factor affects the relative scores across different levels of a first factor, we say that there is an *interaction* between the two factors. (The term "relative scoress" means the difference between the scores.) Suppose, for example, that the mean learning scores for a_1 and a_2 were 12 and 10 under b_1, but 11 and 9 under b_2. Then there is no interaction between A and B. The relative advantage of a_1 over a_2, 2, is the same with b_1 as with b_2. Interactions very often exist in psychological research, however. Usually such interactions change only the size of the relative advantage, but they can also cause a reversal in relative advantage; for example, a reversal such that learning is better for a_1 when b_1 is used, but better for a_2 if b_2 is used. Such interactions can be assessed with a two-factor experiment, but not with a one-factor experiment using A or B, nor with two one-factor experiments using only A and only B, respectively. The multifactor experiment is therefore preferable.

(3). The multifactor experiment is more efficient. The effects of the factors can be detected with fewer total subjects, on the average, by conducting multifactor experiments than by conducting separate experiments for each factor.

Considering the advantages of multifactor experimentation, it is not surprising that such designs are so often used. But even though the number of factors possibly affecting any behavior is very large, experimenters typically use designs with only two or three factors. Why are not more factors used? One reason is that as the number of factors increases, more subject-hours are required to maintain the ability to detect differences among conditions, though not as many more as if separate one-factor experiments were conducted. In addition, as the number of factors increases, preparation and presentation of the stimuli become more complicated. Furthermore, interpretation of the results becomes more difficult, because analysis of variance leads to consideration not only of interactions between pairs of factors, but among triples of factors, quartets, and so on. Although interactions among three or more factors can be interpreted mathematically, such interactions are difficult to understand and use.

Psychologists seem to shun anything more complex than a three-factor interaction. As far as this author is aware, interactions among four or more factors have not played an important role in practical applications, in psychological theory, or in the summarization of empirical findings in psychology. Perhaps interactions among four or more factors are simply of minor importance, even when they exist. Even if they were of considerable importance, however, they would doubtless be ignored because of the limitations of human comprehension.

If an experimenter avoids many-factor designs, he avoids the necessity of trying to interpret interactions among many factors, much as an ostrich can avert his eyes from a disagreeable sight by sticking his head in the sand. This author, however, would prefer that experimenters use multifactor designs so that many variables of possible importance can be studied in an experiment, and then feel free to ignore the possible presence of many-factor interactions in the results. After all, these interactions would of necessity be "ignored" anyway if any of the factors involved in them were omitted from the experiment. By including such factors but ignoring any possible complex interactions involving them, an experimenter can at least gain information about the simpler effects of these factors.

A final reason why designs with more than three factors are infrequently used is that textbooks seldom describe such designs or provide the computational formulas required for the analysis. Most experimenters depend on the texts to provide this information; relatively few venture to construct and use designs without such guidance. An advantage of the present book is that it describes methods that allow an experimenter to construct and analyze any of the many-factor designs belonging to an important class of designs that are most often used in psychological research. Of course, restrictions on the number of subject-hours available and on the experimenter's available time for running them will still discourage the routine use of many-factor designs.

The point here is not that experiments must contain many factors. Rather it is that possibly important factors should not be excluded simply because a familiar textbook does not detail more complicated designs, or because one-sided arguments are advanced in favor of simpler designs.

Because the requirements for subject-hours may become excessive as the number of factors in a design increases, special types of designs have been constructed that reduce the subject-hour requirements—but at a price. The price is a confusion, technically called a *confounding*, between different effects (for example, between different interactions). The confusion is such that an experimenter may detect some effect at work, but cannot be sure if it is due to one effect, another, or both. For example, suppose the AB and ABC interaction effects are confounded. Then the analysis may show that an effect is operating that may be the AB interaction, the ABC interaction, or both. The experimenter who uses such a design may be willing to assume that one of these possibilities is true. He may be willing, for example, to assume that the effect is due to interaction AB, and that interaction ABC is negligible. Although in principle such assumptions might be based on evidence, in practice they are more likely to be based on a leap of faith.

Confounded designs are also viewed with suspicion by the overseers to the gates of publication, as well as by many knowledgeable investigators and research consumers. This suspicion is often well founded. However, confounded designs should not be rejected offhand when planning an experiment. It may be better to include more factors in an experiment at the price of some interpretive uncertainty than to exclude possibly important factors from the experiment.

Nonetheless, this book does not cover confounded designs (at least designs normally considered to involve confounding) in any detail. There is a kind of confounding that occurs in the commonly used nested designs (Section 4.8), but such designs are not normally considered to entail "confounding." The most important confounded designs—the Latin square designs—are considered in Chapter 7. Other confounded designs constitute a topic for advanced study.

1.5 OVERVIEW

This section presents an overview of the forthcoming chapters in this book. Chapter 2 is a review of elementary statistics and notation, including summation notation, normal probability distribution, estimation of parameters, hypothesis testing, and power.

Chapter 3 describes a way of symbolizing designs by combining capital-letter factor symbols with factor-relation symbols. One such symbolization, for example, is $S(A) \times B$. Letters S, A, and B symbolize factors. The $\times$ and

( ) are factor-relation symbols called the *cross* and *nest*, respectively. Chapter 3 discusses how to derive and interpret such symbolizations. The symbolization is useful for communication and comprehension. In addition, from it we can derive the score model, which plays such a central part in the analysis and interpretation of experimental data.

For each experimental design there is a *score model*, which is the topic of Chapter 4. A score model is a set of equations, each equating an observed score X to a set of theoretical terms devised to account for the score. For example, for design $S(A) \times B$, the score model is

$$X_{abs} = \mathbf{m} + \mathbf{a}_a + \mathbf{b}_b + \mathbf{s}_{s(a)} + \mathbf{ab}_{ab} + \mathbf{bs}_{bs(a)} + \mathbf{e}_{abs}$$

This formula can be thought to represent a single equation or a set of equations, one for each combination of values that the variable subscripts a, b, and s may assume. One such equation, for example, might be

$$X_{214} = \mathbf{m} + \mathbf{a}_2 + \mathbf{b}_1 + \mathbf{s}_{4(2)} + \mathbf{ab}_{21} + \mathbf{bs}_{14(2)} + \mathbf{e}_{214}$$

Such an equation asserts that a score, X_{214} in this case, can be interpreted as the sum of theoretical terms. Term **m** is the *population mean*. It represents the average scoring tendency across all conditions and across all persons in the population from which the experimental subjects were drawn. Term $\mathbf{e}_{abs}$ is the *error term*, and represents a random score component that cannot be accounted for by the condition to which the subject is exposed.

The other terms are called *main* and *interaction terms*. Their purpose is to account for the true (nonrandom) deviations of scores from the population mean, **m**. Such deviations are attributed to the different factorial levels to which different subjects are exposed. A term (other than **m** or **e**) represented by just one bold-faced letter (for example, $\mathbf{a}_a$) is called a *main term*, or a *main effect*. A main term, such as $\mathbf{a}_2$, is some quantity, such as 2.6, that contributes to the score of any subject exposed to the corresponding factor level (a_2 in this example). An interaction term with two bold-faced letters (for example, $\mathbf{ab}_{ab}$) is called a *two-factor interaction term* or *effect*. (There are also nested interaction terms, such as $\mathbf{bs}_{bs(a)}$.) Interaction terms involving three, four, or more factors can and do occur, but no such term appears in the score model shown. A two-factor interaction term, such as $\mathbf{ab}_{21}$, is some quantity, such as -4.3, which contributes to the score of any subject exposed to both of the corresponding factor levels (a_2 and b_1 in this example). The score model is important in experimentation because the major reason for analyzing the scores (that is, the data) of an experiment is to draw conclusions about the main and interaction terms.

Chapters 5 and 6 concern the analysis of variance of the scores obtained from experimentation. Analysis of variance commences with the calculation of *sums of squares*, the topic of Chapter 5. A sum of squares is a quantity, and there is such a quantity for each term of the score model, except for **m**. Sums of squares are of little direct interest for the interpretation of the experiment; they are only a way station. From them we calculate mean squares simply by dividing each sum of squares by its appropriate *degrees of freedom*. (Degrees of freedom and mean squares are discussed in Chapter 6.)

For most main and interaction terms of a score model, we calculate an F-ratio, which is the ratio of two mean squares. If the F-ratio exceeds a critical value, we conclude that the main effect or interaction term associated with the numerator of the F-ratio contributes significantly to the scores observed. If the F-ratio does not exceed the critical value, the term under consideration is *nonsignificant*. A nonsignificant term is one deemed to make no contribution to the scores; in effect, such a term can be omitted altogether from the score model without affecting its adequacy. The procedure just described, called *significance testing*, is also discussed in Chapter 6.

Chapter 7 concerns a special type of confounded design, the Latin square designs. Such designs require special consideration; they cannot be handled directly by the procedures given in Chapters 3 to 6.

Chapter 8 discusses the various considerations the investigator should keep in mind when planning his experiment. For example, if a subject will be exposed to several conditions, and a score is to be derived from each, the experimenter must minimize the distortions in the results due to the effects on the later scores of earlier exposures.

The probability that an experiment will show some experimental factor to have a significant effect on the scores depends in part on the design used. Generally it is preferable that this probability be as high as possible, given the practical constraints occurring with any investigation. Chapter 8 considers and compares the various alternatives available to the experimenter for making his design sensitive to experimental effects. The chapter also discusses the advantages and disadvantages of running subjects in groups rather than individually.

Chapter 9 concerns various considerations that arise during the data analysis. For example, what should the investigator do if, for one reason or another, certain of the scores required by the design are missing, or if certain assumptions about the scores that are required for the analysis appear to be untrue? Chapter 9 also discusses some common types of data analysis not previously considered, such as *trend analysis* and *multiple comparisons*.

CHAPTER TWO

Review of Elementary Statistics

2.1 SUMMATION NOTATION

We will often be adding a set of quantities together, and will use the special symbolizations described in the following paragraphs to indicate such summations succinctly.

2.1.1 One-Way Arrays of Numbers

Suppose there are four numbers, 3, 8, 4, and 5. These numbers are symbolized by X_1, X_2, X_3, and X_4, respectively; that is, $X_1 = 3$, $X_2 = 8$, $X_3 = 4$, and $X_4 = 5$. The term X_i symbolizes an arbitrary one of these quantities; i is a variable subscript ranging from 1 to I. Capital Greek sigma, $\sum$, is used to symbolize the sum of several, or all of the X_i's, as follows:

$$\sum_{i=1}^{3} X_i = X_1 + X_2 + X_3 = 3 + 8 + 4 = 15$$

The summation designation may also be written $\sum_{i=1}^{3} X_i$ so that it will fit within a normal line of print.

The number beneath the $\sum$ sign equaling i is the value of i for the first X_i of the sum. By convention, the subscript increases by 1 for each successive term of the summation. The subscript value for the final term of the summation is the number written above the $\sum$; it is 3 in the preceding example. Had we wished to sum all four numbers in the array, we would have written 4 above the $\sum$.

Usually, the variable subscript starts at 1 and ends at its largest possible value for the set of numbers under consideration. This is true of most of the summations in this book. Therefore, we normally omit the lower and upper subscript limits from $\sum$, agreeing in such circumstances that 1 is the initial value and that the upper limit is the largest possible value for the available data. In other words, we may write $\sum_i X_i$ in place of $\sum_{i=1}^{4} X_i$. Furthermore, we often will omit all appendages to $\sum$ altogether, and simply write, for example, $\sum X_i$. It is understood, in such an example, that the subscript on X varies from 1 to its maximum value for the available data.

2.1.2 Two-Way Arrays of Numbers

Rather than to arrange all numbers in a single series, it is often natural and convenient to tabulate them in a two-way array, as shown in Table 2.1-1. An X symbol for a specific number now has two subscripts. We let X_{ij} be a variable, which, when specific values are assigned to i and j, equals one or another of the numbers of the array.

We can represent the sum of all numbers in the array in $\sum$-notation as $\sum_{i=1}^{3} \sum_{j=1}^{4} X_{ij} = 67$. We often employ the simpler notation $\sum X_{ij}$, with the implicit assumption that the summation proceeds over all values of both subscripts beginning with 1. At other times, for clarity, we use the notation $\sum_{ij} X_{ij}$, to emphasize that the summation is over both subscripts.

Using the $\sum$-notation for two-way arrays, we can indicate the sums for individual rows and columns, as well as for the entire array. For example, $\sum X_{2j} = X_{21} + X_{22} + X_{23} + X_{24} = 21$. When a subscript on X is a specific number rather than a variable, then that subscript remains fixed—the summation is only over subscript variables. In other words, in finding $\sum X_{2j}$ we summed over the different values of j from 1 to 4, but i remained fixed at 2. Likewise, $\sum X_{i3}$ stands for the sum of the third column of the array as i goes from 1 to 3, but j remains fixed at 3.

By $\sum_i X_{ij}$ we specify that the summation is only over i, but not over j. Such a notation means "column sum," but does not stand for a specific quantity until j is specified.

Table 2.1-1 A two-way array of numbers and the X_{ij} symbols for them

5	7	4	9
2	6	5	8
9	1	7	4

	$j = 1$	$j = 2$	$j = 3$	$j = 4$
$i = 1$	X_{11}	X_{12}	X_{13}	X_{14}
$i = 2$	X_{21}	X_{22}	X_{23}	X_{24}
$i = 3$	X_{31}	X_{32}	X_{33}	X_{34}

In summing over two subscripts we can first sum over columns, for specific rows, and then add these row sums together, or vice versa. When the two-$\sum$ notation is used, the order of writing of $\sum_i$ and $\sum_j$ implies the order for summing. If we write $\sum_i\sum_j X_{ij}$, the implication is that we sum by rows; that is, the subscript of the inner $\sum$ varies faster than the subscript for a $\sum$ more to the left. If we write $\sum_j\sum_i X_{ij}$, the implication is that i changes faster than j; that is, we sum down the columns. The two ways of summing must give the same result, since, of course, the sum of all numbers in the array is the same regardless of the order in which they are summed. In other words, $\sum_j\sum_i X_{ij} = \sum_i\sum_j X_{ij}$. In terms of the simplified notation we shall use, $\sum_{ji} X_{ij} = \sum_{ij} X_{ij}$. Note that, although we may change the order of the subscripts appended to $\sum$, the order of the subscripts appended to X is never altered. We may have arrays of greater dimensionality than two, and for these arrays, also, all the different ways of ordering the $\sum$ subscripts yield the same sum.

2.1.3 Summing Functions of X

A summation sign can apply not only to X, but to functions of X. For example, we often wish to deal with the summation of the squares of the array quantities, so we have forms such as $\sum X_{ij}^2$. If we wish to sum the numbers first, and then to square the sum, we write $(\sum X_{ij})^2$. Care must be taken not to confuse these two operations.

When each quantity in an array is multiplied by a constant before summing, the final sum is the same as if we summed the array first, then multiplied this result by the constant. In other words, if W is a constant (the same) for each member of an array X_{ij}, then $\sum WX_{ij} = W\sum X_{ij}$. As with the squaring, an operation on X written to the right of the summation sign should, in general, be carried out on the original array entries before summing. When the operation (multiplying by W in this example) is indicated at the left of the $\sum$ sign, the summation on the array is to be carried out first, and then the indicated operation is performed on the sum. A multiplier may be a constant with respect to one summation sign, but variable with respect to another. Let W_i be a variable as i changes, but the same for each j, when i stays the same. Then $\sum_i\sum_j W_i X_{ij} = \sum_i W_i \sum_j X_{ij}$; that is, we may move a multiplier to the left of any $\sum$ with respect to which it is constant. We must, however, keep it to the right of a $\sum$ with respect to which it is a variable.

2.1.4 Summing a Constant

Let J stand for the upper limit of subscript j; in other words, j goes from 1 to J so that $\sum_j$ implies a sum with J terms: $\sum_j W = W\sum_j 1 = WJ$. *The sum by $\sum$ of a constant is simply that constant times the number of terms in the summation.* Note that if the constant were placed to the left of $\sum$, only a blank would remain; the notation would be meaningless. Here this blank was filled with 1, an implied multiplier of W. The sum of $1 + 1 + \cdots + 1$ for J terms is, of course, J.

Let I stand for the upper limit of subscript i; then $\sum_i 1 = I$, and $\sum_i \sum_j W = W \sum_i \sum_j 1 = WJI$. When 1 is summed over several $\sum$'s, the result is the product of the subscript upper limits. (We assume that the variable subscripts begin at 1.)

We may have $\sum$'s whose subscripts are absent in the symbol being summed. Such an expression is legitimate. For example, $\sum_{ij} Y_j$ is a meaningful expression. It equals $\sum_i \sum_j Y_j$, that is, $\sum_i(\sum_j Y_j)$. The notation $\sum_j Y_j$ stands for a definite quantity, which is a constant relative to variation in subscript i. Since $\sum_i W = WI$, we know that $\sum_i \sum_j Y_j = I \sum_j Y_j$.

2.1.5 Summing a Sum

The symbol to which we apply the $\sum$-notation may itself be written as a sum. For example, besides array X_{ij} we may have another array Y_{ij} of the same size. To indicate the sum of the identically located quantities in the two arrays, we write $X_{ij} + Y_{ij}$. Then $\sum_{ij}(X_{ij} + Y_{ij}) = \sum_{ij} X_{ij} + \sum_{ij} Y_{ij}$. In other words, when one or more $\sum$'s are applied to a summation of terms, we may equivalently apply them to each term separately. We must take care, especially in view of the use of one $\sum$ with multiple subscripts in place of several $\sum$'s, that all $\sum$'s are applied to each component. For example, $\sum_{ij}(X_{ij} + Y_j)$ does not equal $\sum_{ij} X_{ij} + \sum_j Y_j$; it equals $\sum_{ij} X_{ij} + \sum_{ij} Y_j = \sum_{ij} X_{ij} + I \sum_j Y_j$.

2.1.6 Summary of $\sum$ Rules

Following is a summary of some of the important rules for $\sum$ that have been covered thus far.

RULE 1. The order of adjacent $\sum$'s may be rearranged without affecting the final result; for example,

$$\sum_i \sum_j X_{ij} = \sum_j \sum_i X_{ij}$$

RULE 2. A multiplier may be moved to the left of any $\sum$ with respect to which it is a constant; for example,

$$\sum_j W X_j = W \sum_j X_j$$

RULE 3. A set of $\sum$'s applied to a sum of terms may equivalently be applied separately to each term; for example,

$$\sum_{ij}(X_{ij} + Y_{ij}) = \sum_{ij} X_{ij} + \sum_{ij} Y_{ij}$$

2.1.7 Dot Notation

We often use dot notation, which is an even more compact way of representing sums than the $\sum$-notation. If we have a subscripted symbol such as X_{ij} representing an array of numbers, then we can indicate the summation over that subscript by replacing a subscript with a dot. For example:

$$X_{..} = \sum_{ij} X_{ij}$$
$$X_{.j} = \sum_{i} X_{ij}$$
$$X_{i.} = \sum_{j} X_{ij}$$

We use the dot notation not only to indicate sums but also to indicate mean values of array quantities, by adding a bar over a dot-subscripted symbol. The averaging is only over subscripts that have been replaced with dots.

A mean is simply a sum divided by the number of quantities summed. Therefore,

$$\bar{X}_{..} = \frac{X_{..}}{IJ} = \frac{\sum_{ij} X_{ij}}{IJ}$$

$$\bar{X}_{.j} = \frac{X_{.j}}{I} = \frac{\sum_{i} X_{ij}}{I}$$

$$\bar{X}_{i.} = \frac{X_{i.}}{J} = \frac{\sum_{j} X_{ij}}{J}$$

At times we indicate the mean of a set of numbers simply as $\bar{X}$, without dots, with the understanding that the summation is over all numbers in the set.

2.2 PROBABILITY DISTRIBUTIONS

2.2.1 Discrete and Continuous Distributions

A *random variable* is a symbol (such as X) that, on each of a series of *trials*, assumes one or another of a set or *range* of possible values; the value assumed by X varies randomly from trial to trial. If an infinite series of X's were observed, we would typically find that some values occurred relatively more frequently, and some occurred relatively less frequently. A *probability distribution* is a description of the range of values that X can assume, together with the relative frequencies for each value that would obtain over an infinite series.

If the random variable X can take on only distinct values, each value being an isolated point on the real number continuum, its probability distribution is said to be *discrete*. For example, the sum of the pips appearing face up when two dice are thrown is a random variable. The sum can assume only one of 11 different values—the integers 2 to 12. The probability distribution for this random variable is discrete.

If the random variable X can take on any of a continuum of adjacent values, each value lying next to others infinitesimally close, its probability distribution is said to be *continuous*. The best-known continuous probability distribution is the bell-shaped *normal distribution*.

A graphic portrayal of the discrete and continuous probability distributions mentioned here is given in Figure 2.2-1. In graphs of probability distributions, the various values that the random variable can assume are represented along

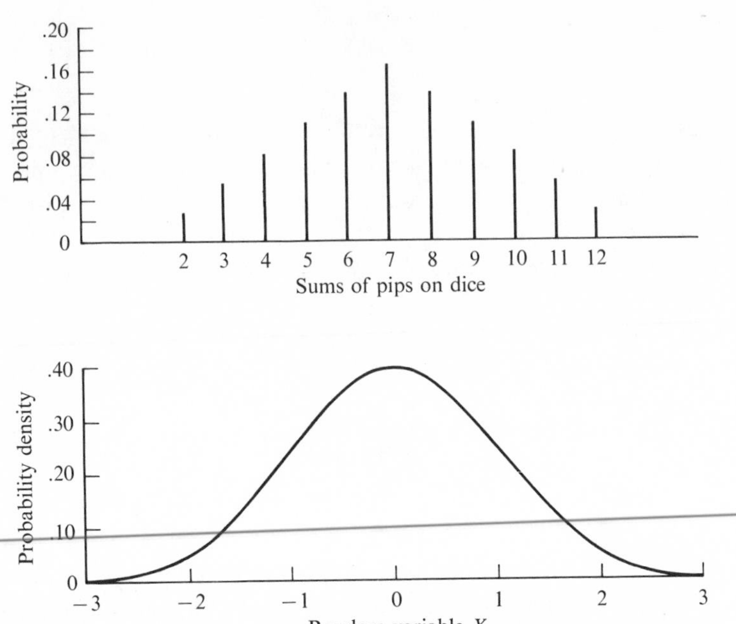

Figure 2.2-1. *Top:* The discrete probability distribution for the random variable, the sum of pips on two dice. *Bottom:* The standard normal probability distribution, a continuous distribution.

the *abscissa* (that is, horizontally). The relative frequencies of occurrence are represented along the *ordinate* (that is, vertically).

The normal distribution is actually a family of distributions that differ according to the value of X at the distribution center and the relative concentration of the distribution around its center. In other words, distributions may be tall and narrow, or short and broad, or anything in between. The distribution illustrated in Figure 2.2-1 is the *standard normal distribution*, which centers at $X = 0$ and includes 68% of its area between $+1$ and -1 on the abscissa. The normal distribution extends infinitely above and below its center, though the height of the curve becomes extremely small far from the center. The value of X at the center of the normal distribution is symbolized by **m**.

The relative frequency differs for discrete and continuous distributions. For discrete distributions, a relative frequency is a *probability;* the ordinate is so labeled in Figure 2.2-1. For continuous distributions, the ordinate is a *probability density;* the ordinate of the normal distribution in Figure 2.2-1 is so labeled. The probabilities for the different X values of a discrete distribution sum to 1, as the probabilities for a set of mutually exclusive, exhaustive events must do, according to the usual assumptions of probability theory.

(The members of set of events are *mutually exclusive* if only one of the events can occur on a trial; a set is *exhaustive* if one of the events must occur on a trial.)

Finite probability values cannot be attached to each X of a continuous distribution, because with an infinite number of different X's, the sum of the probabilities for an exclusive, exhaustive set of events is infinite, whereas probability theory requires that the sum be 1. For this reason, continuous distributions have probability densities that are associated with each X, rather than probabilities. Probabilities per se are associated not with specific X values, but with intervals along X. We can think of the probability densities at two locations of X as specifying the relative probabilities for very narrow intervals located around those two X's.

There is a probability that an observation will lie between any two X values X_1 and X_2; it is the area under the curve and between X_1 and X_2. The total area under a continuous distribution is 1, corresponding to a probability of 1 that some value of X will occur on a trial. The area to which we refer is not measured in square inches, square feet, or any such common units. It is determined by multiplying the units of the ordinate and abscissa.

2.2.2 The Central Tendency of a Probability Distribution

The *mean* of a finite set of numbers is simply the sum of the numbers divided by N, the number of numbers: $\bar{X} = \sum X/N$. The mean of a set of numbers is a useful summary that usually gives us an idea of the central tendency of a set of numbers, that is, where the "middle" of the set is. The mode and the median are two other important measures of central tendency. The *mode* is the number in a set that occurs most frequently. The *median* is that value of X below and above which half the numbers fall.

A probability distribution also has a mean, a mode, and a median, and although their definitions differ somewhat from those applying simply to a set of numbers, the mean, mode, and median are measures of central tendency of the probability distributions.

The mean of a probability distribution is also called the *expectation* (or sometimes the *expected value*). It is symbolized by $\mathrm{E}(X)$, rather than by $\bar{X}$. The terms "mean" and "expectation" may equivalently be applied to the random variable or to the probability distribution; that is, we may speak of the expectation of a random variable or the expectation of the probability distribution describing that random variable. Either term refers to the same quantity.

First we consider the mean for a discrete distribution. Let $p(X)$ be the probability that X occurs. Then the mean is

$$\mathrm{E}(X) = \sum X p(X),$$

where the summation is over each distinct value of X that occurs. Suppose, for example, that we flip a pair of coins and observe how many heads land face up, this value being a random variable X. The possible values for X are 0, 1, and 2, and these values have probabilities of 1/4, 1/2, and 1/4, respectively. Then $\overline{X} = (1/4)\cdot 0 + (1/2)\cdot 1 + (1/4)\cdot 2 = 1$.

The mean for a discrete probability distribution is comparable to the simple mean of a finite set of numbers, although the mathematical definitions are different. If a finite set of values of a random variable occurred in the exact proportions specified by the probabilities, then the mean of that set would equal the mean of the distribution. Normally, however, a series of observations of a random variable will not occur in such exact proportions. If we flipped the pair of coins four times, we might or might not get no heads (two tails) once, one head and one tail twice, and two heads once. Because of the random nature of the observations, the mean of a series of N observations of a random variable usually differs from series to series. The mean of a set of N observations of a random variable is itself a random variable based on repetitions of N observations.

The mean (or expectation) of a continuous probability distribution is very similar in concept to the mean of a discrete distribution, but the definitions differ in form. The definition of the mean of a continuous probability distribution requires the symbols of calculus, so that students lacking calculus will necessarily be a bit mystified. However, the concept of mean should nevertheless be understandable.

The expectation (or mean) of a continuous probability distribution is defined to be

$$\mathrm{E}(X) = \int_{-\infty}^{+\infty} X\cdot f(X)\cdot dX,$$

where $f(X)$ is the probability density of the distribution at X. This definition is analogous to the definition of a discrete distribution, with $f(X)\cdot dX$ substituted for $p(X)$ and the integration symbol $\int$ substituted for the summation sign $\sum$. (The integration sign indicates summation in calculus just as the summation sign indicates summation for finite quantities.) We can think of dX as a very narrow interval along the X continuum—so small that $f(X)$ can be considered constant throughout the range. Then $f(X)\cdot dX$ is the probability that an observation will fall in the interval dX, since the area under the distribution at X for a very narrow interval dX is approximately the area of a rectangle $f(X)$ tall and dX wide. (The area of a rectangle is, of course, its height times its width.)

The *mode* of a discrete distribution is that value of X having the largest $p(X)$; for a continuous distribution it is that value of X having the largest $f(X)$. The *median* of a discrete distribution is that value of X both above and below which the $p(X)$'s sum to 0.5. (The $p(X)$ for the median itself, if the median is a possible observation, is split among the "below" and "above" sums.) The median of a continuous distribution is that value of X both above and below which lie one-half the total area of the distribution. In any distribution symmetric around the center, such as the two distributions shown in Figure 2.2-1, the mean and median are equal; each lies at the distribution's center. (In Figure 2.2-1, the mode is the same as the mean and median.)

2.2.3 The Variance of a Probability Distribution

The preceding subsection concerned measures of central tendency for a set of numbers or a probability distribution. Another important characteristic of a set of numbers or a distribution is the degree of dispersion (spread) around the central tendency. The degree of dispersion tells us whether X's lie relatively close to the center, or whether they disperse widely to either side.

The most important measure of dispersion is the variance, which is symbolized as σ^2. (σ is lower-case Greek sigma.) The *variance* of a set of numbers is defined to be

$$\sigma^2 = \frac{\sum(X - \bar{X})^2}{N}$$

The summation is over each number in the set. If the same number is repeated several times in the set, $X - \bar{X}$ must be entered in the formula for each repetition.

The variance for a discrete probability distribution is defined as

$$\sigma^2 = \sum(X - \mathrm{E}(X))^2 p(X)$$

The definition of the variance is the same as the definition of the mean of X, $\mathrm{E}(X)$, except that $(X - \mathrm{E}(X))^2$ of the present formula replaces X of the former one. In essence, then, σ^2 is the mean of the random variable $(X - \mathrm{E}(X))^2$, the squared deviation of X from the mean value of X. For a continuous distribution, the definition is analogous, but again, calculus is required:

$$\sigma^2 = \int_{-\infty}^{+\infty} (X - \mathrm{E}(X))^2 f(X) \cdot dX$$

Again, we can think of σ^2 as the mean of $(X - \mathrm{E}(X))^2$.

Suppose that $\mathrm{E}(X) = 0$. Then the formula for σ^2 would be

$$\sigma^2 = \int_{-\infty}^{+\infty} X^2 f(X) \cdot dX$$

As previously stated,

$$\mathrm{E}(X) = \int_{-\infty}^{+\infty} X \cdot f(X) \cdot dX$$

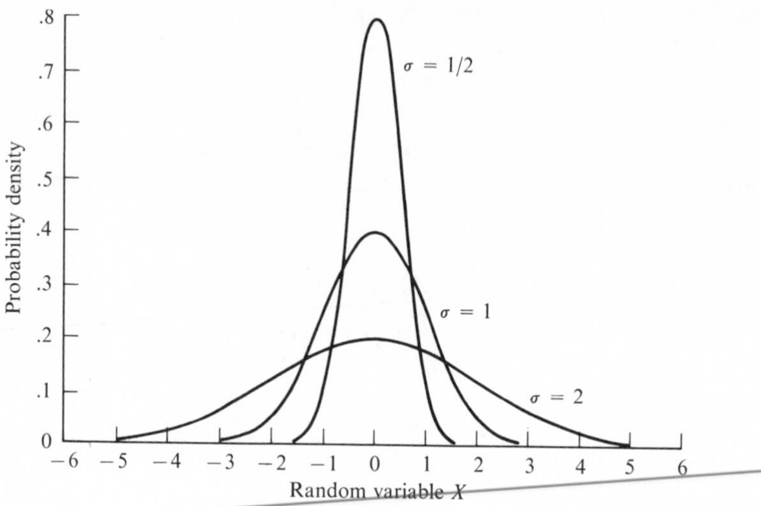

Figure 2.2-2. Three normal distributions differing in standard deviation (σ) and variance (σ^2). The mean (**m**) is 0 for all three distributions.

If we wanted the expectation of X^2 instead of X, we would replace X by X^2 and write

$$\mathrm{E}(X^2) = \int_{-\infty}^{+\infty} X^2 f(X) \cdot dX$$

The expression for $\mathrm{E}(X^2)$ is the same as the expression for σ^2 when $\mathrm{E}(X) = 0$; hence, when $\mathrm{E}(X) = 0$, $\sigma^2 = \mathrm{E}(X^2)$. The same result holds whether X is continuous or discrete.

The variance of the standard normal distribution equals 1. The variance of a normal distribution that is more tall and narrow than the standard distribution is less than 1. The variance of a normal distribution that is shorter and broader than the standard distribution is greater than 1.

The square root of the variance is called the *standard deviation*, and it is symbolized, logically enough, σ. Figure 2.2-2 shows three normal distributions, all centered at 0, but having standard deviations of 1/2, 1, and 2. The area under each of the three curves is 1, as is the area under any normal probability distribution. A tall, narrow distribution (small σ) compensates in height for what it lacks in breadth. A short, broad distribution (large σ) compensates in breadth for what it lacks in height.

2.2.4 Using Tables of the Normal Distribution

Table A.3 of the Appendix provides important data on the standard normal distribution. To use it we must understand the meaning of *percentile* and *percentile score*. A particular value X_P of a probability distribution is said to have a *percentile score* of P if exactly $P\%$ of the distribution's area lies below

X_P. In other words, the probability that an observation X will fall below X_P is $P/100$. We then refer to X_P as the P*th percentile* of the distribution. It is very easy to confuse percentile scores and percentiles: a particular X value, X_P, *has* a percentile score of P, meaning that $P\%$ of the distribution's area falls below X_P; but though X_P *has* a percentile score, it *is* a percentile, the Pth percentile. The 50th percentile of a distribution is the median, and is symbolized X_{50}. The 50th percentile X_{50} equals zero for the standard normal distribution. The percentile score of the median is 50. Percentile scores fall between 0 and 100, regardless of the shape of the distribution. Percentiles cover the same range as random variable X.

The abscissa of the standard normal distribution (Section 2.2.1) is usually symbolized by z. Table A.3 lists percentile scores P, beginning with 50, together with the corresponding percentiles z. The table shows, for example, that the z below which 60% of the area lies is .25. The table does not show percentile scores for each .01 increment in z. If the values shown are not close enough, interpolation can be used. Because of the symmetry of the normal distribution, P's below 50 need not be listed. The percentage of area below a z of $-.25$ is exactly the same as the percentage above a z of .25, namely, $100\% - 60\% = 40\%$.

Table A.3 can be used for nonstandard distributions as well—that is, those having means other than 0 and standard deviations other than 1. To find the percentile score for an X, we first convert that X to a standard score z by the formula

$$z = \frac{X - \mathbf{m}}{\sigma}$$

We then use Table A.3 to find the percentile score for the z calculated. It is the percentile score desired. To convert a percentile score to a percentile for a nonstandard distribution, we use the table as before to find the percentile on the z scale; we then convert to the X scale by the formula

$$X = \sigma z + \mathbf{m}$$

which is simply a rearrangement of the preceding formula.

2.3 ESTIMATION OF PARAMETERS

2.3.1 Populations and Samples

In this book a probability distribution is generally considered to be derived from an infinite *population* of experimental subjects, each of which has an associated score X. Certain quantities, such as the mean ($\mathbf{m}$) and variance (σ^2), characterize the population. These are called *parameters* of the population. We never observe the infinity of scores, and so we never learn exactly what the parameters equal. We only observe a finite number of randomly chosen scores, called a *sample*. A number characterizing the sample is called a *statistic*. The word "statistic" is also used more abstractly, to refer to the method or formula used in the calculation. Although we never know exactly what the parameters equal, we can use sample statistics to *estimate* the parameters.

There are two important types of estimates: point and interval. A *point estimate* is a single number expressing a best guess for the parameter. An *interval estimate* is an interval (range) of numbers within which the parameter is believed to lie.

2.3.2 Estimators and Estimates

It is worthwhile to distinguish between *estimators* and *estimates*. An *estimator* is a method or formula for computing an estimate using whatever scores might be sampled. An *estimate* is a particular value computed using a particular sample. The sample mean, $\overline{X}$, conceived as a computational formula applicable to any sample that might occur, is a *point estimator* of the population mean, $\mathbf{m}$ (though it is not the only estimator). For a sample of three scores, say, 10, 6, and 2, we could calculate an *estimate* of $\mathbf{m}$: $\overline{X} = (10 + 6 + 2)/3 = 6$.

It is, however, extremely difficult to maintain purity of usage for the two words in exposition. For example, do we refer to the "sample mean" as an "estimate" or an "estimator"? In many contexts it is not clear that one meaning or the other is, or should be, exclusively intended. Because of this ambiguity, one author may use only "estimate" for both meanings, whereas another will use both words almost as synonyms. In this book, both "estimate" and "estimator" are used in a manner that is as consistent as possible with the definitions given, although it is acknowledged that, in many contexts, the choice of words could be disputed, or that both meanings might be appropriate.

Although the definition of "statistic" given here implies that a statistic might be an estimate, but not an estimator, in fact in some parts of the discussion a statistic is described in abstract terms, and divorced from a *particular* sample; in such situations the statistic can be thought of as an estimator.

2.3.3 Point Estimates

Given a random sample of I scores from a normal population, there are innumerable statistics that might be used to estimate a particular parameter such as $\mathbf{m}$. For example, we might use the sample mean, or the sample median. How do we know which statistic to use? There are various criteria by which we can compare different possible estimators. Two important ones are expectation and variability.

Expectation. Suppose we draw a sample of size I from a population, calculate a statistic, draw a new random sample, calculate the same kind of statistic (whose value, however, varies from sample to sample), and so forth, for an indefinitely long series of samples. Each sampling and statistic calculation is called a *replication*. The statistic is a random variable over replications; it has a probability distribution and a mean (expectation). One desirable property of a statistic as an estimator is that its expectation equal the parameter being estimated. If it does, the statistic is said to be *unbiased;* otherwise, it is *biased*. Both the mean and the median are unbiased estimators of **m** for a normal distribution.

The preferred estimator for the population mean **m** is the sample mean $\bar{X}$. Such simplistic and obvious analogies as that between the population mean and the sample mean cannot always be relied on to determine the appropriate statistic. For example, in Section 2.2.3 it was noted that the population variance is defined as the mean of $(X - \mathrm{E}(X))^2$ over the infinity of X scores composing the population. The analogous definition for the sample would be the mean of $(X - \bar{X})^2$, that is, $(1/I)\sum_{i=1}^{I}(X_i - \bar{X})^2$. The expectation of this formula, however, does not equal σ^2. The preferred estimator of σ^2 is, instead, $(1/(I-1))\sum_{i=1}^{I}(X_i - \bar{X})^2$, which is unbiased. This latter formula differs from the former only by having $I - 1$ rather than I, the sample size, as a divisor.

The symbol for a parameter with a "circumflex" ($\hat{\ }$) over it refers to a point estimate of that parameter. For example, $\hat{\mathbf{m}}$ refers to an estimate of **m**, and $\hat{\sigma}^2$ refers to an estimate of σ^2. The symbol $\hat{\mathbf{m}}$ refers to an estimate equaling the mean $\bar{X}$ of a sample of scores; $\hat{\sigma}^2$ always refers to the unbiased estimate. The symbol $\hat{\sigma}$ refers to $\sqrt{\hat{\sigma}^2}$; it is not exactly an unbiased estimate, but it is routinely used as the estimate of σ.

If purity in the usage of "estimate" and "estimator" is to be maintained, the ubiquitous phrase "unbiased estimate" should be avoided. Strictly speaking, the concept of "bias" should apply only to estimators, not to estimates; but in practice, it is applied to both terms. When the student sees the phrase "unbiased estimate" in this book or others, he should translate it to mean "an estimate calculated with an unbiased estimator."

Variability. Successive replications involve separate samplings, and therefore separate X's. It follows that an estimator such as $\bar{X}$, which depends on the sample X's, varies from replication to replication. Indeed, an estimator is a random variable over replications, and the estimator has a probability distribution over an infinity of replications, just as X itself does.

The probability distribution for the estimator has both an expectation and a variance. It is desirable that the variance be as small as possible. The smaller the variance, the greater the *precision* of the estimator. Two estimators of **m** may both be unbiased, but if one has high variability, we have little confidence that the estimate for a particular single sampling is close to the population value. If the estimator is precise, however, chances are better that the estimate for a particular replication is close to the parameter. Although both the sample mean and the sample median are unbiased estimators of the normal distribution mean, the sample mean is more precise—its variability around **m** is less over replications than the variability of the sample median. For this reason, the sample mean is preferred to the sample median as an estimator of **m**.

The sample mean $\bar{X}$ from a normal distribution is, itself, normally distributed over replications. The variance of $\bar{X}$, $\sigma_{\bar{X}}^2$, has a simple relationship to σ^2 of the original distribution:

$$\sigma_{\bar{X}}^2 = \frac{1}{I}\sigma^2$$

The *sum* of the X's of a sample has a greater variance over replications than the variance of the X's taken singly; the increment is directly proportional to sample size I.

2.3.4 Interval Estimates

Instead of using a single number to estimate the value of a parameter, we can use two numbers to delimit a range of values within which the parameter is thought to lie. Generally speaking, it is not possible to specify a finite interval, however wide, that contains the parameter with a probability of 1. There will be some probability that the parameter falls outside the interval specified. Given that the interval is judiciously located, the wider the interval, the greater the probability that the parameter falls within that interval.

Since many different intervals can be specified, it is necessary to state on which probability of parameter inclusion the interval estimate is based. Normally, this probability is multiplied by 100 and referred to as a percentage. The interval is then referred to as a P *percent confidence interval*. Typically, a 95% or a 99% confidence interval is reported. Many different intervals are possible, all of which have the same probability of including the parameter. The smallest interval for a specified probability is normally the preferred one.

Psychologists use point estimates much more frequently than confidence intervals. It is usually more convenient to have a single best guess than a range of possibilities. If an experimenter has a confidence interval in addition to the point estimate, however, he has some idea about how accurate the point estimate is. If the confidence interval around the point estimate is small, the error in the point estimate is probably small, but if the interval is large, the error is probably large. The point estimate is often, but not invariably, in the center of the confidence interval.

2.4 HYPOTHESIS TESTING

2.4.1 General Principles

Whereas the statistical theory of estimation concerns making guesses about the value of population parameters on the basis of samples of data from the population, the theory of hypothesis testing concerns the acceptance or rejection of specific hypotheses about those parameters. For example, we may propose the hypothesis that $\mathbf{m} = 7$ or some other constant. To test the hypothesis, we perform various calculations on the data sampled, compare a resultant statistic with tabled values for some theoretical probability distribution, and, on the basis of the comparison, either (1) accept the hypothesis that $\mathbf{m} = 7$, that is, we conclude that the data confirm, or are at least consistent with, the hypothesis, or (2) reject the hypothesis, that is, we conclude that $\mathbf{m} \neq 7$. If we reject our hypothesis that $\mathbf{m} = 7$, the hypothesis testing procedure does not yield a specific alternative, such as $\mathbf{m} = 12$, that we can accept. We simply conclude that $\mathbf{m}$ has some unspecified value other than 7. Of course, we could then proceed to estimate $\mathbf{m}$.

This book is not concerned with testing whether a parameter equals a specific value, but rather whether comparable parameters for two or more populations equal each other. In its simplest form, the problem is to test the hypothesis that two normal distributions have the same mean. In other words, if $\mathbf{m}_1$ and $\mathbf{m}_2$ are the means of populations (distributions) 1 and 2, then our hypothesis would be that $\mathbf{m}_1 = \mathbf{m}_2$. In a typical example, by two "populations" we would mean not two different classes of subjects, but two different distributions of scores for subjects randomly chosen from the same source, who were then exposed to two different experimental treatments. Usually, an experimenter hopes to find differences between "populations" that were created by different experimental conditions. However, he begins by postulating the lack of a difference, that is, by postulating that $\mathbf{m}_1 = \mathbf{m}_2$. A hypothesis so assumed for the purpose of testing its validity is usually called a *null hypothesis*, symbolized by H_N. After performing his computations, the experimenter either accepts this null hypothesis or rejects it. If he rejects H_N, he accepts the *alternative hypothesis*, symbolized by H_A, which merely asserts that a circumstance other than H_N is true.

Hypothesis testing always involves the following:

(1). There is some *test statistic* that can be calculated for the available data (the data of a sample).

(2). By making certain assumptions, including a null hypothesis, mathematical statisticians demonstrate that over innumerable replications, this statistic will have a particular probability distribution, called the *sampling distribution of the statistic.*

(3). The abscissa of the sampling distribution is divided into *acceptance* and *rejection regions*. If the statistic for the sample lies in the acceptance region, the H_N is accepted; if it lies in a rejection region, H_N is rejected (and, perforce, H_A is accepted).

The sampling distribution typically has a central region of higher probabilities, with regions of decreasing probability to either side, comparable to the normal distribution. If H_N were true, we would expect the sample statistic to fall somewhere around the central region of the sampling distribution, for under H_N this is the region where the statistic has greatest probability. If H_N were false, however, another sampling distribution would actually be appropriate for the statistic at hand. Depending on the particular test, failure of H_N would make the sample statistic more likely to fall either in the region of one tail (extremity) only of the null sampling distribution, or in one of the two tails of the null distribution, with neither possibility excluded.

Null sampling distributions typically encompass any value of the test statistic that might occur. Therefore, we cannot reject a null hypothesis merely on the basis of impossibility that the test statistic could occur were it true. Instead, when the test statistic occurs in a region of low probability far from the center of the null distribution, the conclusion that the statistic derived from a distribution based on some H_A is more credible than the conclusion that it derived from a distribution based on some H_N.

The region of the null distribution that determines rejection of the null hypothesis is the region at one or both extremes (tails) that includes $100\alpha\%$ of the area of the distribution. In other words, if the null hypothesis is true, it will be incorrectly rejected with probability α.

The probability α is called the *significance level* of the test. It is usually set at .05, but many experimenters prefer to see rejection at the .01 level before they feel very confident in accepting H_A. Although .05 is, by convention, the significance level used most frequently in hypothesis testing, it is not, in fact, a value for which there is particular mathematical justification. The smaller the α, the less likely is the null hypothesis to be falsely rejected. Of course, we wish to avoid falsely rejecting true null hypotheses. Therefore, we are motivated to set α as small as possible. The problem is that the smaller we make α, the smaller is the region on the abscissa for rejecting H_N, so when H_N is false, the smaller we make α, the more likely we are to err by accepting H_N. It is concern for this error, accepting H_N when it is false, that keeps us from making α smaller than we do.

Therefore, it is obvious that in setting α, we have to balance the chances for making two types of errors: rejecting H_N when it is true (equivalently, accepting H_A when it is false), called a *type 1 error*, and rejecting H_A when it is true (equivalently, accepting H_N when it is false), called a *type 2 error*. The probability of a type 1 error, when H_N is true, is α, the significance level. The probability of a type 2 error when H_A is true is symbolized by β.

Table 2.4-1 The four possible results of accepting or rejecting a null hypothesis

Hypothesis accepted	True hypothesis	
	H_N	H_A
H_N	Correct acceptance of H_N (Given H_N, probability $= 1 - \alpha$)	Type 2 error (Given H_A, probability $= \beta$)
H_A	Type 1 error (Given H_N, probability $= \alpha$)	Correct acceptance of H_A (Given H_A, probability $= 1 - \beta$)

By this analysis, there are two possible true conditions: null hypothesis true (therefore, alternative hypothesis false), and alternative hypothesis true (therefore, null hypothesis false). Similarly, two conclusions are possible: null hypothesis true, and alternative hypothesis true. Thus there are four possible resultant situations, shown in Table 2.4-1. For two situations the conclusion is correct, and for two it is not.

2.4.2 Testing H_N that $\mathbf{m} = W$

As mentioned earlier, this book is not concerned with testing hypotheses of the type that $\mathbf{m} = W$, where W is some constant. Nonetheless, this very simple example is useful for exemplifying the fundamental steps in hypothesis testing. Let us follow the steps invoved in testing H_N that $\mathbf{m} = 7$ for some population.

We observe I scores randomly drawn from the population. We have to consider (1) the test statistic, (2) the sampling distribution of the statistic, and (3) regions of acceptance and rejection for H_N along the sampling distribution abscissa.

(1) **Test statistic.** The test statistic used in this example is:

$$t = \frac{\bar{X} - \mathbf{m}_N}{\dfrac{\hat{\sigma}}{\sqrt{I}}}$$

The symbols are: $\mathbf{m}_N$, the population mean according to H_N; $\hat{\sigma}$, the square root of $\hat{\sigma}^2$, the unbiased estimate of σ^2; and I, the number of scores (X's) sampled from the distribution. As was explained in Section 2.3.3,

$$\hat{\sigma} = \sqrt{\frac{\sum_{i=1}^{I} (X_i - \bar{X})^2}{I - 1}}$$

(2) **Sampling distribution.** The sampling distribution of t is a special and important distribution called the t-*distribution*; hence our designation of the statistic as t. Actually, there is not one but a family of t-distributions. The

exact distribution depends on a quantity called the *degrees of freedom* (abbreviated df). The df for the problem being analyzed is $I - 1$.

The graph of the t-distribution is similar to the normal distribution; it is bell-shaped and symmetric around a mean of zero. As the df becomes larger, the t-distribution becomes more like the standard normal distribution. The approximation is, in fact, quite accurate for an I as small as 30.

If σ were a known quantity, we would use σ instead of $\hat{\sigma}$ in the formula for our test statistic, and the sampling distribution would then be the standard normal distribution regardless of sample size. In practice, however, σ is typically unknown and must be estimated from the same sample data as are used to compute $\bar{X}$.

(3) **Acceptance and rejection regions.** In general, H_N could fail because $\mathbf{m}$ is either larger than or smaller than 7. Therefore, we set up rejection regions in both tails of the null distribution. In other words, we perform what is called a *two-tailed test*.

We want the rejection region to include the proportion α of the null distribution, and we want one-half of this proportion to fall in each tail. (It has been suggested that, in a two-tailed test, equal rejection probabilities need not fall in each tail, but they customarily do.) Therefore, we wish to find that value of the t-distribution above which lies $(1/2)\alpha$ of the area, and that value below which lies $(1/2)\alpha$ of the area. Since the t-distribution is symmetrical around 0, we actually need to find only one value or the other. Let t_c be the value above which lies $(1/2)\alpha$ of the distribution, and let c stand for *critical value*. Then the lower rejection region has an upper limit at $-t_c$.

Table A.4 of the Appendix gives percentiles of t-distributions defined by various df's. Percentiles, symbolized by $t_P(\text{df})$, are given for percentile scores of 90, 95, 97.5, 99, 99.5, and 99.95. The entries in each row refer to the same distribution (the same df). The information on a particular t-distribution given by Table A.4 is, then, much more abbreviated than the information given on the normal distribution. More complete tables for a t-distribution are available, but the information provided in this table is sufficient for significance testing. Let us see how Table A.4 is used in testing the hypothesis that $\mathbf{m} = 7$.

Suppose α is .05, so $(1/2)\alpha$ is .025. Since .025, or 2.5%, of the t-distribution lies above t_c, P must equal $100(1 - (1/2)\alpha) = 97.5$. If $I = 15$, df $= 14$; and from Table A.4 we find that

$$t_c = t_P(\text{df}) = t_{97.5}(14) = 2.14$$

Therefore,

$$\text{Acceptance region: } -2.14 \leq t \leq 2.14$$
$$\text{Rejection region: } t < -2.14, \text{ and } t > 2.14$$

Note that we include the boundary values, ± 2.14, in the acceptance region.

It makes little difference whether these points are said to be in the acceptance or rejection region, but some convention must be agreed on.

If the t-statistic calculated for our sample of I scores falls between -2.14 and $+2.14$, we accept the null hypothesis H_N that $\mathbf{m} = 7$. If the t-statistic lies above $+2.14$ or below -2.14, we reject H_N and conclude that $\mathbf{m}$ for the population from which we sampled is a value other than 7. For example, if t were $-.62$, we would accept H_N. If t were 3.94, we would reject H_N.

2.4.3 Testing H_N that $\mathbf{m}_1 = \mathbf{m}_2$

Essentially the same approach is required here as in Section 2.4.2. We have a sample statistic, we have a sampling distribution of the statistic that we believe would describe variations in our sample statistic if we replicated our sampling many times, and we have acceptance and rejection regions along the sampling distribution.

(1) **Test statistic.** In this example also, the test statistic is distributed as t:

$$t = \frac{\bar{X}_1 - \bar{X}_2}{\sqrt{\hat{\sigma}^2\left(\frac{1}{I_1} + \frac{1}{I_2}\right)}}$$

The subscripts 1 and 2 specify the score distribution. As previously, we assume that the variance, which we understand to be common to the two distributions, is unknown. We estimate this common variance, $\hat{\sigma}^2$, as the weighted average of the estimated variances computed from samples 1 and 2:

$$\hat{\sigma}^2 = \frac{(I_1 - 1)\hat{\sigma}_1^2 + (I_2 - 1)\hat{\sigma}_2^2}{(I_1 - 1) + (I_2 - 1)}$$

To compute $\hat{\sigma}_1^2$ and $\hat{\sigma}_2^2$, we use the formula

$$\hat{\sigma}^2 = \frac{\sum_{i=1}^{I}(X_i - \bar{X})^2}{I - 1}$$

separately for each sample.

(2) **Sampling distribution.** As we have already noted, the sampling distribution is t, just as it was for testing H_N that $\mathbf{m} = W$. The df is $I_1 + I_2 - 2$. Since $I_1 + I_2$ equals the total number of scores in both samples, df is determined in the same manner as in Section 2.4.2, except that we subtract 2 instead of 1. The difference in method exists because previously, to calculate the variance estimate, we had first to estimate $\bar{X}$ from the data whose variance we sought; here, we have to calculate $\bar{X}$ for each distribution. The concept of df will be discussed again in Section 6.2. Here it can be said that whenever we estimate an independent parameter, that is, one that cannot be calculated from other estimates, we use up 1 df.

(3) **Acceptance and rejection regions.** The approach here is precisely as in Section 2.4.2. If H_N were false, could it fail in either direction? In other

words, could we have either $\mathbf{m}_1 > \mathbf{m}_2$ or $\mathbf{m}_2 > \mathbf{m}_1$? Unless we have very good grounds indeed for assuming the contrary, we should answer "yes," and do a two-tailed test. If α is .05, then $t_c = t_{97.5}(I_1 + I_2 - 2)$, and the region of acceptance is $-t_c \leq t \leq t_c$. The regions of rejection occur above $+t_c$ and below $-t_c$.

2.4.4 Testing H_N that $\sigma_1^2 = \sigma_2^2$

In testing H_N that $\mathbf{m}_1 = \mathbf{m}_2$, we assumed that $\sigma_1^2 = \sigma_2^2$. Actually, this assumption can be tested with the same sample data with which we tested H_N that $\mathbf{m}_1 = \mathbf{m}_2$.

(1) **Test statistic.** The test statistic for testing H_N that $\sigma_1 = \sigma_2$ is

$$F = \frac{\hat{\sigma}_1^2}{\hat{\sigma}_2^2}$$

(The formula for computing $\hat{\sigma}_1^2$ and $\hat{\sigma}_2^2$ was described in Section 2.4.3.) We call this statistic F because, assuming that we have independent random samples of I_1 and I_2 scores from two normal distributions, the statistic would be distributed over innumerable replications of sampling as the F-distribution.

(2) **Sampling distribution.** The F-distribution is extremely important, since it is used for significance testing for all the experimental designs that are considered in this book. Like the t- and normal distributions, the F-distribution has a central peak and the distribution diminishes monotonically to both sides. Unlike the t- and normal distributions, however, the F-distribution is not symmetrical around its peak. Whereas the upper tail extends indefinitely with an ever-decreasing probability density, the lower tail ends abruptly at a value of 0. It cannot go below 0 into negative values because a variance estimate, being based on the sum of squared differences, cannot be negative. Therefore, neither can the ratio of two variance estimates be negative.

We only have to table one normal distribution—the standard one. However, we have to table different t-distributions for different df's. Similarly, there are many different F-distributions, but they differ according to two different df's. An F-distribution has a numerator df and a denominator df, symbolized as df_1 and df_2, respectively. There is a separate F-distribution for each combination of df_1 and df_2. Table A.5 in the Appendix gives percentiles of F-distributions defined by different df_1's and df_2's. Percentiles, which are symbolized $F_P(\mathrm{df}_1, \mathrm{df}_2)$, are given for percentile scores of 75, 90, 95, 99, and 99.9. As with the t-distributions, when the df's become large the percentile points of F-distributions change very slowly, so that F-distributions need not be tabled for each possible combination of df's. The table distribution with df's that most nearly approximate those for the available test statistic is the one that normally is used. Interpolation in the F-tables is explained in Section 6.5.7.

(3) **Acceptance and rejection regions.** To test H_N that $\sigma_1^2 = \sigma_2^2$, a two-tailed significance test is required, since, a priori, if H_N is false, the reason could be either that $\sigma_1^2 < \sigma_2^2$, which would tend to produce an F-statistic in the lower tail, or that $\sigma_2^2 < \sigma_1^2$, which would tend to produce an F-statistic in the upper tail.

If $\alpha = .05$, the rejection region would consist of $F > F_{97.5}(\mathrm{df}_1, \mathrm{df}_2)$ and $F < F_{2.5}(\mathrm{df}_1, \mathrm{df}_2)$. It is not possible to find $F_{2.5}(\mathrm{df}_1, \mathrm{df}_2)$ in most F-distribution tables because only percentile scores above 50 are listed. In contrast to the t-distribution, the F-distribution is not symmetric around zero, so we cannot set $F_{2.5}$ equal to $-F_{97.5}$. However, we do have the following inverse relation:

$$F_{100(\alpha/2)}(\mathrm{df}_1, \mathrm{df}_2) = \frac{1}{F_{100(1-(\alpha/2))}(\mathrm{df}_2, \mathrm{df}_1)}$$

Note that the order in which the df's are written changes for the two F's.

Such a two-tailed procedure requires that we designate one distribution as "1" and the other as "2" before sampling so that F may end up smaller or larger than 1. An alternative and entirely equivalent procedure is to wait until the sample variances have been estimated and then to use the larger estimate as the numerator, that is, to label the distribution with the larger estimated variance as "1." When such a procedure is used, an extreme value of F must, of course, lie in the upper tail. The critical F, F_c, is $F_{100(1-\alpha)}(\mathrm{df}_1, \mathrm{df}_2)$. We reject H_N only if $F > F_c$. The test is now "one-tailed."

2.5 POWER

The *power* of a statistical test is the probability that H_N will be rejected, given that H_A is true. Let us analyze power considerations for a specific test: for H_N that $\mathbf{m} = 7$. We assume a normally distributed population with unknown σ^2. We also assume that if H_N fails, it fails only in that $\mathbf{m}$ has some value other than 7; the distribution is still assumed to be normal with variance σ^2.

The power of a test is defined only with respect to a specific H_A. Let us assume an H_A that $\mathbf{m} = 4$. Although H_N is that $\mathbf{m} = 7$, and the sampling distribution of our test is based on that assumption, if $\mathbf{m} = 4$ we still have a sample statistic t calculated on the I scores sampled from H_A. We could replicate this sampling innumerable times, and under H_A the sampling distribution would be something different from the one we assumed for H_N. Under H_A, the test statistic would, over innumerable replications, fall into the rejection regions set up under H_N some proportion of such replications. That proportion is different from α; it is the power of the test for H_A. Since the size of the rejection region varies directly with α, the smaller the α, the smaller the power of a test against any specified H_A. Also, the greater the difference between H_A and H_N, the larger the power of the test, for if H_A is far removed from H_N, the test statistic t based on it will tend to fall in one tail or the other of H_N for most replications,

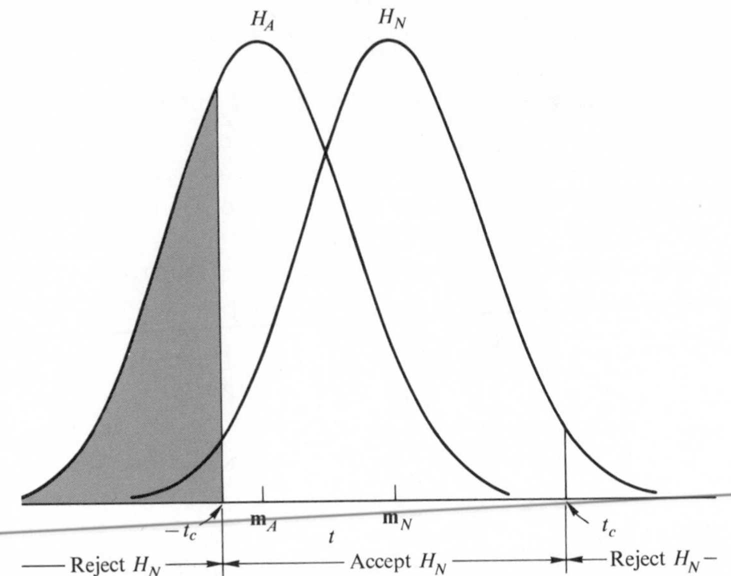

Figure 2.5-1. The t-distribution assuming H_N, with mean $\mathbf{m}_N$, and the t-distribution assuming H_A, with mean $\mathbf{m}_A$. The power of the statistical test against H_A is that proportion of the H_A distribution falling in the "Reject H_N" region. The proportion is shown by the gray area. (A minute proportion of the curve would fall in the upper "reject" region as well.)

Another consideration affecting power is sample size: The larger the sample size, the smaller the variability of the sampling distributions, and the smaller the range of t's falling within the acceptance region.

To summarize, with everything else held constant, the power of a statistical test is larger,

1. the larger the α;
2. the larger the discrepancy between H_A and H_N;
3. the larger the sample size.

Figure 2.5-1 illustrates a sampling distribution of t for H_A and for H_N. The power is the proportion of the H_A distribution falling within the rejection region; this proportion of the H_A distribution is shown by the gray area. The remainder of the H_A distribution falls within the H_N acceptance region. (We ignore the very small proportion of the H_A distribution that falls within the upper rejection region.) The proportion of the H_A distribution falling within the H_N acceptance region is the probability that H_N will be accepted though H_A is true. This is the probability of a type 2 error β. Thus the power of a test equals $1 - \beta$.

It is helpful to imagine how the various considerations affecting the power of a test would alter Figure 2.5-1. If we enlarged α, we would shrink the rejection region, and the gray proportion of the H_A distribution would obviously increase. Similarly, this area would increase if we enlarged the discrepancy between H_A and H_N by making $\mathbf{m}_A$ smaller. Finally, if we increased the sample size I, the variability of both distributions would decrease. The shrinkage of the acceptance region would cause an increase in the proportion of the H_A distribution in the H_N rejection region.

Although our example concerned only testing H_N that $\mathbf{m} = 7$, similar considerations hold for all statistical tests of interest to us. For H_N that $\mathbf{m}_1 = \mathbf{m}_2$, that is, $\mathbf{m}_2 - \mathbf{m}_1 = 0$, we have as H_A that $\mathbf{m}_2 - \mathbf{m}_1$ equals some other constant. Figure 2.5-1 would apply to this situation as well, with $\mathbf{m}_N$ and $\mathbf{m}_A$ equaling differing values of $\mathbf{m}_2 - \mathbf{m}_1$.

With the F-test for variances, we can also increase power by increasing the significance level, the discrepancy between hypotheses, and the sample size, although Figure 2.5-1 does not adequately illustrate the H_N and H_A distributions. First, the F-distribution is not symmetric around its mean. Second, the F-distribution under H_A, known as a *noncentral* F-distribution, is not a simple translation of the (central) F-distribution of H_N; that is, it has a different shape as well as a different mean.

The probability of accepting the null hypothesis when it is true is obvious; it is $1 - \alpha$. The probability of accepting H_A when it is true (that is, the power of a test) is not obvious; it is a much more complicated quantity. Up to now, power has been discussed in relation to a definite H_A. Actually, in an experiment H_A is unknown. Indeed, the experimenter presumably does not even know that H_N is false, or he would not be performing a statistical test based on the assumption of H_N. In considering the power of a test, then, we usually consider a curve giving the power for each possible H_A that could occur. Such power curves are important for deciding among possible test statistics. In the discussion of the various statistical tests, test statistics were given without any explanation of why one particular test statistic is chosen rather than another. However, it is now clear that, of two possible test statistics having corresponding sampling distributions, if the power curve of one is higher than that of the other for all H_A's, then the test statistic with the higher power is the preferred one. This assertion requires that the two tests make equal assumptions about the null and alternative distributions; for example, both might require normality. A test with smaller power is sometimes preferred to one with larger power if the former test requires weaker but more plausible assumptions than the latter. Because nonparametric tests are based on weaker assumptions than the parametric tests that have been discussed, a nonparametric test is frequently used in place of a more powerful parametric one if the experimenter is dubious about the validity for his situation of the assumptions required by the parametric test.

Under the same assumptions, the test statistic t based on $\bar{X}$ is superior to the test statistic based on the median of the X's for testing H_N that $\mathbf{m} = 7$. The variability of the median is greater, and a test statistic based on it would have a sampling distribution with greater variability, and thus with smaller power.

Choosing the test statistic with larger power is very sensible. The probability of correctly accepting H_N is determined by the significance level. Given that significance level, the aim is to maximize the probability of a correct conclusion when H_A is true.

2.6 REMARKS ON THE THEORY AND PRACTICE OF HYPOTHESIS TESTING

Section 2.4 presented the usual standard version of the theory of hypothesis testing. This section discusses some criticisms of that theory and of conventional practices arising from it.

2.6.1 The Null Hypothesis as a Range of Values

One of the more remarkable aspects of the theory is the assumption that, when H_N is one value on a continuous range of alternative hypotheses, H_N could be true. It is absurd to believe that when we are faced with a continuous range of possible values for a parameter, or for a relation between parameters, that one precise point on that range could be accepted as the true value on the basis of sample data. To accept H_N, then, is to say that $\mathbf{m}$ is 7.000 . . . , not 7.001 . . . or 6.998 . . . , and so on. But if a continuum of possibilities is available, we typically have no reason to believe that our exact H_N is any more likely than the infinite number of nearby alternatives.

It would seem that when H_N asserts that some parameter is a point on a continuous range of alternative hypotheses, H_N, taken literally, is false to begin with. There is no need to do a statistical test. We can simply reject H_N, taken literally, and accept H_A—even before we collect any data. Therefore, we can either discontinue using hypothesis-testing procedures of the kind mentioned, or we can construct an alternative rationale for using the same calculational procedures. Many investigators are content to follow the former path, but the majority of them follow the latter. Journal editors and an investigator's colleagues normally expect him to subject his data to hypothesis testing. Therefore, convention, if nothing else, requires that an investigator test hypotheses; so the situation is less awkward if the investigator has a rationale to lend his test results some meaning.

Such an alternative line of reasoning says that H_N does not really postulate a point value for the parameter (or some function of several parameters), but rather a range of values, which, for practical or theoretical purposes, are equivalent. In other words, we should not conclude that H_A is true unless the true value of the parameter differs from the point value given for H_N by at least some minimal amount. Suppose, for example, that an experiment compared some current procedure with an experimental one to see if an improvement could be effected. If both procedures were approximately equally effective, changing to the experimental one would not be worthwhile. We would test the null hypothesis that $\mathbf{m}_1 = \mathbf{m}_2$, but if rejection of H_N meant a change in procedures, we would want to accept H_N if $\mathbf{m}_1$ and $\mathbf{m}_2$ were reasonably close.

This idea that H_N is a range makes sense, particularly in comparison with the idea that it is a point. However, investigators seldom tell us what the range is in a particular case. Their written reports indicate that they apparently do not give the matter much thought. Rather, such reports imply that whenever H_N is rejected, the parameter tested is considered to be outside the range, and whenever H_N is accepted, the parameter is considered to be within the range, whatever the range is. "Range," then, would appear to be more important as a theoretical rationale than as a working concept.

The idea of an H_N range is important for those interested in the power of a test. Experimenters sometimes state a minimal discrepancy between H_N and H_A that they feel would be noteworthy. Then they calculate the power of their test for that discrepancy to see if the power of the test is reasonably high. There is little point of doing an experiment if the discrepancy being investigated can only be detected with a probability of, say, .10. But most investigators forego power calculations, though it would be an eye-opener for many to discover that the H_A that failed to reach statistical significance had only a slight chance of doing so, even if it were true.

Such computations should be done more often, even if they are not done routinely. Ideally, they should be done before an experiment is performed, but the necessary data on sample variability often are not available ahead of time. Then again, it can be perplexing and dangerous to make judgments about "minimally important discrepancies"; such judgments are at best arbitrary. (Power computations for experimental designs are discussed in Section 8.4.)

2.6.2 The Magical Number .05

As previously mentioned, by convention α is usually set at .05. This number has no special meaning, however, from the point of view of either mathematics or logic. Whether a test statistic just falls short of $t_{97.5}$ or just exceeds it has no more particular importance than whether it just misses or just exceeds $t_{98.72}$, so long as the significance level has not yet been set. Once the significance level is set, of course, it determines the location of the acceptance and rejection regions. Once these regions have been established, whether a test statistic falls just to one side of a boundary or the other is important, since it determines the conclusion.

Statisticians sometimes point out that α and β should be adjusted for different experiments, depending on the relative importance of the two types of possible errors. Suppose rejection of H_N implies a costly change of procedure, which, however, would be minimally worthwhile if H_A were true, but a complete waste otherwise. Then the experimenter would particularly wish to avoid a type 1 error (accepting H_A when H_N is true), so a small α would be appropriate. Making a type 2 error would not be very serious, since, considering the cost of changeover, the benefits gained would be small. Therefore, the increase in β resulting from a decrease in α would not be of much concern.

Consider now another experiment in which rejection of H_N implies acceptance of the effectiveness of an inexpensive drug for treating some ailment, when the standard treatment, used as a control condition in the experiment, is virtually useless. Then the experimenter would wish the power of the test to be high, that is, β should be low. One way to lower β is to increase α. In such a situation, an α of .10 or .20 would not be objectionable.

Although in principle, statistical theorists do not give blanket prohibitions against the use of such α's, they offer no objective standards for determining what α should be in particular instances. In practice, experimenters seldom reject H_N's using α's $> .05$, though they do find ways of hinting about the possible presence of effects that approach, but do not reach, significance at the .05 level. If an investigator really believes that it is important not to overlook a possibly true H_A, he will normally have to use techniques other than increasing α above .05 for increasing the power of his test, or his colleagues may reject his conclusions.

2.6.3 Multiple Significance Levels

Theoretical discussions of significance testing proceed on the presumption that each test uses a single significance level, chosen previous to testing. Theorists often assert that the degree of credibility in rejecting H_N is the same, no matter where in the rejection region the test statistic lies. In practice, however, we see something quite different. An investigator usually uses multiple significance levels (for example, .05, .01, and .001) for the same test. The acceptance and rejection regions are determined by the largest level, usually .05. If a test statistic falls in the rejection region, it is reported as significant at the smallest significance level possible, of the several being used. In other words, if the test statistic could indicate rejection of H_N for rejection regions based on an α of .01, but not on an α of .001, then it is reported that the particular H_N reached significance at the .01 level. Under such circumstances, an H_N reported as "significant at the .05 level" could not be rejected at the .01

level or, of course, at the .001 level. Otherwise it would have been reported as significant at the lower level. It is quite apparent that investigators, journal editors, and consumers of psychological research have greater confidence in an H_A, the smaller the significance level associated with it, regardless of any theoretical arguments to the contrary.

2.6.4 Estimation versus Hypothesis Testing

It was mentioned previously that some investigators would almost prefer to forego hypothesis testing altogether. What type of analysis would such investigators provide with their data? Some would provide merely descriptive summary statistics, such as means, standard deviations, and correlation coefficients. Others would provide more elaborate estimation data, such as confidence intervals. There is much truth in the oft-voiced criticism that investigators have generally underrated the importance of estimation and overemphasized hypothesis testing. Such criticism particularly bears on an investigator who reports the results of his hypothesis testing but neglects to provide estimates of parameters. It is all very well to reject the H_N that $\mathbf{m}_2 - \mathbf{m}_1 = 0$, but it seems equally important to provide a best guess about the magnitude of $\mathbf{m}_2 - \mathbf{m}_1$. If, however, H_N is accepted, it seems of less importance to report a best guess for $\mathbf{m}_2 - \mathbf{m}_1$, even though that best guess would not generally equal the value stated in H_N.

Granted that the role of estimation in psychological research has often been underrated, consumers of psychological research probably would not be very happy if estimation displaced hypothesis testing altogether. Many experiments, including the multifactor experiments that are discussed in this book, include the test of many H_N's, with each H_N involving many parameters. The consumer of psychological research would have a difficult time, indeed, trying to wade through estimates and confidence intervals for all design parameters. Hypothesis testing, for all its faults, limitations, and weaknesses, can screen out many of these parameters from consideration, so that we may concentrate on what are, in general, the more important parameters.

CHAPTER THREE

Factors and Designs

3.1 INTRODUCTION

As was pointed out in Section 1.3, the first stage of an experimental investigation is conception, in which an investigator considers various factors and criteria he might include in his design. Stage (2) is design, in which the experimenter decides exactly what factors are to be included and how they relate to each other to form a design. Factors and their relations to form designs are the topic of this chapter.

Many different kinds of factors can enter into a design, and they are included in a design for different reasons. The kinds of factors are discussed in Section 3.2. Given that certain factors are to be included, we must specify how these factors are to relate to each other. This book does not attempt to tell the experimenter what factors should be included in his design, how many and what levels of each should be used, or what the criterion should be. The factors should be related to each other, however, in only a few simple ways. The relation of one factor to another means how the levels of that factor are combined with the levels of another factor. Most designs used in psychological research are based on two simple relations, crossing and nesting, which are explained in Sections 3.3 and 3.4. Less frequently used are designs based on the *Latin square*, a relation among three factors. This chapter concerns only designs based on crossing and nesting. Chapters 4 to 6 focus on the score models and data analysis for such designs. Latin square designs are discussed in Chapter 7.

The present chapter explains how to construct an acceptable experimental design and how to symbolize it. An acceptable design, for our purposes, is simply one of the class of designs with which the system used in this book can deal. Since this class includes the large majority of those described in standard texts and used in practice, the restriction is not great. Nonetheless, the author does not wish to suggest that all reasonable experimental designs are included within the system described here.

The method for symbolizing designs explained in this chapter is useful in and of itself as a concise way of describing different designs. It is also the starting point for the derivation of the mathematical model relating the scores to the factor levels used. (Models are the topic of Chapter 4.) The mathematical model, in turn, is required for the derivation of analysis of variance formulas and as an aid to understanding the results of the analysis.

The use of capital letters A and B as factor symbols has already been explained. Capital letters C, D, and H are also used as general purpose factor symbols. Capital letter S is reserved to symbolize the subject factor and E symbolizes the error factor. The use of F as a factor symbol is avoided, since F is generally used to symbolize a distribution and test statistic. When the design has a group factor, the letter G is used to symbolize it, but otherwise G is used as a general purpose factor symbol.

A level of a factor is symbolized by the corresponding small letter with a numerical subscript to indicate the specific level. When a level of a factor is discussed in general terms, a variable subscript is used that is precisely the same as the level symbol (for example, c_c, d_d, s_s, and so on). The subscripts run from 1 to an upper limit symbolized by the same capital letter as that symbolizing the factor.

3.2 THE KINDS OF FACTORS

Different kinds of factors enter into experimental designs, and they are included in designs for different reasons. Before examining the symbolization of designs, it is well to consider these different kinds of factors. Each factor in an experimental design is one of the following kinds: *treatment*, *blocking*, *group*, *trial*, *unit*, or *error*. Furthermore, each factor is either a *fixed* or a *random* factor, and each is *qualitative* or *quantitative*. In addition, some factors are not included in the design by the investigator, but they nonetheless unobtrusively inject themselves into the experiment, possibly distorting the results and conclusions. In this book, these are referred to as *phantom factors*.

3.2.1 Treatment Factors

The main purpose of an experiment is normally to investigate the effects of the design's treatment factors. A treatment factor is characterized by the following two attributes:

1. An investigator could assign any of his experimental subjects to any one of the levels of the factor.
2. The different levels of the factor consist of explicitly distinguishable stimuli or situations in the environment of the experimental unit.

Concerning attribute (1), the experimenter does not normally assign subjects to levels by whimsy, but uses a random process. With treatment factors, however, there is no reason why any particular subject should not be assigned to any level of the factor by such a random process. The requirement of attribute (2) that stimuli or situations be explicitly distinguishable will be more clear subsequent to the discussion of group factors in Section 3.2.4. Attribute (2) does not apply to group factors, although, with some additional qualification, attribute (1) does. Possible treatment factors might be the number of food pellets given as a reinforcement for a bar press, the temperature of the room in which the subject performs, the number of hours a subject is deprived of sleep before performing some task, and the giving or withholding of knowledge of results to a subject firing a gun at a target.

The term *treatment* refers to a combination of levels, one from each treatment factor in the design. A treatment is not a *condition* of the experiment unless the design consists only of treatment factors, since a condition, by definition, requires one level from every design factor for its specification.

Frequently, the experimenter is not interested or is only slightly interested in the effect of a treatment factor included in the design. The factor is included, however, because it might prove to have importance, or because the experimenter wishes to control the effects of unavoidable variations between experimental sessions by arranging such variations systematically into the design.

For example, if time of day were thought to affect the scores, even though the experimenter had no real interest in studying such effects, he might include "time of day" as a factor in the design so that the effects would not enter in a haphazard way. Such factors, whose purpose is to systematize the possible effects of essentially extraneous variables, are often called *control factors*. By definition, the effects of a control factor are of secondary interest, and therefore whether a factor is conceived as a control factor depends on the interest of the investigator, not on any absolute considerations. For example, some investigations are primarily aimed at studying the effects of time of day on performance. Time of day, in such investigations, would not be a control factor. The results relating to the control factor may turn out to be of greater interest than results relating to the substantive treatment factors in an experiment. Then the experimenter may emphasize the control factor in his written description of the experiment, and, indeed, may no longer call it a control factor. The term "control factor" is frequently applied to blocking factors (Section 3.2.3), as well as to treatment factors.

3.2.2 Trial Factors

If the subjects are to be scored separately for each of a succession of trials in some experiment, and if the scores on separate trials are to be kept distinct for the analysis of variance, then the design has a *trial factor*. The change in scores over trials is attributed to the operation of this trial factor, which has the same number of levels as there are trials in the experiment.

An experiment with such a trial factor is, of course, a repeated measurements experiment, but not all repeated measurements experiments involve an explicit trial factor. If the successive scores of a subject are obtained under different levels of a treatment factor or factors, then the score changes are attributed to the treatment factor(s), not to the number of trials. In such experiments, however, there is always the possibility that a trial effect could be operating also, which could produce misleading results (see Section 8.1).

An experiment in learning need not necessarily involve a trial factor or repeated measurements, as these terms are used in this book. Let us differentiate between learning trials and design trials. An investigator may designate a sequence of stimulus presentation and response as a "trial," but the successive responses may not be considered to be separate scores for the analysis of variance. Instead, such learning trials may be repeated until the subject reaches some standard of performance, called a (*learning*) *criterion* (not to be confused with *criterion* as previously defined). The number of such *learning trials* required to reach the criterion might then be considered a "score" for the analysis of variance. Suppose each subject is given only one series of learning trials leading to one score. Then there is only one *design trial*, there is no trial factor in the design, and the design is classified as "single measurement" instead of "repeated measurements."

An experiment in learning need not involve either learning trials or a trial factor. For example, different subjects might be exposed to two methods for teaching the same topic, and then be given the same examination to see which method produced higher scores. This learning experiment involves neither a series of learning trials nor a trial factor.

At first it might be thought that if there were a trial factor with, say, five levels, then each subject would have five scores. Actually, each subject may have some multiple of five scores, the trial number being reset to one when a change occurs in the level of some other factor. For example, a subject may appear in the laboratory daily for five trials a day. The scores on different days are considered to derive under different levels of a "sessions factor," and the first score on each day is said to be under "trial one."

3.2.3 Blocking Factors

There are various bases by which the subjects in an experiment may be conceived to form different groups. In many experimental designs, such groups constitute the levels of a design factor. *Group factors*, on the one hand, relate to groups arbitrarily formed by random assignment of subjects into groups for temporary experimental purposes. (Group factors are discussed in the succeeding subsection.) *Blocking factors*, on the other hand, relate to groups formed on the basis of pre-experimental subject similarity. The information required to form the blocks may be obvious (as with sex—usually), may be obtained by asking the subjects (as with age and occupation), or may be gained from a psychological test (as with IQ). If psychological test scores are not available for all the subjects before experimentation, the subjects must first be tested. The test may not be a standard one, such as a standard IQ test, but rather may be a special purpose test that is similar to the experimental task.

A blocking factor may be included in an experiment because the experimenter wishes to investigate scoring differences among the blocks, but frequently it is included in the design to increase the power of the design to detect treatment effects, not for its own sake. The use of blocking factors to increase power is considered in more detail in Section 8.3.

A blocking factor may be qualitative or quantitative (see Section 3.2.8). The levels (blocks) of a *quantitative blocking factor* are based on contiguous ranges of measurements along some quantitative variable. Subjects are usually assigned to the blocks of a quantitative blocking factor by the *counting off method*. Each subject has a value along the blocking variable. The subjects are rank-ordered according to these values. Then, if there are to be N subjects per block, the N subjects with the highest scores form one block, the N subjects with the next highest scores form another block, and so on. Age and IQ would be possible quantitative blocking factors. A frequently used synonym for a quantitative blocking factor is concomitant variable.

The levels of a *qualitative blocking factor* are based on nonquantitative groupings, such as sex, occupation, the school a subject attends, and his ethnic background. Some authors restrict the use of the terms "blocking factor" and "block" to quantitative factors, but the terms are used here in relation to qualitative factors as well. The term *control factor* may be applied to blocking factors as well as to treatment factors. Blocking factors are discussed further in Section 3.8.2.

3.2.4 Group Factors

As pointed out previously, a group factor is like a blocking factor in that its levels relate to groupings of subjects. However, a group factor differs from a blocking factor in that with the former, subjects are grouped on an arbitrary, random basis (or at least subjects within some block are so grouped), whereas with the latter, subjects are grouped on some natural, pre-experimental basis.

A design in which subjects are randomly formed into "groups" does not necessarily require a group factor. For example, suppose the design consists only of factor A, whose levels form the different treatments of the design. An equal number of subjects is randomly assigned to each of these treatments. The subjects assigned to the same level might be said to form a treatment group. Yet no group factor is included in this design.

In essence, a group factor is included in a design when it is, or might be, needed to account for systematically different scores among different groups assigned to the same treatment. If such differences were ignored, the analysis of variance would be in error. In the preceding example using design factor A, no group factor is required because systematic score differences among groups are attributed to the different levels of factor A operating on these groups. This explanation is sufficient. If the subjects within one level of A were randomized into separate groups, however, systematic score differences among such groups could not be attributed to the level of A, which would be the same for all such groups. A group factor would then be necessary.

Different groups exposed to the same treatment could have systematically different scores because each group is, or might be, exposed to unique influences affecting their scores. The unique influences might come about because the subjects are formed into functioning groups, designed as such, as in social psychology experiments. For example, a design might involve one treatment factor A, whose levels specify, say, different instructions. Subjects are randomly assigned to the levels of A. The subjects within each level of A are randomly formed into groups; there might be 12 subjects in each level, divided into 3 groups of 4 subjects each. Each group is run separately. The members hold a discussion with each other, and then a score is obtained for each subject—perhaps a rating of his satisfaction with the group. The scores for the subjects in a group are affected by unique group influences. One group, for example, may include a particularly talented individual who is primarily responsible for the success of the group. The satisfaction of all group members is thereby increased. Another group may include a particularly disruptive and uncooperative subject, who deleteriously affects every subject in the group. Such unique group influences can be accounted for by postulating a group factor, each level of which is unique to one group.

The inclusion of a group factor may be justified even if the subjects do not form groups on the formal sense previously considered. For example, an experimenter's laboratory may be set up to run four subjects at the same time, but in individual cubicles. Four subjects appear at the same hour and are given their instructions as a group. The four subjects who are run at the same time are exposed to common influences. For example, although the instructions are supposed to be the same at all times, the exact way they are spoken will vary from group to group, and a question may arise in one group but not in another. Also, such conditions as the time of day, the weather, and the experimenter's clothing vary from group to group. Although none of these may affect the scores, a group factor can be included as a precaution.

A group factor is not as likely to be included in experiments with such informal groupings as are used in social psychology experiments. One reason is that, in the latter, group size is an important variable and must be controlled. When subjects are simply run at the same time and in the same place, groups frequently vary in size because of scheduling difficulties or "no-show" subjects who were scheduled. Variable group size is inconsistent with the previously stated requirement for an equal number of scores per condition. A second reason for exclusion of the group factor is that when subjects are run individually, group influences on scores are likely to be of minor importance, even if subjects come at the same time and receive instructions together. Scores collected under such circumstances are normally analyzed as if obtained from a completely randomized design. With a completely randomized design, however, subjects ideally should be run on a completely individual basis; they should come at different times and should receive instructions separately (see Section 8.5 for further discussion of individual and group experiments).

A group factor is like a treatment factor in that an investigator could assign any of his experimental subjects to any one of the groups. (But if arbitrary groups are formed within a block, arbitrary assignments are possible only within the block.) However, a group factor does not fit the second attribute of a treatment factor (Section 3.2.1). Since subjects from a common subject pool are arbitrarily assigned to groups (or to groups within blocks), systematic differences among groups within a treatment cannot be attributed to the subjects in a group. (Such differences among blocks can be attributed to the subjects.) Neither can the systematic differences be attributed to explicitly distinguishable stimuli or situations in the environment, since the treatment is the same for the different groups under consideration. Instead, the group factor serves as a kind of catchall to account for the effects of the host of inevitable differences in the situations that the different groups face.

Capital letter *G* is used to symbolize the group factor when it occurs in a design, but the letter *G* is not reserved for this purpose exclusively; it is also used as a general purpose factor symbol.

3.2.5 Unit Factors

Generally speaking, in experimental design the term *unit* refers to that entity from which we take a score (or several scores in a repeated measurements design). For psychologists, the "units" are usually human or animal experimental subjects, so this discussion will proceed in terms of a subject factor. The unit is sometimes a group of subjects. The "score" for such a "group unit" might be the mean of measures on subjects composing the group.

In repeated measurements experiments, and only in repeated measurements experiments, it can be noted that subjects exposed to the same treatment have systematically different mean scores, even if these subjects belong to the same level of a blocking factor. These differences are too large to be accounted for by the random error components of the score. The differences are attributed instead to the action of a *subject factor* (symbolized by capital letter S), which has one level for each subject in the design.

The subject factor is similar to a blocking factor in that the level to which a subject belongs is inherent in the subject—the level is not arbitrarily assignable by the experimenter. With the blocking factor, however, the experimenter must decide what groups to use, and he must make some observation on each subject in order to categorize him into his proper group. With the subject factor, as with the group factor, however, no such steps are required. Also, whereas the experimenter may be very interested in analyzing and commenting on the effects of a blocking factor, the subject factor, like the group factor, is included only as a formal necessity.

Two subjects whom the experimenter "exposes" to the same treatments have entirely different conditions if S is a design factor, since each has a distinct level of factor S. The experimenter, however, hardly "exposed" the subjects to different levels of S. Each subject arrived with his own level of S, and the experimenter has no hand in the matter.

As previously noted, a *condition* of a design is specified by naming one level from each design factor. Group, trial, and unit factors, as well as treatment factors, are included when enumerating design factors. (The error factor is not included however, as will be explained later.) In Chapter 1 the meaning of the word "condition" in some contexts is necessarily imprecise. (For example, the experimenter was said to expose the subjects to different conditions.) However, now that the different kinds of factors that a design can have have been described, in subsequent chapters a differentiation will be made between treatments and conditions.

A differentiation is also made between a *cell* and a *condition*. A *cell* of a design is specified by naming one level from each design factor except the subject factor (or, more generally, the unit factor), if there is one. In single measurement designs, the cells and the conditions are the same, since there is no subject factor. In repeated measurements experiments, however, a cell includes several conditions. In other published works, the term "cell" is used variously, sometimes meaning "condition" and sometimes meaning "cell" as defined here.

3.2.6 Error Factors

Error is not usually conceived as a factor of an experiment. Indeed, in this book the error factor is not included as a design factor. When "the factors of a design" are referred to, the error factor is not included among them, unless specifically mentioned. When a three-factor experiment is referred to, the error factor is not included among the three. (Unlike many other sources, however, the present book includes all other kinds of factors in the count, not excepting unit, group, category, and trial factors.) Nonetheless, it is sometimes useful to conceive the error component of a score as being caused by an *error factor*, symbolized by capital letter *E*.)

3.2.7 Fixed versus Random Factors

The results reported for any experiment are meant to be more than a record of what was observed at a particular time under particular circumstances. The results are intended to indicate discoveries of lawful regularities that will appear in other, similar experiments. In other words, investigators and consumers of psychological research *generalize* the conclusions of an experiment to situations that differ more or less from the experimental situation.

For our purposes, such generalization can be classified as *statistical* or *subjective*. The statistical theory underlying the analysis of variance provides a kind of objective justification for some of the generalizations—those that this book refers to as statistical. The classification of design factors as random or fixed relates to this theory for statistical generalization. So that we may proceed with the analysis of variance, each design factor must be classified as random or fixed. Such a classification is supplemental to the breakdown given in the preceding six subsections; that is, each of the kinds of factors previously discussed—treatment, trial, blocking, group, unit, and error—must also be categorized as random or fixed.

A factor is classified as *random* under the following circumstances:

1. The experimenter intends that the conclusions from his experiment should apply to a larger population of levels of this factor than those used in the experiment, and he wishes to use the statistical theory of analysis of variance to justify such a generalization.
2. The factor levels used in the experiment are determined by random choice from a very large population of levels to which the generalization applies.
3. It is understood that, in any replication of the experiment, the levels to be used for that factor should be determined by a new random selection.

A factor is classified as *fixed* under the following circumstances:

1. The experimenter may or may not wish to generalize his conclusions to a larger population of levels of this factor than those used in the experi-

ment, but any such generalizations will be subjective and not justified by the statistical analysis of the experimental scores.

2. The factor levels used in the experiment can be determined by any procedure whatsoever.
3. It is understood that, in any replication of the experiment, the levels to be used for that factor should be precisely the same as in the original experiment.

Oftentimes a design factor may not exactly qualify as random or fixed, but a decision must be made one way or the other. The subject factor S is always classified as random, since experimenters intend that their conclusions should apply to some population from which the subjects were selected. In practice, however, the requirement of random selection is generally not met. Indeed, it is normally unclear just what the subject population is, but when college students are used as subjects, the population is likely to be considered "college students in the country." Certainly laboratory experiments never use a random method of selection of college students across the country, and only very rarely do they use a random selection from the one campus.

The error factor E is always classified as random, since, by assumption, the error components of scores are random samples from a probability distribution. In replicating an experiment, each score is conceived to have a new, random error component.

Treatment factors are usually considered to be fixed, but they are sometimes random. For example, one factor in an experiment might have as levels five visual patterns selected at random from a large population of patterns. The experimenter may wish to see if subjects respond differently to different members of this population. If he wishes his conclusions to apply to the population and wishes to justify his generalization with his analysis of variance, the pattern factor must be interpreted as random. If his conclusions are meant to apply only to the specific five patterns used in the experiment, the pattern factor must be interpreted as fixed, and any replication should use the same five patterns.

The trial factor is always fixed. The group factor is always random. Sometimes a blocking factor is fixed and sometimes it is random. When blocks are formed by the counting off method (Section 3.2.3), the blocking factor does not precisely fit the definition for either a fixed or a random factor. In a replication the blocks entail different ranges along the blocking variable, but for a fixed factor the ranges defining the levels should be constant. If the distribution of measures along the blocking variable were known for the subject population of interest, rather than using the counting off method, the experimenter could set up ranges along the blocking variable and include an equal number of subjects within each range. Subjects for an experimental block would then be chosen randomly only from subjects within the range. A replication would then involve exactly the same ranges and the blocking could more reasonably be considered fixed. In actuality, however, the counting off method is used most of the time, since it is much more practical; the blocking factor is considered to be a fixed factor. In general, when an experimenter is in doubt, he probably should consider a factor to be fixed, for the analysis may then be simpler. "Doubts," however, should not lead him to classify factor S as fixed.

Given that a factor is random, the statistical treatment of the scores depends on the number of levels in the population of levels. There obviously must be more levels in the population than the number of levels required by the experiment; otherwise, the levels could hardly be selected randomly—but there can be an infinite number more. Actually, the number of levels for the most common random factor in psychological experiments, S, cannot be specified with any precision. The number of students in the class from which subjects were chosen might be considered to be the appropriate number. In fact, however, an experimenter is loath to take such a small group to be the population to which his results apply.

In practice, statistical analysis in psychology typically proceeds as if the population from which subjects are drawn is very large—strictly speaking, as if it were infinite. Henceforth, the discussion in this book will proceed on the same basis. Indeed, the population for any random factor will be assumed to be "very large." When the number of population levels for a random factor is not much greater than the number of levels used in the experiment, the methods presented in this book are somewhat in error. As a general rule, however, experimenters assume that the number of population levels is very large and use the methods presented here. Experimental analyses that assume a small number of population levels for a random factor are rare.

3.2.8 Quantitative versus Qualitative Factors

Besides being classified as random or fixed, factors can be classified as quantitative or qualitative. A *quantitative factor* is one whose levels can be described by numerical values on some scale. Strength of drug dosage (say, in milligrams), number of learning trials, and sound level in decibels are some examples of possible quantitative factors.

The levels of a *qualitative factor* can be distinguished from each other, but not simply on the basis of a numerical value on some scale. Sex of the subject would be a qualitative factor with two levels. The levels of a "pattern" factor might simply be five different visual patterns that cannot be differentiated adequately by numerical scale values. An experiment might utilize three strains of mice, and "strain" would then be a qualitative factor.

The score models discussed in Chapter 4 and the analysis of variance techniques described in Chapters 5 and 6 apply regardless of which factors are quantitative and which are qualitative. In essence, the model allows us to proceed without considering the descriptive possibilities inherent in the numerical description of quantitative factors. It is as if we interpreted the quantitative factor as a qualitative factor, as, indeed, we may do. We may ignore the mathematical possibilities that inhere in quantitative factors versus the qualitative, and simply consider the levels of the quantitative factors to be different from one another, no more and no less than the levels of qualitative factors are different. The numerical values of the quantitative factors simply serve as convenient labels to distinguish the levels. Initially, then, all factors will be treated in this book as if they were qualitative. Section 9.5 discusses how the numerical values of quantitative factors can give rise to additional analytical possibilities—in particular, the possibility of finding a mathematical function that relates the criterion to the values of a quantitative factor.

3.2.9 Constant and Phantom Factors

Undoubtedly, innumerable potential factors could affect virtually any criterion an investigator may choose to examine, if only to a small degree. Only one or a few of these potential factors are included as design factors in an experiment. Potential factors not included as design factors are divided into *constant factors* and *phantom factors*.

Only one level of a *constant factor* applies throughout the experiment. The room in which the experiment is conducted might constitute such a level. By using only one room, the experimenter assures himself that score differences among treatment groups cannot be attributed to room differences. On the other hand, the experimenter cannot use his statistical analysis to justify the generalization of his results to any other possible levels to the constant factor. Generalization, of course, occurs. The investigator does not restrict his conclusions to experiments carried out in his room. But such generalizations are subjective, not statistical. This is not to say that the experimental room would affect the results of many experiments; but many constant levels are associated with an experiment, and many of them must change when another investigator attempts to duplicate the experiment. Sometimes the change in the constant level may account for a failure to duplicate results. In recent years, for example, there has been much discussion about the effect that an experimenter (that is, the one who runs the subjects) can have on the results. It is now recognized that two experimenters may conduct the same experiment and obtain different results, even though nothing else about the experiments changes. Formerly, this possibility was usually overlooked.

The levels of *phantom factors*, in contrast to those of constant factors, vary from condition to condition in an experiment. In this respect, they are like the design factors. Unlike the design factors, however, they are likely to be overlooked altogether. Innumerable changes exist among conditions, in addition to those described by the design factors. An experiment may be described as testing different teaching methods; but teachers, as well as methods, may change. Instructions are described as constant, but the way they are presented may change over the course of the investigation, and these changes may affect the scores. The danger of phantom factors is that the score changes they bring about may be incorrectly attributed to the design factors. There are two general methods an investigator may use to protect himself against this danger:

1. He may set out to *control* the potential phantom factor.
2. He may *randomize* the presentation of treatments to minimize the threat posed by phantom factors.

There are two general ways of controlling a potential phantom factor:

1. Convert it into a constant factor, if possible.
2. Convert it into a design factor.

In either case a potential phantom factor, once controlled, does not appear in the experiment as a phantom factor. In an earlier section it was mentioned that "time of day" for running a subject can be a phantom factor (though the term "phantom factor" was not used). One way to control the phantom factor is to make time of day a constant factor, that is, to run all subjects at the same hour of the day. Another way is to convert time of day into a "control factor" of the design, say, by planning systematically to run subjects at 10 A.M., 12 noon, 2 P.M., and 4 P.M.

Rather than setting out to control time of day, the investigator might be content to randomize the order of presentation of treatments. Such randomization is necessary anyway as a protection against the innumerable other phantom factors that will be operating. Since randomization is a protection against phantom factors, and since randomization will characterize the experiment in any event, why worry about controlling this or that phantom factor? One reason is that controlling phantom factors increases the power of the design to detect treatment effects. Another is that, when a phantom factor is converted to a control factor, the investigator can learn just how important that factor is in affecting his scores. A disadvantage of converting phantom factors to constant factors is that the generality of the experimental results is narrowed. The results then only apply, strictly speaking, to the constant levels used, instead of to the range of levels the phantom factors would otherwise assume.

3.3 THE CROSSING RELATION

Having considered the kinds of factors that enter into experimental designs, we now examine how these factors are related to each other in designs. This section concerns the crossing relation; Section 3.4 concerns the nesting relation.

3.3.1 Crossing between Two Factors

Two factors in a design, say, A and B, are *completely crossed* (or simply *crossed*) if each level of each factor appears, in some condition, with each level of the other factor. For example, if A has two levels and B has three levels, all of the following six combinations would occur in the design: ab_{11}, ab_{12}, ab_{13}, ab_{21}, ab_{22}, ab_{23}. If A and B are the only design factors, there would be exactly six conditions in the design, namely, those six combinations just listed. If the design has other factors, ab_{ab} by itself would not denote a condition, because, as we have noted, a condition denotation requires that one level from each design factor be specified. (Recall that the error factor is excluded from this requirement.) Let $A \times B$ mean that factors A and B are (completely) crossed. (The symbol $\times$ is called a *cross*.) If $A \times B$, then equivalently $B \times A$; that is, the meaning is the same regardless of the order of writing.

3.3.2 Crossing among Several Factors

The method for design symbolization developed in this book requires that each factor symbol appear once, and that the relations among factors be shown. We therefore have to consider not only relations between pairs of factors, but relations among many factors. Suppose we have a third and final factor C, for which $B \times C$ and $A \times C$. (Equivalently, $C \times B$ and $C \times A$.) We might then include the three factors in one symbolization, $A \times B \times C$. Crossing has previously been defined as a relation between pairs of factors. What then is the meaning of $A \times B \times C$, a relation among three factors? It is not simply a shorthand way of saying that $A \times B$, $B \times C$, and $A \times C$, because to specify $A \times B$, $B \times C$, and $A \times C$ is not to specify how the three factors fit together.

Whenever three or more factors are combined with crosses, it means that all combinations of all the levels occur. For example, $A \times B \times C$ implies that all $A \cdot B \cdot C$ different possible combinations occur. For $A \times B \times C \times D$ we have $A \cdot B \cdot C \cdot D$ different combinations. The Latin square design, discussed in Chapter 7, shows how it is possible to have $A \times B$, $B \times C$, and $A \times C$ while having fewer than $A \cdot B \cdot C$ different combinations.

3.4 THE NESTING RELATION BETWEEN FACTORS

3.4.1 Nesting between Two Factors

Factor B is nested within factor A if each meaningful level of factor B occurs in conjunction with only one level of factor A. The relation is symbolized $B(A)$. Factor B is called the *nested factor* and A is called the *nest factor*. For

$B(A)$, if the combination ab_{23} appeared, then b_3 would not occur with any other a_a level except a_2 in any condition in the design.

We noted that $A \times B$ and $B \times A$ are equivalent ways of saying the same thing, and either form is acceptable. There is no analogous equivalent for $B(A)$. The form $A(B)$ is *not* equivalent, though it is the proper symbolization for "factor A nested within factor B." The form $(A)B$, with the nest factor appearing first, is not legitimate and is never used in this book.

3.4.2 Nested Factor S

The factor that most commonly appears as a nested factor is S, the subject factor. Factor S frequently appears as a nested factor in repeated measurements designs, though it may appear crossed as well. For example, consider the three-factor design shown in the top part of Table 3.4-1. In such a tabular representation of a design, the conditions are not listed in compact form (such as acs_{122}). Instead, a condition is inferred from the individual factor

Table 3.4-1 Design $S(A) \times C$ using literal labeling (*top*) and initialized labeling (*bottom*)

		c_1	c_2	c_3	c_4
a_1	s_1 s_2 s_3				
a_2	s_4 s_5 s_6				
a_3	s_7 s_8 s_9				

		c_1	c_2	c_3	c_4
a_1	s_1 s_2 s_3				
a_2	s_1 s_2 s_3				
a_3	s_1 s_2 s_3				

levels relating to each cell of the table. Factor S is nested within A, since each level of S (for example, s_5) appears with only one level of A, namely, a_2. However, S is crossed with C, and so is A. The design symbolization is $S(A) \times C$.

The tabular representation of $S(A) \times C$ in the top part of Table 3.4-1 shows that there are nine subjects, $s_1, s_2, \ldots, s_9$, three at each level of A. Actually, it is more convenient, for reasons that will later become clear, to begin the indexing of the subject levels anew for each level of the nest factor A. Therefore, the tabular representation and labeling of levels for $S(A) \times C$ would normally be as shown in the bottom part of Table 3.4-1. This alternate labeling method has advantages, but can lead to ambiguities.

At first, this part of the table would appear to represent not $S(A) \times C$, but $S \times A \times C$, for "s_1" occurs with every combination ac_{ac}. But the s_1 subjects under a_1, a_2, and a_3 are unrelated; they are actually three arbitrarily chosen subjects of a total of nine subjects. It is in recognition of the confusion that the common level symbol could create that the nesting relation is defined in this book in terms of the "meaningful" levels of the nested factor. The s_1's under a_1, a_2, and a_3 are meaningfully different, in spite of the common level symbol. Therefore, S is not crossed with A; it is nested within it.

The design of the bottom part of Table 3.4-1 could represent $S \times A \times C$ if s_1 were the exact same subject under a_2 and a_3 as under a_1 (and the same statement applied to s_2 and s_3). However, when the diagram represents $S(A) \times C$, each of nine subjects in the experiment would be exposed to four treatments and would have four scores. If the bottom part of Table 3.4-1 represented $S \times A \times C$, each of the three subjects in the experiment would be exposed to twelve treatments and would have twelve scores.

The labeling of the top part of Table 3.4-1, in which each distinct subject has a distinct level symbol, is called *literal labeling*. The labeling of the bottom part of Table 3.4-1, in which the numbering of subjects begins anew within each nest factor level, is called *initialized labeling*.

We are already familiar with the number of levels of a factor. For a nested factor we can distinguish between the number of literal levels and the number of nominal levels. The number of *literal levels* is the number of meaningfully different levels. For design $S(A) \times C$ of Table 3.4-1, the number of literal levels is nine, there being nine distinct subjects. The number of *nominal levels* is the upper limit of the subscript when initialized labeling is used, that is, the number of levels within a nest. For nested factors, the factor symbol is used to represent the number of nominal levels, not the number of literal levels. In other words, for $S(A) \times C$ of Table 3.4-1, we let $S = 3$; we do not let $S = 9$. When we wish to indicate the number of literal levels, we use an L subscript; for example, we write $S_L = 9$.

3.4.3 Nesting among Several Factors

Crossing was defined as a relation between two factors, but then it was shown how the cross could be used to symbolize designs with more than two factors. Nesting can also relate more than two factors; for example, we might have $S(A(B))$. This designation means that $S(A)$ and $A(B)$. But $S(A(B))$ by itself does not symbolize a design (see Section 3.4.4).

It necessarily follows that if $S(A)$ and $A(B)$, then $S(B)$; that is, the nesting relation is *transitive*. On the other hand, the crossing relation is not transitive, since $A \times B$ and $B \times C$ allows, but does not ensure, that $A \times C$. For example, in design $S(A) \times C$, shown in Table 3.4-1, $S \times C$ and $C \times A$, but S and A are not crossed. However, in $S \times A \times C$ (an alternative possible interpretation for the bottom part of Table 3.4-1), $S \times C$ and $C \times A$ are consistent with $S \times A$.

Table 3.4-2 illustrates design $S(A(B)) \times C$, using initialized labeling for S and A. Factors S, A, and B are each crossed with C. With a literal interpretation of all factor level labels, the design illustrated would be interpreted as $S \times A \times B \times C$. However, eighteen different subjects are used. Each subject receives four scores, one at each of the four C levels, and each subject

Table 3.4-2 Design $S(A(B)) \times C$ using initialized labeling for S and A

			c_1	c_2	c_3	c_4
b_1	a_1	s_1				
		s_2				
		s_3				
	a_2	s_1				
		s_2				
		s_3				
b_2	a_1	s_1				
		s_2				
		s_3				
	a_2	s_1				
		s_2				
		s_3				
b_3	a_1	s_1				
		s_2				
		s_3				
	a_2	s_1				
		s_2				
		s_3				

appears at only one level of A and at one level of B. Likewise, the a_1's and a_2's are different; that is, there are really six levels of A, but labeling is begun with a_1 within each B level. Factor A might be schools, and factor B might be cities. Three subjects (students) are chosen from each of two schools in each of three cities. We assume that the two schools were chosen arbitrarily from each city, and that the labeling of the particular school by a_1 under each B level is arbitrary. Under these circumstances, the proper symbolization for Table 3.4-2 is $S(A(B)) \times C$.

3.4.4 Crossing in Nested Designs

The student has perhaps wondered why all the examples of designs with nesting have also included crossing. The reason is that the most common nesting relation has S as a nested factor, but there is no design that is symbolized by $S(A)$ alone. If factor C of $S(A) \times C$ (Table 3.4-1) were eliminated so that only one score were obtained from each subject, the resulting design would be simply the one-factor design A, not a two-factor design $S(A)$.

Recall that in Section 3.2 it was said that an S factor is necessary to account for systematic differences among subjects in repeated measurements designs. Without C, $S(A) \times C$ is not a repeated measurements design. Each subject has only one score, observed under some single treatment a_a. Differences among the scores of subjects that are exposed to the same treatment are attributed to the error term of the score model. Further explanation of factor S will be postponed until Chapter 4, which discusses the score model.

To repeat, by eliminating C in $S(A) \times C$, we end up not with $S(A)$, but with a one-factor design symbolized as A. What would be the result of eliminating C in $S(A(B)) \times C$ (Table 3.4-2)? Again, there would be no repeated measurements, so factor S would be eliminated. We would go from the four-factor design to the two-factor design $A(B)$. It is possible, then, to have designs involving only nesting that do not include factor S (or, more generally, that do not include a unit factor).

3.4.5 Error as a Nested Factor

We can think of the error factor E as a factor nested within all other factors in the design. A specific, "meaningful" level of the error factor appears only once in an experiment, since all the error terms for the different scores are conceived to be entirely independent and unrelated to each other. If a level of the error factor appears only once in an experiment, it must of necessity appear only in conjunction with one level of each of the other design factors, and therefore, by definition, must be nested within each of the other design factors.

Normally, however, E is omitted when symbolizing designs, just as it is omitted when enumerating and listing the design factors. If we wished to include it, however, the design would be symbolized $E(Z)$, where Z is the design symbolization excluding E. (More generally, Z is used to designate a factor symbol or some legitimate combination of them using nests and crosses.)

3.5 COMPACT DESIGNS

In this section some concepts are developed for working with design symbolizations and a class of designs is specified that is covered by the symbolization introduced thus far. In this text this class is referred to as *compact designs* because such designs can be symbolized so concisely. The symbolization for a compact design includes exactly one occurrence of each design factor. First we examine the structure of this symbolization, and then we consider number of conditions and subjects that this symbolization must represent.

3.5.1 Structure

A compact design symbolization has either 0, 1, 2, 3, or some larger integral number of crosses that exist within no nests. These are called *first-order crosses*. The symbolizations joined by the first-order crosses are called *first-order modules*. If there are n first-order crosses, then there are $n + 1$ first-order modules. There can be only two kinds of first-order modules, of the types A and $B(Z)$, where Z stands for an acceptable symbolization composed of one or more factors. These are called *simple modules* and *nested modules*, respectively. Each nested first-order module, in turn, contains one or more second-order modules. If a cross (or a module or a factor) is enclosed within n sets of parentheses (nests) in the design symbolization, it is a cross (or a module or factor) or order $n + 1$. The terms *set of crosses*, *set of modules*, and *set of factors*, refer, respectively, to those crosses, modules, and factors enclosed by precisely the same nests.

Letter Z might symbolize a single factor, an entire design composed of many factors, or a part of a multifactor design including several factors. If Z contains several factors, it is called a *complex factor*. We can speak of the number of levels of a complex factor, just as we can speak of the number of levels of a *simple factor* such as A or B. In general, we can substitute Z for a simple factor in a design symbolization, and still have an acceptable symbolization representing a compact design, except that we cannot, in general, replace the nested factor of a nested module by an arbitrary Z. For example, we cannot replace A of module $A(Z)$ by Z', where Z' is $C \times D$.

Let us see how the breakdown into modules applies to the following design symbolization:

$$S(A \times B(C)) \times D(G) \times H$$

There are three crosses. Two of them—the last two—are first-order crosses.

There are, then, three first-order modules: $S(A \times B(C))$, $D(G)$, and H. The first two are nested modules; the third is a simple module.

There are two sets of second-order modules. The first set has two second-order modules, A and $B(C)$, separated, of course, by a second-order cross; the second set has only a simple module, G. There is only one set of third-order modules, and it has only a single member, C. Naturally, with only one third-order module there are no third-order crosses.

Note that a breakdown of a design symbolization from lower- to higher-order modules always ends with simple factors. Each set of modules must consist of $n + 1$ modules, each of form A or $B(Z)$, joined by n crosses. By definition, any "compact" design symbolization has the structure that has been described in this section.

3.5.2 Number of Conditions and Subjects

For a nested factor such as $S(A)$, we have distinguished between the number of literal levels and the number of nominal levels. We also speak of modules as having levels. A nested module has the same number of literal levels as its nested factor. Of course, the number of levels of a simple-factor module requires no additional explanation. Now a further requirement can be stated for compact designs and their symbolizations. By definition, for compact designs, crossing is *exhaustive* for each set of modules of any order; that is, the design must include all possible combinations of the levels of that set of modules (literal levels for nested modules). The number of such combinations for first-order modules equals the number of conditions in the experiment.

For a nested first-order module, such as $S(Z)$, the number of literal levels for S, S_L, is obtained as by multiplying the number of nominal levels S by the number of combinations for Z, which are called the number of literal levels for Z. For example, a design symbolized $S(A \times B \times C) \times D$ is presumed to be compact by its structure, but a further requirement is that S have $A \cdot B \cdot C \cdot S$ literal levels; that is, that there be $A \cdot B \cdot C \cdot S$ different subjects.

It is useful to be able to determine the number of conditions, scores, and subjects for a symbolized design. These numbers can be checked against the numbers determined from a tabular representation to see if the same designs are represented. Also, it is important to know early in the experiment the number of subjects a design would require, since the number of available subjects is usually limited. In repeated measurements designs, it is important to keep the number of scores per subject in mind during the design stage, since there are limits on a subject's endurance and available time.

In a single measurement experiment, the number of subjects (or scores) equals the number of conditions (cells) times E, the number of subjects per condition. In a repeated measurements experiment, the number of conditions equals the number of scores. If the repeated measurements design is symbolized $S(Z) \times Z'$, then the number of subjects equals the number of literal levels for S, which is the number of levels for Z times S, the number of subjects per level. If the number of subjects per level differs depending on the level, then the total number of subjects must be determined by adding the number of subjects in the different levels. The number of different conditions for a single subject, that is, the number of scores per subject, is determined by eliminating $S(Z)$ from the symbolization, and then determining the number of conditions for the design Z' remaining.

If a repeated measurements design is symbolized by $S \times Z$, then the number of subjects equals S, the number of levels for factor S. The number of scores per subject (conditions per subject) equals the number of conditions in the design symbolized by Z.

Let us determine the number of conditions and subjects for several designs. For a compact design involving only crossing, such as $A \times B \times C$, the number of conditions is simply the product of the number of levels for the factors. If $A = 2$, $B = 4$ and $C = 3$, then the number of conditions equals $2 \cdot 4 \cdot 3 = 24$. Presuming that this is a single measurement design, the number of cells also equals 24. The number of subjects, which equals the number of scores in this example, is $24 \cdot E$, where E is the number of subjects per cell.

Next consider the repeated measurements design $S(A) \times B$. Suppose $A = 3$, $B = 2$, and $S = 4$. The number of conditions equals $S_L \cdot B$, where S_L is the number of literal levels for factor S. Since S_L equals the number of levels of A times the number of subjects per level S, $S_L = 3 \cdot 4 = 12$, which is also the number of different subjects required for the design. The number of conditions, then, is $12 \cdot 2 = 24$. The number of conditions per subject (scores per subject) is 2, since if $S(A)$ is removed from the design symbolization, only B remains.

Now consider a more complicated design of the same type: $S(G(A)) \times B \times C$, where $A = 2$, $B = 3$, $C = 3$, $G = 4$, and $S = 3$. The total number of conditions equals $S_L \cdot B \cdot C$. To find S_L, we need first to find G_L, which equals $A \cdot G = 2 \cdot 4 = 8$. Then the total number of subjects $S_L = 8 \cdot 3 = 24$, and the number of conditions equals $24 \cdot 3 \cdot 3 = 216$. The number of conditions (scores) per subject equals $B \cdot C = 3 \cdot 3 = 9$.

For design $S \times A \times B$, where $A = 3$, $B = 4$, and $S = 6$, the number of subjects equals 6 and the number of conditions (scores) equals 72. The number of scores per subject equals $A \cdot B = 12$.

3.6 THE TREE DIAGRAM OF A DESIGN

The symbolization for a compact design can be shown in an upside-down *tree diagram*, with the branches heading downward. The tree diagram for $S(A \times B(C)) \times D(G) \times H$ is:

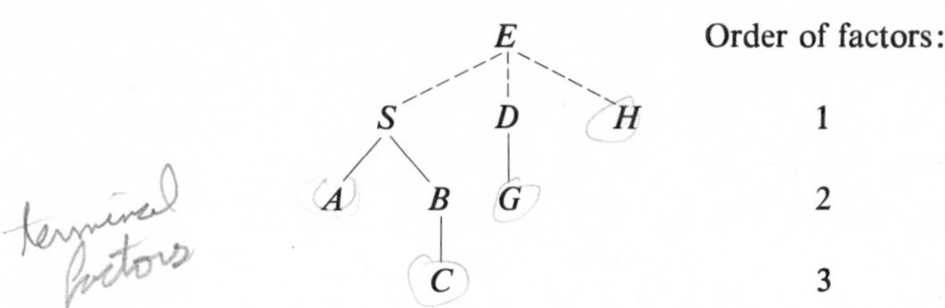

The usual symbolization, printed on one line, is sometimes called the *linear design symbolization* in order to distinguish it from the tree diagram.

Although factor E is included in this diagram so that the branching relation holds throughout, factor E is otherwise ignored. Three branches lead off from E, one for each first-order module. These branches are dashed as a reminder that E is not actually considered part of the symbolization; the design is not symbolized by $E(Z)$, nor is the nest of $E(Z)$ counted in ascertaining the order of modules.

A branch leading down to a simple factor ends there. A simple factor at the lower end of a branch is called a *terminal factor*. Terminal factors are simply those that are not nested within any other design factors, though other factors may be nested within them. In the preceding design, only factors A, C, G, and H are terminal factors. If a branch leads to a nested factor, there are additional branches extending from it—one for each module within the nest.

The relation between two factors can easily be determined from the tree diagram. If two factors are related by nesting, we can trace along the tree diagram from one factor to the other, going only up or only down branches but never reversing from up to down. A factor that is higher on the branch is nested within any factor below it along the branch. Any two factors not located along the same branch are crossed. Thus, from the tree diagram it is apparent that S is nested within A, B, and C; B is nested within C; D is nested with G. All other pairs of factors (excluding E) are crossed.

When we write the score model for the design, the tree diagram will simplify the task (see Section 4.10). The relations between all pairs of factors can be observed quite readily from the linear design symbolization as well as from the tree diagram. The tree diagram presents the organization in a somewhat clearer fashion, however, so the relations are somewhat easier to observe. Similarly, the score model can be written from the design symbolization, but the tree diagram simplifies the task. Given a tree diagram, the student should be able to write the design symbolization.

3.7 THE FACTOR RELATION TABLE

If we know only the pairwise relations between all factors in a design, we can write the design symbolization quite easily. We assume that the design is, indeed, from the class of designs to which we have restricted ourselves. If it is not, that fact will probably become apparent when the procedure described in this section fails in some way. Or, once a design symbolization has been derived, we may realize that the design symbolization does not conform to the tabular or other basic description of the design that we started with. We may, for example, check to see if the number of conditions implied by the design symbolization corresponds to the number in our plan, if we have made such a detailed plan.

We begin by constructing a *factor relation table*, which gives the relations—nesting or crossing—between all pairs of factors in the design. All factors in a design are listed along the top and along the side, as shown in Table 3.7-1. If two factors are related by crossing, a cross is placed in the appropriate location. Nesting relations must be entered with factor symbols to specify which factor is nested. The entries along the positive diagonal of the table are dashed out, since these concern only a single factor, not a pair. The entries above the diagonal are redundant with those below it, but both are included. From the factor relation table we can derive the design symbolization. We begin by extracting first-order simple modules. A first-order simple module has only ×'s (plus a dash) in its column. From Table 3.7-1 we see that *H* is the only such factor. All other factors are part of one or more first-order nested modules.

The column entries for a first-order nested factor show that factor nested in another factor or factors, while having no other factors nested in it. From Table 3.7-1, observe that, although factor *B* is nested in *C*, it is also a nest

Table 3.7-1 Factor relation table

	A	*B*	*C*	*D*	*G*	*H*	*S*
A	–	×	×	×	×	×	*S*(*A*)
B	×	–	*B*(*C*)	×	×	×	*S*(*B*)
C	×	*B*(*C*)	–	×	×	×	*S*(*C*)
D	×	×	×	–	*D*(*G*)	×	×
G	×	×	×	*D*(*G*)	–	×	×
H	×	×	×	×	×	–	×
S	*S*(*A*)	*S*(*B*)	*S*(*C*)	×	×	×	–

factor for S, so it fails the test. Only D and S both pass the test. There are, then, exactly three first-order modules, one simple and two nested.

The design symbolization has the form $H \times D(Z_1) \times S(Z_2)$. The factors composing Z_1 are simply all factors in which D is nested, and the factors composing Z_2 are simply all factors in which S is nested. To find the Z_1 factors, we simply examine column D in Table 3.7-1, and we find that Z_1 is composed of a single factor, G. One first-order module, then, is simply $D(G)$.

From the S column of Table 3.7-1 we determine that Z_2 is composed of three factors, A, B, and C. But what is the form of Z_2? To determine this, we imagine that A, B, and C are the only factors in the design. We ignore all rows and columns in Table 3.7-1 that are associated with other factors, and proceed as we did when we began to extract first-order simple modules. Since, in a factor relation table that applies only to factors A, B, and C, the A column would have only crosses, A is a simple module of Z_2. We can also determine that B is the basis of a nested module, $B(C)$. We have thus determined that Z_2 is $A \times B(C)$.

We have now placed every factor of the design. The design symbolization is $H \times D(G) \times S(A \times B(C))$, the same design for which a tree diagram was shown in Section 3.6. (The arrangement of the first-order modules is different, but the arrangement is arbitrary.)

3.8 CLASSES OF DESIGNS

It is convenient to subdivide experimental designs into four classes, which we label I, II, III, and IV. Normally, the experimental unit is the "subject," but as previously mentioned, it may also be something else (for example, a group of subjects). We assume that the units are individual subjects. (The same principles would hold for any kind of experimental unit, however.)

Table 3.8-1 diagrams the four classes. Classes I and II are single measurement designs; only one score has been taken on each unit. Classes III and IV are repeated measurements designs.

Table 3.8-1 Four classes of designs

Class			
I	II	III	IV
Completely randomized designs	Randomized blocks designs (partially randomized designs)	Partially repeated measurements designs	Completely repeated measurements designs
Single measurement designs		Repeated measurements designs	

3.8.1 Class I Designs

In Class I, *completely randomized designs*, each subject is randomly assigned to one of the conditions (cells) of the design, usually with the restriction that an equal number of subjects should appear in each treatment. (Some variations from this restriction are discussed in Sections 5.11 and 9.1.) A priori, any subject could have been assigned to any condition of the design, the one he ended up in being simply a matter of happenstance. A Class I design may include nesting relations as well as crossing. Among the design factors, there must be one or more treatment factors and there may be a group factor.

If the design has but one factor A, a treatment factor, it is sometimes called a *simple randomized design*. The symbolization $A \times B$, where A and B are treatment factors, can represent a two-factor completely randomized design. If the subjects within each level of the single treatment factor A are randomly divided into groups, the design is symbolized $G(A)$. If the subjects within each combination ab_{ab} of $A \times B$ are so divided, the design is symbolized by $G(A \times B)$.

Completely randomized designs are appealing because of their simplicity and because, by using them, an experimenter may avoid making certain dubious assumptions. On the other hand, Class I designs normally require more subjects than other designs for detecting treatment effects; that is, power is normally lower for Class I designs unless many more subjects are used than are used with other designs. For this reason, Class I designs are often inadvisable.

3.8.2 Class II Designs

Class II designs are single measurement designs with a blocking factor. They are called *randomized blocks designs* because a subject is randomly assigned to a condition (cell) only within his characteristic block, usually with the restriction that an equal number of subjects should appear in each condition. The number of subjects in a block sometimes equals the number of treatments, and then the design has only one subject per condition. Randomized blocks designs have also been called *treatments* × *levels*, *treatments* × *blocks*, and *stratified designs*.

With a Class I design, before assignment of subjects to conditions, it is conceivable that any particular subject could be assigned to any cell of the design. In contrast, with a Class II design, before assignment of subjects to cells, we know that a particular subject will be assigned to one of the cells having his block level and cannot possibly be assigned to any other cell of the design. If the Class I designs are called completely randomized, then a parallel designation for Class II designs would be *partially randomized designs*.

As mentioned in Section 3.2.3, a major purpose of blocking is to increase the power of the experiment for detecting treatment effects, although determination of block differences per se may also be of interest to the experimenter. Blocking is an effective means of increasing design power only if the blocking variable correlates with the criterion, that is, if mean scores differ among blocks. Blocks need not be based on a quantitative variable to increase design power, as long as the correlation holds. For example, natural qualitative groupings, such as occupation, race, sex, the school the subject attends, and his place of employment, could also increase design power. However, the purpose of including such groupings in an experiment is more likely to be to study group differences than to increase test power. One reason is that these qualitative blocks would normally not be as effective as a well-chosen quantitative variable for increasing design power.

We have already noted that the blocking of subjects along a quantitative blocking variable can increase design power relative to the completely randomized design. Can we do better yet by adding a second blocking factor that is crossed with the first? This approach is not promising, for several reasons. First, the number of subjects required may increase, since there would be more blocks, and the number of subjects required is at least equal to the number of treatments times the number of blocks. Second, it would be difficult to place subjects along two blocking variables simultaneously so that an equal number of subjects would appear in each category. As an example, suppose that the two factors are IQ and grade-point average in school. First we divide the subjects equally into two IQ blocks, high and low. Then we divide the subjects into two grade-point-average blocks, high and low. Each subject is high or low in IQ and high or low in grade-point average, and on this basis he would fit into one of four categories. We would find, however, that, because of the well-known correlation between IQ and grade-point average, the high-high and low-low categories contain many more subjects than do the other two categories. The situation can be illustrated as follows:

		Grade-point average	
		High	Low
IQ	High	Many	Few
	Low	Few	Many

In view of the way we wish to analyze the data, there is no satisfactory resolution to this problem. All things considered, it is best to use only one blocking variable.

3.8.3 Class III and IV Designs

In contrast to Class I and II designs, Class III and IV designs involve repeated measurements; that is, each subject yields more than one score for the analysis of variance. In Class IV designs, each subject contributes one score to each cell of the design. Recall that a cell is defined as a combination of one level from each design factor except S. Design $S \times A \times B$ is a Class IV design. Each subject has a score for each ab_{ab} combination; that is, each subject is exposed to each ab_{ab} combination. Because of this, we call Class IV designs *completely repeated measurements designs.*

In a Class IV design, factor S is crossed with every other design factor. In Class III designs, however, S is nested within at least one design factor. That one factor may be a treatment, block, or group factor. A subject, by definition, only appears at one level of the factor(s) within which S is nested. Hence a subject, in a Class III design, cannot have a score for each cell. Therefore, Class III designs are only *partially repeated measurements designs.*

The design symbolization for a Class III or IV design always includes symbol S (assuming that the subject is the experimental unit). A Class IV design is symbolized by $S \times Z$; for example, $S \times A$ or $S \times A \times B$. A Class III design is symbolized by $S(Z) \times Z'$; for example, $S(A) \times B$ or $S(A \times B) \times C$. When a blocking factor appears in a Class III design, S must be nested within it, since a subject cannot appear at every block, but only at his one characteristic one. It follows, also, that a blocking factor cannot appear in a Class IV design.

A group factor G may appear in repeated measurements designs as well as in single measurement designs. If a treatment factor B is appended to the single measurement design $G(A)$ in such a way that each subject is scored at each level of B, then the enlarged design is symbolized by $S(G(A)) \times B$. When B is appended, the design becomes a repeated measurements design, so factor S now appears in the symbolization as well as factor B.

A major purpose of the repeated measurements designs is to increase the power of the experiment for detecting treatment effects in comparison with the completely randomized designs. In this respect, repeated measurements designs are like randomized blocks (Class II) designs. The subject is comparable to a "block" of similar subjects, where there is one subject per treatment. Indeed, what this book refers to as Class IV designs are, in some sources, included within the designation "randomized blocks designs," but here the latter designation is restricted to single measurement (Class II) designs.

The effectiveness of a randomized blocks design for increasing design power depends on the extent to which subjects in a block tend to respond similarly on the criterion, relative to all subjects in the experiment. Repeated measurements designs are even more effective than randomized blocks designs for increasing design power, since a single subject is more homogeneous across measurements than the merely "similar" subjects in a randomized blocks design. The repeated measurements design, however, has disadvantages from which the randomized block design is free:

1. Many treatments must be applied to the same subject, which may be too time-consuming or fatiguing for the subject.
2. There is a considerable possibility that *carry-over effects* may occur, that is, that the scores may be affected by preceding treatments as well as by the current treatment. Carry-over effects may lead to misinterpretation of the results.

A Class III design, in comparison with a completely randomized design, has greater power to detect treatment differences only with regard to factors with which a subject is crossed. There is no advantage for detecting effects involving only the subject's nest factor or factors. Therefore, other things being equal, it is better to cross S with the factors whose effects are most important to detect and to nest S within the factor or factors whose effects are of less interest. With a Class IV design, S is crossed with all other design factors, so the power is high throughout. Because of time limitations, fatigue, or other considerations, however, it may be impossible or unwise to require each subject to submit to each treatment.

Class III designs are used quite commonly in learning experiments with a trial factor. A subject receives every trial, and therefore S is crossed with trials. Performance over trials can be compared for groups of subjects having different treatments or belonging to different blocks. Such a design would be symbolized by $S(A) \times B$, where B is the trial factor.

We have noted that the blocking of subjects can increase design power, and that the use of repeated measurements can do the same. Would using both techniques together (that is, combining repeated measurements and blocking) produce yet more powerful tests? No, not for tests already benefiting from the repeated measurements. The subject is, in essence, already matched with himself, and blocking adds nothing further to test power.

As we have noted, however, some tests in a Class III design—tests for effects involving only nest factors—do not have enhanced power. Blocking could enhance the power for such tests. For example, to design $S(A) \times B$, having treatment factors A and B, we might add a blocking factor C, yielding $S(A \times C) \times B$. Such a change could enhance test power for factor A, but there would be no improvement for factor B.

We might want to take a Class IV design and block the subjects according to some quantitative or qualitative blocking factor, turning it into a Class III design, with S nested within the blocking factor. Such a design would allow us to investigate scoring differences among the groups, but it would not improve test power for detecting treatment effects.

3.8.4 Miscellaneous Terminology

Designs with nesting. Sometimes designs involving nesting are called *hierarchical* (or *hierarchic*) designs. If crossing appears as well, the design is sometimes called *partially hierarchical.* Designs having both nesting and crossing are also referred to as *split-plot* or *mixed* designs. Such designs may have single or repeated measurements.

The definitions of these terms sometimes depend on the kinds of factors included in the design. For example, design $S(A) \times B$ might be called split-plot if both A and B were treatment factors, but (partially) hierarchical if A were instead a blocking factor.

The preceding terms will not be employed in this book. When the student encounters them elsewhere it would be as well to ignore them; they are used in such variable ways that they seldom enlighten. Instead, he should determine the proper symbolism for the design, the class as defined previously, the kinds of factors involved, and whether they are random or fixed. Then he will understand the design that was used.

Replications and replicates. The term "replication" has been used thus far to mean the repetition of an experiment under identical conditions, insofar as is possible, except that the levels of random factors should be determined anew for each replication by random draws from the same populations. In practice, what is called a replication would usually depart considerably from this ideal. Many experimental details change between replications, and subjects are normally not even derived from the same institution.

In some statistics books, "replication" may have a somewhat different meaning from that stated here. Oftentimes the term refers to a repetition of the basic experimental layout within what we consider one experiment. For example, an investigator may plan to run essentially the same experiment in teaching methods at different schools. Although the experiment is planned as a whole and the data are to be analyzed as a whole, the repetitions at the various schools may be called replications.

"Replication" is sometimes used in the following way also. If we have $A \times B$ with ten subjects per condition, all to be run at the same location under the same arrangements, we can think of the experiment as ten replications of $A \times B$ with one subject per condition; we speak of ten replicates. Likewise, if we have $S \times A \times B$ with ten subjects, each of the ten may be called a replicate, though all are part of one experiment.

CHAPTER FOUR

The Score Model

4.1 INTRODUCTION

Each score of any design is set equal to a sum of theoretical component terms. Any such equation or the entire set of them for a design is called the *score model.* Chapter 4 is devoted to score models.

4.1.1 The Role of Score Models

Almost every component term of a score model relates to some design factor or combination of design factors. It is by means of these models and their components that we interpret the effects of the various design factors on the scores. The conclusions reached from the analysis of variance are, basically, statements about the model terms. Therefore, an understanding of the model is fundamental to an understanding of the results of an experiment. Furthermore, the score model is the starting point for the derivation of the calculational formulas required to perform the analysis of variance. Clearly, then, a student of experimental design should understand and be able to work with score models.

4.1.2 Terminology and Symbolism of Score Models

The terminology for score models can be illustrated using single measurement design $A \times B$, which can be diagrammed as follows, if factor A has 2 levels, factor B has 3 levels, and there are 4 subjects per condition.

	b_1	b_2	b_3
a_1	X_{111} X_{112} X_{113} X_{114}	X_{121} X_{122} X_{123} X_{124}	X_{131} X_{132} X_{133} X_{134}
a_2	X_{211} X_{212} X_{213} X_{214}	X_{221} X_{222} X_{223} X_{224}	X_{231} X_{232} X_{233} X_{234}

The design calls for $A \cdot B \cdot E = 2 \cdot 3 \cdot 4 = 24$ subjects; in a single measurement design each subject has one score, represented by X with appropriate subscripts, so the table has 24 different X's. Each X stands for some number such as 12.0, 8.2, 15.3, or -7.0, which is some subject's score along the criterion.

The score model for design $A \times B$ is

$$X_{abe} = \mathbf{m} + \mathbf{a}_a + \mathbf{b}_b + \mathbf{ab}_{ab} + \mathbf{e}_{e(ab)}$$

A score model is written with a subscripted X to the left of an equals sign and a sum of theoretical terms to the right. The subscripts of X are always in alphabetical order, except that e always comes last. The theoretical terms are unknown and unobservable, and will remain so to the end, although these terms can be estimated. The scores (X's), on the other hand, are not theoretical terms. They are observed quantities, or are derived from direct observations by objective, describable methods. It is from the scores that the theoretical terms are estimated.

The preceding score model (equation) may be interpreted as a single equation or as a set of equations. If we substitute specific values for the subscripts, we get the score model for a specific score in the experiment. For example, we might have

$$X_{231} = \mathbf{m} + \mathbf{a}_2 + \mathbf{b}_3 + \mathbf{ab}_{23} + \mathbf{e}_{1(23)}$$

Suppose A has 2 levels, B has 3 levels, and there are 4 subjects per condition ($E = 4$). Then there would be $A \cdot B \cdot E = 2 \cdot 3 \cdot 4 = 24$ scores from the experiment, and 24 equations like the exemplary one (except that each equation would have a different pattern of subscripts).

There are four kinds of theoretical terms in a score model:

1. *Population mean*, symbolized by $\mathbf{m}$. Each score model contains the term $\mathbf{m}$.
2. *Main term* (or *main effect*), symbolized by a single bold-faced letter (except $\mathbf{e}$) with one or more subscripts. In our example, there are two main terms, $\mathbf{a}_a$ and $\mathbf{b}_b$. When there is more than one subscript, the rest are nest subscripts, and we speak of a *nested main term* (or *nested main effect*).

An example of a nested main term is $\mathbf{s}_{s(a)}$; here subscript s is a *nested subscript* and a is a *nest subscript*. There is a main term (possibly a nested one) for each design factor.

3. *Interaction term* (or *interaction effect*), symbolized by two or more bold-faced letters with two or more subscripts. When there are more subscripts than letters, the excess subscripts appear within parentheses and we have a *nested interaction term*. In our example there is only one interaction term, a *two-factor interaction term*, $\mathbf{ab}_{ab}$. An example of a *nested two-factor interaction term* is $\mathbf{bs}_{bs(a)}$.
4. *Error term*, symbolized either by $\mathbf{e}_{e(\ldots)}$, with a subscript for each design factor within the parentheses (the *subscript nest*), as in the preceding model, or by $\mathbf{e}$ with a subscript for each design factor, but no subscript e. For example, we might have an error term $\mathbf{e}_{abs}$.

We always write the nested subscripts first, and then we write the nest subscripts within the subscript nest; that is, we never see a form such as $\mathbf{as}_{a(b)s}$. Also, although we may have several tiers of nests when we symbolize a design, such as $S(A(B))$, when we write nested terms we use only one set of parentheses at most. In other words, we never see a form such as $\mathbf{as}_{as(b(d))}$.

According to the symbolism employed in this book, the letters of the model terms indicate the factors to which the terms relate. For example, $\mathbf{c}_c$ is the main term corresponding to factor C, and $\mathbf{ab}_{ab}$ is the two-factor interaction term corresponding to factors A and B.

As stated previously, the score model may be interpreted as a single equation or as a set of equations. Normally, the alternative possibilities cause no confusion. Similarly, a model term may be thought of as a single number or as a set of numbers. For example, when we write $\mathbf{a}_a$ we may have in mind a specific, single number, though we do not care which one it is, so we use a variable subscript instead of a specific numerical subscript. It is in such a specific sense that we interpret $\mathbf{a}_a$ when we view the score model as a single equation. On the other hand, when we write $\mathbf{a}_a$, we may refer to all its values for all values of subscript a. The two interpretations of $\mathbf{a}_a$ and other terms can sometimes be confusing. When we wish to emphasize the general interpretation, we speak of a *key term*. When we wish to emphasize the specific interpretation, we speak of a *parameter*. Thus we say that the parameters of key term $\mathbf{a}_a$ are $\mathbf{a}_1, \mathbf{a}_2, \ldots, \mathbf{a}_a, \ldots, \mathbf{a}_A$. When we say that key term $\mathbf{a}_a = 0$, we mean that parameter $\mathbf{a}_a = 0$ for each value of subscript a. Likewise, we sometimes refer to *key equations* or *parameter equations*.

The subscripts on a model term correspond exactly to the subscripts for factor levels or combinations of levels. Just as factor A has A levels $a_1, a_2, \ldots, a_a, \ldots, a_A$, it also has A parameters $\mathbf{a}_1, \mathbf{a}_2, \ldots, \mathbf{a}_a, \ldots, \mathbf{a}_A$. There is a direct relationship between a level, such as a_3, and the corresponding parameter $\mathbf{a}_3$.

Parameter $\mathbf{a}_3$, for example, is a quantity expressing the influence of level a_3 on any score deriving from a condition that includes a_3. Parameter $\mathbf{ab}_{23}$ is a quantity expressing the influence of the combination of levels, a_2 and b_3, on any score deriving from a condition that includes both a_2 and b_3. The presence of $\mathbf{ab}_{23}$ in a score model does not obviate the requirement that $\mathbf{a}_2$ and $\mathbf{b}_3$ be included as separate main terms in the model. An interaction term supplements main terms; it does not replace them. Likewise, higher-order interaction terms supplement lower-order interaction terms and main terms.

4.1.3 Side Conditions

A *side condition* is an equation or set of equations, each specifying a sum for parameters of a key term. (The concept of side condition is entirely distinct from that of "condition," a combination of one level from each design factor.) The side conditions differ depending on whether the various factors are interpreted as fixed or random, and it is through the side conditions that the classification of factors as random or fixed affects the analysis. This section discusses only the side conditions for main terms. Side conditions for interaction terms are discussed in subsequent sections.

If the factor is fixed, the sum of all parameters of a key term applying for a specific experiment equals zero. For factor A fixed, the side condition is

$$\sum_{a}^{A} \mathbf{a}_a = 0$$

or, in dot notation (Section 2.1.7), $\mathbf{a}_. = 0$. In particular, when $A = 2$, that is, when factor A has two levels, the side condition is $\mathbf{a}_1 + \mathbf{a}_2 = 0$. Since $\bar{\mathbf{a}}_.$, the mean of the parameters, equals $\mathbf{a}_./A$, then whenever $\mathbf{a}_. = 0$, likewise $\bar{\mathbf{a}}_. = 0$.

If factor A is random, the sum of all $\mathbf{a}_a$ parameters for the population of all a_a levels equals zero. Recall from Section 3.2.7 that, for a random factor, the factor levels used in an experiment are supposedly drawn at random from a very large population of possible levels. For a random factor, in contrast to a fixed factor, the summation to zero is for all $\mathbf{a}_a$'s of this population of possible levels. Then what is the sum of the $\mathbf{a}_a$'s used in the experiment when A is random? The sum is no particular quantity that we can specify. The sum would vary randomly from replication to replication, since new levels of factor A would be randomly chosen each time. Therefore, the sum of the $\mathbf{a}_a$'s is a *random variable* over replications when A is a random factor. Although we would call our levels a_1, a_2, and so on, for different replications "a_1" would refer to a different level for each replication, as would a_2, a_3, and so on.

Since subscript limit A equals the number of levels in the experiment, not in the population, we write for random factor A that

$$\sum_{a}^{\infty} \mathbf{a}_a = 0$$

Dot notation implies summing and averaging over levels used in the experiment, not over all population levels in the case of random factors; therefore, if A is a random factor, we write

$$\mathbf{a}_{.} \text{ is r.v.,}$$

where r.v. stands for random variable over replications. Naturally, if $\mathbf{a}_{.}$ is r.v., then $\bar{\mathbf{a}}_{.}$ is r.v. also.

With a fixed factor, exactly the same levels would be used over successive replications. Level a_1 is not only labeled the same, but is the same over replications, and likewise for a_2, a_3, and so on. Therefore, the $\mathbf{a}_a$'s are the same, and the sum of the $\mathbf{a}_a$'s for the experiment is zero for each replication, not a random variable.

4.1.4 Estimation of Parameters

Considerable attention is focused on the estimation of model parameters throughout this chapter. Formulas are presented for calculating these estimates from the observed scores. The chapter also explains how an estimate of a model parameter can be equated to a sum of quantities based on various model parameters and averages across them.

Two kinds of formulas are introduced for purposes of calculation. One type allows calculation of estimates from other, simpler estimates that have previously been calculated. Two-factor interaction parameters, for example, are estimated using previously estimated main term and population mean parameters. Using such formulas, we say we estimate by the *method of successive extraction*. An alternative type of estimation equation is based on a sum of various averages across the score subscripts. These averages are called $\bar{X}$ ("X bar") quantities, and this method of estimation is referred to as the *method of* $\bar{X}$*'s*. The dot notation is used to symbolize such averages. For example, if the score is symbolized by X_{abe}, then $\bar{X}_{ab.} = \sum_e X_{abe}/E$, $\bar{X}_{a..} = \sum_{be} X_{abe}/BE$, and so on. The formulas for estimating parameters from $\bar{X}$ quantities can be derived from the formulas for successive extraction. Derivations of this kind are explained later in the chapter, but in Section 4.12 simple rules are given for writing the $\bar{X}$ estimation formulas directly, so that complicated derivations can be circumvented.

This chapter also shows how parameter estimates can be equated to a sum of quantities based on various model parameters and averages across them. For each $\bar{X}$ quantity in an estimation formula, we can also take the mean across the model terms for the X's being averaged. If we then substitute these theoretical equivalents of the $\bar{X}$ quantities in the computational formulas, we can equate each parameter estimate to a sum of model terms and averages thereof. Such formulas are quite useless for computation, of course, since we do not know what the model terms equal. However, the formulas are very important for understanding sums of squares, expected mean squares, and significance testing, the topics of Chapters 5 and 6.

4.1.5 Confounded Terms

Two terms are said to be *confounded* if the respective parameters of the two terms are paired in such a way that one parameter is a component of a score when and only when its mate is also a component. This chapter will be particularly concerned with confounding between error and an interaction term, which always occurs when there is only one score per condition. Each error parameter in any experiment is conceptually distinct; that is to say, a particular parameter appears in but one score model. Its mate, an interaction parameter, likewise appears but once in the same model, when there is only one score per condition. Under such a circumstance, we can calculate no separate estimate for an error parameter, that is, an estimate unaffected by the values of all nonerror parameters. If, however, we assume that the confounded interaction term is zero, then a formula originally derived for estimating the interaction serves instead to compute an error estimate.

In general, confounded terms are not carried as such (that is, with separate symbols) either within the area of experimental design or within the field of science as a whole. As we shall see, however, certain statistical tests require that the confounded interaction term be zero, but other tests do not. If we assume that the interaction term is zero to begin with, we shall be unable to determine for which tests the assumption was necessary and for which tests it was unnecessary. Furthermore, in general, our approach has greater uniformity and simplicity if the terms for a score model do not depend on the number of scores per condition. Therefore, both the confounded interaction term and error term are kept in a model for a design with one score per condition. We omit the *e* subscript for such models and the subscript nest on term $\mathbf{e}$; since there is only one score per condition and hence one error parameter per condition, there is no need for an *e* subscript in the score model, either on X or on $\mathbf{e}$.

4.1.6 Overview of Chapter 4

Sections 4.2 to 4.4 consider the models for the single measurement designs A, $A \times B$, and $A \times B \times C$, respectively. For each model the estimation of model terms from the scores is also discussed. The model for a single-factor design such as A contains only $\mathbf{m}$, a main term $\mathbf{a}_a$, and an error term $\mathbf{e}_{e(a)}$. It has no interaction term. Discussion of the model for design A, then, largely concerns the nature of main and error terms. Section 4.3 discusses design $A \times B$, whose model does have a two-factor interaction term, as well as main and error terms. This section, then, especially concerns the nature and role of a two-factor interaction term. Section 4.4 discusses design $A \times B \times C$, whose model has a three-factor interaction term in addition to the kinds of terms associated with $A \times B$, so this section especially concerns the nature and role of a three-factor interaction term.

The models discussed through Section 4.4 are for single measurement designs involving only crossing relations. Sections 4.5 and 4.6 consider models for repeated measurements designs involving only crossing relations—Class IV designs. As we shall see, the nature of the design model depends basically on the form of the design symbolization, not on the kinds of factors included in the design. By this reasoning, the models for designs $S \times A$ and $A \times B$ would be analogous, differing only in symbolism. Matters are not quite that simple, however, because the number of scores per condition is generally greater than one for single measurement designs, but normally equals one for a repeated measurements design. When there is only one score per condition, the interaction term having subscripts for all design factors is confounded with the error term. The score model for design $S \times A$, then, normally differs from the model for $A \times B$ not only because of a substitution in the symbolism, but because of the absence of subscript e in the former model. However, this difference does not occur if $A \times B$ has only one score per condition, as sometimes is the case, for then the model for $A \times B$ lacks subscript e as well.

Models for designs with nesting include some special features—namely, nested main and interaction terms. Models for nested designs are described and explained in Sections 4.7 to 4.9. In Section 4.2 through Section 4.9, the score models for a variety of common designs are described and their terms are explained, but derivation of the models is not discussed until Section 4.10, where general rules are presented that will enable the student to derive the score model for any compact design, regardless of the number of factors it includes. In Section 4.11 simple rules are given for the derivation of calculational formulas for estimating model parameters.

4.2 DESIGN A

4.2.1 Introduction

First we shall consider the score model for one-factor single measurement design A, which, for $A = 3$ and $E = 5$, can be diagrammed as follows:

a_1	a_2	a_3
X_{11}	X_{21}	X_{31}
X_{12}	X_{22}	X_{32}
X_{13}	X_{23}	X_{33}
X_{14}	X_{24}	X_{34}
X_{15}	X_{25}	X_{35}

In this example, A symbolizes both the design and the single factor of the design. Normally, A would be a treatment factor, so to be specific let us consider it as such. The score model for design A would be the same, however, if A were a blocking factor. (Factor A would be neither a trial factor nor a subject factor, since these appear only in repeated measurements designs. It could possibly be a group factor, but a group factor would very seldom, if ever, appear by itself in a design.)

There are A different levels of factor A, and each constitutes a treatment (as well as a condition and a cell) of the design. An equal number E of subjects are randomly assigned to each of the A treatments, so there are $A \cdot E$ subjects altogether. Since each subject has one score, the design also has $A \cdot E$ scores. Note that the number of subjects in a condition is expressed by E, which is also the number of error parameters in a condition. Each subject in a condition has one score, and that score has one error parameter; hence the correspondence. A different notation is used for repeated measurements designs.

Section 4.2.2 presents the score model for design A. Section 4.2.3 explains the nature of the error term of the model. Section 4.2.4 explains why subject factor S is not included in the design, even though S might be considered to be nested within A. Section 4.2.5 presents the score expectation model. In Section 4.2.6, score expectations are used to calculate the values of model terms. Since the expectations are never known exactly, however, the model terms cannot be calculated exactly. The calculations, nonetheless, add to our understanding of model terms. Furthermore, if the estimates of the score expectations are substituted for the expectations, the same calculations provide estimates of the model terms (Section 4.2.7). Through Section 4.2.7, factor A is assumed to be fixed. Section 4.2.8 considers the differences that would obtain if factor A were random.

4.2.2 The Score Model

The score model for design A is

$$X_{ae} = \mathbf{m} + \mathbf{a}_a + \mathbf{e}_{e(a)}$$

Suppose $A = 2$ and $E = 3$. (For simplicity, atypically small subscript limits are used in the examples.) The score model would then represent $A \cdot E = 2 \cdot 3 = 6$ equations, as follows:

$$\begin{aligned}
X_{11} &= \mathbf{m} + \mathbf{a}_1 + \mathbf{e}_{1(1)} \\
X_{12} &= \mathbf{m} + \mathbf{a}_1 + \mathbf{e}_{2(1)} \\
X_{13} &= \mathbf{m} + \mathbf{a}_1 + \mathbf{e}_{3(1)} \\
X_{21} &= \mathbf{m} + \mathbf{a}_2 + \mathbf{e}_{1(2)} \\
X_{22} &= \mathbf{m} + \mathbf{a}_2 + \mathbf{e}_{2(2)} \\
X_{23} &= \mathbf{m} + \mathbf{a}_2 + \mathbf{e}_{3(2)}
\end{aligned}$$

The model has two key terms, $\mathbf{a}_a$ and $\mathbf{e}_{e(a)}$, and has no interaction term.

4.2.3 The Error Term

Note that the terms summing to give X_{11}, X_{12}, and X_{13} are the same except for the error term. These three scores derive from the same treatment, a_1. Likewise, the scores from a_2 differ only because of the error terms.

Although conditions may influence scores, they do not completely determine them. Typically, scores deriving from exactly the same condition exhibit a spread of values. An adequate score model, then, must allow for such differences in scores. The "error" term has just this function in single measurement designs. It is simply the catchall required to account for such differences. The error term for a single measurement design is quite different from the error term for a repeated measurements design, so the use of "error term" for both is misleading. The error term for a single measurement design can be conceived as having two components:

1. a systematic component that would be constant for a particular subject if the subject were rescored under the same condition, and
2. a variable component that would fluctuate randomly for each such rescoring.

The score model, and in particular the error term, can be better understood if we compare and contrast it with the equation for a subject's score that is used in psychological test theory—an equation the student has probably seen previously:

$$\text{Observed score } X = \text{true score} + \text{error}$$

The observed test score the subject receives is conceived to consist of the true score, which remains the same each time the subject takes the test, plus an error component, which varies randomly across such repeated testings. The difference in scores a subject typically receives on the same test on different occasions is attributed to fluctuations in the error term. (The true score could change over testings as a result of maturational or learning processes, but if the time between testings is short—say, one or a few weeks—such changes would be relatively small for most tests of ability or personality.)

We can break the true score down into two components: *true population mean* plus *true deviation* from population mean for the particular subject. The true population mean is the true mean score for the population of subjects, for the situation under which the test is given. The true deviation represents the difference between the subject's true score and the population's true score.

The test score, then, can be written

$$X = \text{mean} + \text{deviation} + \text{error}$$

The mean can be compared with $\mathbf{m} + \mathbf{a}_a$ of the score model for design A, since $\mathbf{m} + \mathbf{a}_a$ is the true score for the population of subjects under some given situation a_a. Normally, subjects taking a test may be considered to be under one and the same situation. The "deviation + error" term is comparable with $\mathbf{e}_{e(a)}$ of the score model for A; that is, $\mathbf{e}_{e(a)}$ has the systematic and variable components mentioned previously. Only the variable component corresponds to the "error" term of test theory. In repeated measurements designs, however, the error term $\mathbf{e}$ does correspond exactly to the "error" term of test theory.

Why is the score model for design A not written with separate subject deviation and variable error components? Because in a single measurement design, we have no way of dissecting these two components—they are confounded. We sometimes carry confounded terms in score models. Specifically, when a design has only one score per condition, we carry both an error term and an interaction term confounded with it, for special reasons that were mentioned in Section 4.1.5. In general, however, in both the area of experimental design and the field of science as a whole, confounded terms are not carried separately in equations. To do so would be to violate two canons of the scientific method, parsimony and operationalism. In repeated measurements designs it is possible to dissect the two components, because when two or more scores are available for the same subject, his systematic deviation component can be estimated. Hence the error $\mathbf{e}$ of repeated measurements score models is a purely variable component that corresponds exactly to the "error term" of test theory.

Because the error parameters of a single measurement design have systematic as well as variable components, they are, in general, larger than the repeated measurements error parameters for the same criterion. As we shall see, the larger the error parameters, the smaller the power of the design for detecting treatment differences. This is the reason why single measurement designs generally have less power than repeated measurements designs having the same treatment factors and criterion.

Although conceptually, the $\mathbf{e}$ of single measurement designs has a systematic component that is associated with a subject's presumed tendency toward consistent responding, this systematic component in no way detracts from our conception of the $\mathbf{e}$ for a particular design position as being random and uncorrelated over replications. The replication involves entirely new, randomly chosen subjects, and hence there is random variation in the systematic component of error as well as in the variable component.

4.2.4 Why S Is Not a Nested Factor

With some justification, we could argue that, in the design being discussed, we should include a subject factor S nested within A. After all, each subject occurs in conjunction with exactly one level of A. Under such circumstances, in some designs we would consider each subject to be a level of a subject factor and write $S(A)$; but we do not do so here because of the confounding of the systematic and variable components of error, mentioned previously. Even if we wrote the systematic and variable components separately in our score model, in subsequent calculations we could only treat their sum as a totality. Under such circumstances, as we noted previously, we prefer to have only one term. This term is conventionally referred to as the "error term" and is symbolized accordingly.

Another point concerns the design symbolization and the score model; they are so intimately related that, as we shall see, given the symbolization, we can derive the model. Each factor in the symbolization implies a term or terms in the score model. To maintain the parallelism between the symbolization and the score model, we do not want factors in the design symbolization that have no corresponding terms in the score model. Because we have no term with **s** in the score model, we prefer to have no S in the design symbolization.

We could symbolize our design as $E(A)$, since a specific error parameter applies within only one level of A. The **e** of our model would develop naturally from the E of the symbolization. The E is normally excluded from design symbolizations, however, because it is excess baggage and because it is customarily not considered to be a design factor. It is superfluous to write $E(Z)$ for each design symbolization, because we can always assume that $E(Z)$, where Z is the design symbolization without E.

When S is a design factor, its upper limit S is used to symbolize the number of subjects in a cell. However, when S is not a design factor, its related symbolism is avoided in discussing the design and writing the score model. Hence, as mentioned previously, E is used to symbolize the number of subjects in a cell.

4.2.5 The Score Expectation Model

Suppose the experiment were replicated an infinite number of times. The mean of score X_{ae} across all such replications is called the *score expectation*. The expectation of a quantity X is symbolized by $\mathrm{E}(X)$. To differentiate E of error from E of expectation, the former will always be printed in italics.

There are a few simple but important rules to remember about expectations:

1. The expectation of a constant equals the constant. For example, if W is a constant, then $\mathrm{E}(W) = W$.
2. The expectation of a constant times a random variable equals the constant times the expectation of the random variable. For example, if W is a constant and X is a random variable, then $\mathrm{E}(W \cdot X) = W\mathrm{E}(X)$.
3. The expectation of a sum equals the sum of the expectations of the individual terms summed. For example, $\mathrm{E}(X + Y + Z) = \mathrm{E}(X) + \mathrm{E}(Y) + \mathrm{E}(Z)$.

These rules parallel rules given for the summation sign $\sum$ (Section 2.1), as well they should, for an expectation is, in essence, a summation divided by the number of cases summed. The rules given here can be combined, just as the rules for the summation sign can be combined, so that, for example, if W_1, W_2, and W_3 are constants, and X and Y are variables, then

$$\begin{aligned} \mathrm{E}(W_1 + W_2X + W_3Y) &= \mathrm{E}(W_1) + \mathrm{E}(W_2X) + \mathrm{E}(W_3Y) \\ &= W_1 + W_2\mathrm{E}(X) + W_3\mathrm{E}(Y) \end{aligned}$$

Let us consider the expectation for a specific score, for example, X_{23}. Since

$$X_{23} = \mathbf{m} + \mathbf{a}_2 + \mathbf{e}_{3(2)},$$

we know that

$$\begin{aligned} \mathrm{E}(X_{23}) &= \mathrm{E}(\mathbf{m} + \mathbf{a}_2 + \mathbf{e}_{3(2)}) \\ &= \mathrm{E}(\mathbf{m}) + \mathrm{E}(\mathbf{a}_2) + \mathrm{E}(\mathbf{e}_{3(2)}) \end{aligned}$$

The value of **m** is a constant over replications, as is $\mathbf{a}_2$ when factor A is fixed, as we now assume it to be. Therefore, $\mathrm{E}(\mathbf{m}) = \mathbf{m}$ and $\mathrm{E}(\mathbf{a}_2) = \mathbf{a}_2$. For any $\mathbf{e}_{e(a)}$, $\mathrm{E}(\mathbf{e}_{e(a)}) = 0$, by definition. It follows, then, that

$$\mathrm{E}(X_{23}) = \mathbf{m} + \mathbf{a}_2$$

According to the score model, scores for the same condition (that is, for the same $\mathbf{a}_a$) differ only because of the error parameters. No error parameters are present in the score expectation model, however, and all scores for the same condition have the same expectation. Therefore, the subscript e on X is usually omitted when we take the expectation, and $\mathrm{E}(X_a)$ is written instead of $\mathrm{E}(X_{ae})$. The score expectation model for design A, then, is:

$$\mathrm{E}(X_a) = \mathbf{m} + \mathbf{a}_a$$

Looking at errors and expectations another way, error can be defined as the discrepancy between a subject's score and the expectation for his treatment; that is,

$$\mathbf{e}_{e(a)} = X_{ae} - \mathrm{E}(X_a)$$

Taking the expectation of both sides of the equation yields

$$\begin{aligned} \mathrm{E}(\mathbf{e}_{e(a)}) &= \mathrm{E}(X_{ae} - \mathrm{E}(X_a)) \\ &= \mathrm{E}(X_{ae}) - \mathrm{E}(\mathrm{E}(X_a)) \end{aligned}$$

Since $\mathrm{E}(X_a)$ is a constant over replications (even though X_a per se is not), $\mathrm{E}(\mathrm{E}(X_a)) = \mathrm{E}(X_a)$, and since $\mathrm{E}(X_{ae}) = \mathrm{E}(X_a)$,

$$\mathrm{E}(\mathbf{e}_{e(a)}) = \mathrm{E}(X_a) - \mathrm{E}(X_a) = 0$$

Thus, from the definition of $\mathbf{e}_{e(a)}$ as the difference between a score and its expectation, it follows that $\mathrm{E}(\mathbf{e}_{e(a)}) = 0$. However, $\mathbf{e}_{e(a)}$ is not defined as $X_{ae} - \mathrm{E}(X_a)$ except when A is a fixed factor.

The expectations are never known exactly, though estimates of them can be calculated from the scores. Nonetheless, it is useful for an understanding of the model and its parameters to start out by assuming that the expectations are known. We can thereby see what the model parameters signify in relation to what the average scores for the different treatments would be if the experiment were replicated innumerable times. In the forthcoming subsection, then, the expectations are assumed to be known and formulas are derived for calculating the model parameters from the expectations. To estimate the parameters, we use exactly the same formulas except that we substitute estimates of the expectations derived from the scores.

4.2.6 Solving Linear Equations

In the example for design A of Section 4.2.2, $A = 2$, so the score expectation model has two equations. Assuming A to be a fixed factor, these equations are:

$$\mathrm{E}(X_1) = \mathbf{m} + \mathbf{a}_1$$
$$\mathrm{E}(X_2) = \mathbf{m} + \mathbf{a}_2$$

The equations of the expectation model constitute a set of *linear equations.* In each side of a linear equation, one or more terms are added together, each term being either a known constant, or an unknown multiplied by a known constant. In the example, the $\mathrm{E}(X)$'s are taken to be knowns, and $\mathbf{m}$, $\mathbf{a}_1$, and $\mathbf{a}_2$ are taken to be unknowns multiplied by the constant 1. (The constant 1 need not be indicated explicitly in the equations.)

The student may recall from the study of algebra that there is an important relationship between the solvability of a set of linear equations (that is, the possibility of determining the values of the unknowns), the total number of equations, and the total number of unknowns. It is assumed that the linear equations are *independent*, that is, none of the equations can be derived by performing multiplicative and additive operations on one or more of the remaining equations. A simple example of two linear equations that are not independent is:

$$10 = 2x + 3y$$
$$20 = 4x + 6y$$

The second equation can be obtained simply by multiplying the first equation by 2.

With regard to any set of linear equations, one of three possible situations may obtain:

1. There is no solution; that is, there is no possible set of values for the unknowns that satisfies all the equations.
2. There is exactly one value for each unknown, consistent with the equations.
3. There are many sets of values that the unknowns may assume, each consistent with the equations.

For independent linear equations, situation (1) holds if and only if there are more equations than unknowns; situation (2) holds if and only if the number of equations equals the number of unknowns; and situation (3) holds if and only if there are more equations than unknowns.

In the preceding example with $E(X_1)$ and $E(X_2)$, there are two equations and three unknowns, so situation (3) holds. There are many sets of values for $\mathbf{m}$, $\mathbf{a}_1$, and $\mathbf{a}_2$ that could satisfy these equations (regardless of the values of $E(X_1)$ and $E(X_2)$). This is an awkward situation. It is simpler to have only one solution, and this can be achieved by adding one more equation to the set without including any more unknowns. The additional equation is the side condition for factor A,

$$\mathbf{a}_1 + \mathbf{a}_2 = 0$$

This additional equation contains no new unknowns. By adding this equation to the original set of equations, we have three equations and three unknowns. Now situation (2) holds. There is exactly one solution to the set of equations; that is, assuming the $E(X)$'s to be known constants, there is exactly one value for each unknown parameter for which the equations hold.

Let us write the three equations of the set together:

(1) $$E(X_1) = \mathbf{m} + \mathbf{a}_1$$

(2) $$E(X_2) = \mathbf{m} + \mathbf{a}_2$$

(3) $$0 = \mathbf{a}_1 + \mathbf{a}_2$$

First let us solve for $\mathbf{m}$. We add the first two equations together:

$$E(X_1) + E(X_2) = 2\mathbf{m} + \mathbf{a}_1 + \mathbf{a}_2$$

But by equation (3), $\mathbf{a}_1 + \mathbf{a}_2 = 0$, so we have $E(X_1) + E(X_2) = 2\mathbf{m}$, or

$$\mathbf{m} = \frac{E(X_1) + E(X_2)}{2}$$

The population mean $\mathbf{m}$ is simply the mean of the expectations for the different treatments in the design. If there were more treatments, $\mathbf{m}$ would still be the mean of all treatments in the design. The value of $\mathbf{m}$ depends on more than the population of subjects and the measuring instrument. Obviously, it also depends on which of the possible levels of factor A are included in an experiment.

Once $\mathbf{m}$ has been found, we can easily find $\mathbf{a}_1$ from equation (1) and $\mathbf{a}_2$ from equation (2):

$$\mathbf{a}_1 = E(X_1) - \mathbf{m}$$
$$\mathbf{a}_2 = E(X_2) - \mathbf{m}$$

The main terms equal simply the deviations of the treatment expectations from the population mean. This is also true for designs with more treatments.

Expressed more generally,

$$\mathbf{m} = \frac{\sum_a \mathrm{E}(X_a)}{A}$$

$$\mathbf{a}_a = \mathrm{E}(X_a) - \mathbf{m}$$

In dot notation,

$$\mathbf{m} = \mathrm{E}(\bar{X}_.)$$

Suppose $A = 3$ and we have $\mathrm{E}(X_1) = 5$, $\mathrm{E}(X_2) = 3$, and $\mathrm{E}(X_3) = 10$. Then the expectation model with side condition would be

$$\begin{aligned} 5 &= \mathbf{m} + \mathbf{a}_1 \\ 3 &= \mathbf{m} + \mathbf{a}_2 \\ 10 &= \mathbf{m} + \mathbf{a}_3 \\ 0 &= \mathbf{a}_1 + \mathbf{a}_2 + \mathbf{a}_3 \end{aligned}$$

Thus

$$\begin{aligned} \mathbf{m} &= \mathrm{E}(\bar{X}_.) \\ &= \frac{5 + 3 + 10}{3} = \frac{18}{3} = 6 \\ \mathbf{a}_a &= \mathrm{E}(X_a) - \mathbf{m} \\ \mathbf{a}_1 &= 5 - 6 = -1 \\ \mathbf{a}_2 &= 3 - 6 = -3 \\ \mathbf{a}_3 &= 10 - 6 = 4 \end{aligned}$$

4.2.7 Estimation of Parameters

In actuality, the expectations are not available for an experiment, and so model parameters cannot be found from the calculations indicated. However, if we substitute estimated expectations for the expectations, the preceding formulas yield estimates of the parameters instead of the parameters per se. How do we estimate a score expectation? It is simply the mean of the scores having the same expectation. To differentiate an estimate of a parameter from the parameter itself, a circumflex (ˆ) is placed over the symbol. Thus $\hat{\mathrm{E}}(X_a)$ refers to the estimate of $\mathrm{E}(X_a)$, and

$$\hat{\mathrm{E}}(X_a) = \frac{\sum_a X_{ae}}{E} = \bar{X}_{a.}$$

Therefore,

$$\begin{aligned} \hat{\mathbf{m}} &= \hat{\mathrm{E}}(\bar{X}_.) = \bar{X}_{..} \\ \hat{\mathbf{a}}_a &= \hat{\mathrm{E}}(X_a) - \hat{\mathbf{m}} \\ &= \bar{X}_{a.} - \bar{X}_{..} \end{aligned}$$

Note that to translate a term that contains $\hat{\mathrm{E}}(\quad)$ to one that contains only score means, we average the X quantity in parentheses over the subscript e;

that is, $\hat{E}(X_a) = \bar{X}_{a.}$, and $\hat{E}(\bar{X}_.) = \bar{X}_{..}$. It is assumed here, as elsewhere unless otherwise stated, that the number of scores is the same for each cell of the design.

In Section 4.1.4 it was mentioned that there are two computational forms for estimation equations: one based on other estimates (the method of successive extraction), and one based on $\bar{X}$ quantities (the method of $\bar{X}$'s). The two equations for $\hat{\mathbf{a}}_a$ just given illustrate these two kinds of formulas. The first formula enables us to calculate $\hat{\mathbf{a}}_a$ in terms of a previous estimate, $\hat{\mathbf{m}}$. The second formula is written in terms of $\bar{X}$ quantities. The correspondence between the two formulas is direct and simple here, but, as we shall see later, in more complicated situations the distinction can be considerable.

It was also stated in Section 4.1.4 that we can derive the equivalents of the $\bar{X}$ quantities in model terms, and thus express a parameter estimate by means of such theoretical variables. For example, for $\hat{\mathbf{m}}$ and $\hat{\mathbf{a}}_a$,

$$X_{ae} = \mathbf{m} + \mathbf{a}_a + \mathbf{e}_{e(a)},$$

so

$$\begin{aligned}
\bar{X}_{..} &= \frac{\sum_a \sum_e X_{ae}}{AE} \\
&= \frac{\sum_a \sum_e (\mathbf{m} + \mathbf{a}_a + \mathbf{e}_{e(a)})}{AE} \\
&= \frac{1}{AE}\sum_a \sum_e \mathbf{m} + \frac{1}{AE}\sum_a \sum_e \mathbf{a}_a + \frac{1}{AE}\sum_a \sum_e \mathbf{e}_{e(a)} \\
&= \mathbf{m} + \bar{\mathbf{a}}_. + \bar{\mathbf{e}}_{.(.)}
\end{aligned}$$

Because of the side condition on A, $\bar{\mathbf{a}}_. = 0$, so

$$\bar{X}_{..} = \mathbf{m} + \bar{\mathbf{e}}_{.(.)}$$

To follow the derivation the student must be familiar with the rules of the $\sum$-notation and dot notation given in Section 2.1, so he should review that section if necessary.

Since $\hat{\mathbf{m}} = \bar{X}_{..}$, we see from the preceding equation that the estimate of $\mathbf{m}$, $\hat{\mathbf{m}}$, differs from the true value of $\mathbf{m}$ because of $\bar{\mathbf{e}}_{.(.)}$, the mean of the error parameters for all $A \cdot E$ scores in the design, but it is unaffected by the magnitudes of the $\mathbf{a}_a$'s (when A is fixed). On the average, the larger the value of E, the smaller the discrepancy between $\hat{\mathbf{m}}$ and $\mathbf{m}$ will probably be, since the greater the number of error parameters, the more their average will tend to equal their expectation, which is zero. In equation form,

$$\begin{aligned}
\bar{X}_{a.} &= \frac{1}{E}\sum_e (\mathbf{m} + \mathbf{a}_a + \mathbf{e}_{e(a)}) \\
&= \mathbf{m} + \mathbf{a}_a + \bar{\mathbf{e}}_{.(a)}
\end{aligned}$$

Therefore,

$$\begin{aligned}\hat{\mathbf{a}}_a &= \bar{X}_{a.} - \bar{X}_{..}\\ &= \mathbf{m} + \mathbf{a}_a + \bar{\mathbf{e}}_{.(a)} - (\mathbf{m} + \bar{\mathbf{e}}_{.(.)})\\ &= \mathbf{a}_a + \bar{\mathbf{e}}_{.(a)} - \bar{\mathbf{e}}_{.(.)}\end{aligned}$$

The "error of estimation for $\hat{\mathbf{a}}_a$" means the difference between the estimate $\hat{\mathbf{a}}_a$ and the true value, that is, $\hat{\mathbf{a}}_a - \mathbf{a}_a$. From the preceding equation we can see that the error of estimation for $\hat{\mathbf{a}}_a$ equals $\bar{\mathbf{e}}_{.(a)} - \bar{\mathbf{e}}_{.(.)}$, the difference between the mean of the error parameters for treatment a and the mean of the error parameters for all scores in the experiment. The larger the variance of the errors, the larger this difference will probably be. Note that the error of estimation for $\hat{\mathbf{a}}_a$ does not depend on the magnitude of **m**.

Besides estimating **m** and $\mathbf{a}_a$, we can estimate $\mathbf{e}_{e(a)}$, by

$$\begin{aligned}\hat{\mathbf{e}}_{e(a)} &= X_{ae} - \hat{\mathrm{E}}(X_a)\\ &= X_{ae} - \bar{X}_{a.}\end{aligned}$$

Its theoretical components are

$$\begin{aligned}\hat{\mathbf{e}}_{e(a)} &= \mathbf{m} + \mathbf{a}_a + \mathbf{e}_{e(a)} - (\mathbf{m} + \mathbf{a}_a + \bar{\mathbf{e}}_{.(a)})\\ &= \mathbf{e}_{e(a)} - \bar{\mathbf{e}}_{.(a)}\end{aligned}$$

We see that the estimate of error is not affected by the magnitudes of the $\mathbf{a}_a$'s or of **m**.

4.2.8 Factor *A* Random

If factor A were random, the score model would not change. However, the score expectations would change. Consider the score $X_{11} = \mathbf{m} + \mathbf{a}_1 + \mathbf{e}_{1(1)}$. Expectation $\mathrm{E}(X_1) = \mathrm{E}(\mathbf{m}) + \mathrm{E}(\mathbf{a}_1) + \mathrm{E}(\mathbf{e}_{1(1)})$. For factor A fixed, $\mathbf{a}_1$ is a constant, so $\mathrm{E}(\mathbf{a}_1) = \mathbf{a}_1$. For factor A random, however, "$\mathbf{a}_1$" is simply a label for any randomly chosen parameter from a population of parameters, since, over replications, any parameter is as likely to be chosen for a_1 as another. By the side condition, the sum of these parameters over innumerable replications is zero. Therefore, over an infinity of replications, the mean of the parameters labeled "$\mathbf{a}_1$" would be zero also; that is, $\mathrm{E}(\mathbf{a}_1) = 0$. By the same reasoning, $\mathrm{E}(\mathbf{a}_a) = 0$ for any subscript a. As before, $\mathrm{E}(\mathbf{e}_{e(a)}) = 0$. Then $\mathrm{E}(X_{ae}) = \mathbf{m}$ for each score. To be given the score expectations is to be given one number, which equals **m**. Given the score expectations, then, we cannot solve for the $\mathbf{a}_a$'s. However, we can still estimate the $\mathbf{a}_a$'s appearing in the one experiment by the same formula that is used when factor A is fixed:

$$\begin{aligned}\hat{\mathbf{a}}_a &= \hat{\mathrm{E}}(X_a) - \hat{\mathbf{m}}\\ &= X_{a.} - \bar{X}_{..}\end{aligned}$$

Even though the formulas are the same for calculating estimates when design factors are random as when they are fixed, the theoretical components of the estimates change. Because we no longer have $\sum_a^A \mathbf{a}_a = \mathbf{a}_. = 0$, the $\mathbf{a}_a$'s do not cancel out as when A was fixed, and they contribute to the error of the estimates. For example, if $A = 2$ and $E = 3$, then

$$\bar{X}_{..} = \frac{6\mathbf{m} + 3(\mathbf{a}_1 + \mathbf{a}_2) + \mathbf{e}_{.(.)}}{6}$$

$$= \mathbf{m} + \bar{\mathbf{a}}_. + \bar{\mathbf{e}}_{.(.)}$$

just as when A was fixed. Now, however, the $\mathbf{a}_a$'s do not cancel out. The sum $\mathbf{a}_1 + \mathbf{a}_2$, and thus the average $\bar{\mathbf{a}}_.$, is unknown but cannot be assumed to be zero.

$$\bar{X}_{a.} = \frac{3\mathbf{m} + 3\mathbf{a}_a + \mathbf{e}_{.(a)}}{3}$$

$$= \mathbf{m} + \mathbf{a}_a + \bar{\mathbf{e}}_{.(a)}$$

Then, since $\hat{\mathbf{a}}_a = \bar{X}_a - \bar{X}_{..}$,

$$\hat{\mathbf{a}}_1 = \mathbf{a}_1 - \bar{\mathbf{a}}_. + \bar{\mathbf{e}}_{.(1)} - \bar{\mathbf{e}}_{.(.)}$$
$$\hat{\mathbf{a}}_2 = \mathbf{a}_2 - \bar{\mathbf{a}}_. + \bar{\mathbf{e}}_{.(2)} - \bar{\mathbf{e}}_{.(.)}$$

In general, for any size of limit A,

$$\hat{\mathbf{a}}_a = \mathbf{a}_a - \bar{\mathbf{a}}_. + \bar{\mathbf{e}}_{.(a)} - \bar{\mathbf{e}}_{.(.)}$$

The error of estimation is

$$\hat{\mathbf{a}}_a - \mathbf{a}_a = -\bar{\mathbf{a}}_. + \bar{\mathbf{e}}_{.(a)} - \bar{\mathbf{e}}_{.(.)}$$

In contrast to the situation that exists when factor A is fixed, the error of estimation for $\mathbf{a}_a$ depends on all the $\mathbf{a}_a$ parameters operating for the experiment as well as the $\mathbf{e}_{e(a)}$ parameters.

Now consider the estimation of error:

$$\begin{aligned}\hat{\mathbf{e}}_{e(a)} &= X_{ae} - \hat{\mathrm{E}}(X_a) \\ &= X_{ae} - \bar{X}_{a.} \\ &= \mathbf{m} + \mathbf{a}_a + \mathbf{e}_{e(a)} - (\mathbf{m} + \mathbf{a}_a + \bar{\mathbf{e}}_{.(a)}) \\ &= \mathbf{e}_{e(a)} - \bar{\mathbf{e}}_{.(a)}\end{aligned}$$

The estimate of $\mathbf{e}_{e(a)}$ has the same components as it has when factor A is fixed. The error of estimation for $\hat{\mathbf{e}}_{e(a)}$, $\hat{\mathbf{e}}_{e(a)} - \mathbf{e}_{e(a)}$, is independent of the $\mathbf{a}_a$'s, as it was when A was fixed.

As will become more clear in succeeding chapters, the determination of the key term components contributing to parameter estimates plays an important part in the theory and procedures of the analysis of variance. For example, we cannot understand the analysis of variance without considering the effect of the interpretation of a factor as random or fixed on the presence or absence of such components. When we attend to such components in the present chapter then, it is to lay the groundwork for understanding the analysis of variance, and not simply an abstract mathematical exercise.

4.3 DESIGN $A \times B$

4.3.1 Introduction

We shall now consider the model for the two-factor design $A \times B$. This single measurement design might be from either Class I or Class II. If A and B are treatment factors, the design would be from Class I; if B, say, is a blocking factor, the design would be from Class II. The class of the design (I or II) has no effect on its score model, the estimation of model parameters, or the procedures for analysis of variance.

The two-factor score model, in contrast to the one-factor model, includes an interaction term, $\mathbf{ab}_{ab}$. Explanation of this two-factor interaction constitutes an important part of this section. There are a total of $A \cdot B$ different conditions (cells). Each subject in the experiment is exposed to only a single condition. Initially, we assume that there are E subjects per condition, or $A \cdot B \cdot E$ subjects altogether, where $E > 1$. Each subject has one score, so the design yields $A \cdot B \cdot E$ scores also. Section 4.3.9 discusses the situation where E equals 1.

4.3.2 The Score Model

A score for design $A \times B$ is symbolized by X_{abe}. The score model for $A \times B$ is

$$X_{abe} = \mathbf{m} + \mathbf{a}_a + \mathbf{b}_b + \mathbf{ab}_{ab} + \mathbf{e}_{e(ab)}$$

This one key equation symbolizes $A \cdot B \cdot E$ different parameter equations, one for each score in the experiment. Error $\mathbf{e}_{e(ab)}$, as in design A, consists in part of a true score deviation for subject e plus a variable component, combined in unknown proportions.

4.3.3 The Score Expectation Model and Side Conditions

As with design A, we begin by assuming that the expectations are known and calculate the parameters therefrom. Estimates of the parameters can be obtained by substituting estimated expectations for the true expectations in these formulas.

As with design A, the expectation of the error term is zero; $\mathrm{E}(\mathbf{e}_{e(ab)}) = 0$. Hence the expectation for any score from the same condition is the same, and an expectation of X_{abe} is symbolized by $\mathrm{E}(X_{ab})$. For factors A and B fixed, any parameter $\mathbf{a}_a$, $\mathbf{b}_b$, or $\mathbf{ab}_{ab}$ is a constant over replications, so

$$\begin{aligned} \mathrm{E}(X_{ab}) &= \mathrm{E}(\mathbf{m} + \mathbf{a}_a + \mathbf{b}_b + \mathbf{ab}_{ab} + \mathbf{e}_{e(ab)}) \\ &= \mathbf{m} + \mathbf{a}_a + \mathbf{b}_b + \mathbf{ab}_{ab} \end{aligned}$$

There are $A \cdot B$ such parameter equations, and there are $1 + A + B + A \cdot B$ unknown parameters. With $A \cdot B$ linear equations and more than $A \cdot B$ unknowns, the equations have many solutions. When we include the side conditions, however, there are an equal number of unknowns and independent equations. Assuming both A and B to be fixed factors, the side conditions for design $A \times B$ are:

$$\sum_{a}^{A} \mathbf{a}_a = \mathbf{a}_. = 0$$

$$\sum_{b}^{B} \mathbf{b}_b = \mathbf{b}_. = 0$$

$$\sum_{a}^{A} \mathbf{ab}_{ab} = \mathbf{ab}_{.b} = 0 \quad \text{(for any constant value of subscript } b\text{)}$$

$$\sum_{b}^{B} \mathbf{ab}_{ab} = \mathbf{ab}_{a.} = 0 \quad \text{(for any constant value of subscript } a\text{)}$$

If the sums equal zero, it follows that the means $\bar{\mathbf{a}}_.$, $\bar{\mathbf{b}}_.$, $\overline{\mathbf{ab}}_{.b}$, and $\overline{\mathbf{ab}}_{a.}$ must also be zero. The two key side conditions for the interaction term represent B and A different equations, respectively. Altogether, then, there are $B + A + 2$ parametric side conditions.

We learned earlier that the expectation model consists of $A \cdot B$ equations, which, with the side conditions just mentioned, make $A \cdot B + A + B + 2$ equations. We also noted earlier that there are $A \cdot B + A + B + 1$ unknown quantities in the expectation model. The side conditions introduce no additional unknowns. It appears, then, that there are more equations than unknowns. If these equations were all independent, there would be no solution. However, all the side conditions for the interactions are not independent. One of the equations can be derived from the remainder.

For example, suppose that $A = 2$ and $B = 3$. Suppose also that all the side conditions mentioned previously are given except the following one:

$$\mathbf{ab}_{21} + \mathbf{ab}_{22} + \mathbf{ab}_{23} = 0$$

We can show that this equation is dependent on the others by deriving it from them, as follows. The other side conditions on the interactions are:

(1) $$\mathbf{ab}_{11} + \mathbf{ab}_{21} = 0$$

(2) $$\mathbf{ab}_{12} + \mathbf{ab}_{22} = 0$$

(3) $$\mathbf{ab}_{13} + \mathbf{ab}_{23} = 0$$

(4) $$\mathbf{ab}_{11} + \mathbf{ab}_{12} + \mathbf{ab}_{13} = 0$$

Adding equations (1), (2), and (3) yields

(5) $$\mathbf{ab}_{11} + \mathbf{ab}_{12} + \mathbf{ab}_{13} + \mathbf{ab}_{21} + \mathbf{ab}_{22} + \mathbf{ab}_{23} = 0$$

But equation (4) says that the first three terms of equation (5) sum to zero. Therefore, the last three terms, $\mathbf{ab}_{21}$, $\mathbf{ab}_{22}$, and $\mathbf{ab}_{23}$, must also sum to zero if equation (5) holds. In other words, $\mathbf{ab}_{21} + \mathbf{ab}_{22} + \mathbf{ab}_{23} = 0$, the remaining side condition, can be derived from the others.

We have demonstrated that one side condition is not independent of the others. Therefore, we have one less independent equation in the side conditions than would appear at first. For $A \times B$, in general, for the expectation model with side conditions, there are then $A \cdot B + A + B + 1$ independent equations and the same number of unknowns. There is exactly one solution for the unknown parameters, which we now derive.

4.3.4 Solving for the Parameters

We shall now find formulas for the parameters, assuming that the expectations, the $\mathrm{E}(X_{ab})$'s, are known quantities. Although actually the $\mathrm{E}(X_{ab})$'s are never known (and thus the parameters can never be calculated exactly), by substituting estimated expectations into the same formulas, we can calculate parameter estimates. Since $\mathrm{E}(X_{ab}) = \mathbf{m} + \mathbf{a}_a + \mathbf{b}_b + \mathbf{ab}_{ab}$,

$$\begin{aligned} \frac{1}{AB}\sum_{ab} \mathrm{E}(X_{ab}) &= \frac{1}{AB}\sum_{ab} (\mathbf{m} + \mathbf{a}_a + \mathbf{b}_b + \mathbf{ab}_{ab}) \\ &= \mathbf{m} + \bar{\mathbf{a}}_. + \bar{\mathbf{b}}_. + \overline{\mathbf{ab}}_{..} \\ &= \mathbf{m} \end{aligned}$$

Because of the side conditions, $\bar{\mathbf{a}}_. = 0$, $\bar{\mathbf{b}}_. = 0$, and $\overline{\mathbf{ab}}_{..} = 0$. Rearranging the preceding equation, we have

$$\mathbf{m} = \frac{1}{AB}\sum \mathrm{E}(X_{ab}) = \mathrm{E}(\bar{X}_{..})$$

The population mean is the mean of the expectations for all conditions (just as it is with design A).

$$\begin{aligned} \mathrm{E}(\bar{X}_{a.}) &= \frac{1}{B}\sum_{b} \mathrm{E}(X_{ab}) \\ &= \frac{1}{B}\sum_{b} (\mathbf{m} + \mathbf{a}_a + \mathbf{b}_b + \mathbf{ab}_{ab}) \\ &= \mathbf{m} + \mathbf{a}_a + \bar{\mathbf{b}}_. + \overline{\mathbf{ab}}_{a.} \\ &= \mathbf{m} + \mathbf{a}_a \end{aligned}$$

Because of the side conditions, $\bar{\mathbf{b}}_. = 0$ and $\overline{\mathbf{ab}}_{a.} = 0$. Note that $\sum_b \mathbf{a}_a/B = \mathbf{a}_a \sum_b 1/B$; since any $\mathbf{a}_a$ is a constant with respect to changes in subscript b, it can come to the left of the summation sign. Of course, $\sum_b 1/B = B/B = 1$. By rearranging $\mathrm{E}(\bar{X}_{a.}) = \mathbf{m} + \mathbf{a}_a$, we find that

$$\begin{aligned} \mathbf{a}_a &= \mathrm{E}(\bar{X}_{a.}) - \mathbf{m} \\ &= \mathrm{E}(\bar{X}_{a.}) - \mathrm{E}(\bar{X}_{..}) \end{aligned}$$

The calculation of main term $\mathbf{a}_a$ is also similar to the calculation for design A. In design A, however, a particular main term, such as $\mathbf{a}_2$, appeared in only one condition, and it was calculated from the single condition expectation $\mathrm{E}(X_a)$. In $A \times B$ a parameter such as $\mathbf{a}_2$ appears in B different conditions and it is calculated from an average over these conditions. In both cases, however, a main term $\mathbf{a}_a$ is the mean deviation from $\mathbf{m}$ of the expectations having $\mathbf{a}_a$ as a component. In other words, $\mathbf{a}_2$ is the mean deviation of all expectations having $\mathbf{a}_2$ (instead of another $\mathbf{a}_a$) from $\mathbf{m}$, the mean expectation for all conditions.

The main term $\mathbf{b}_b$ is analogous:

$$\mathbf{b}_b = \mathrm{E}(\bar{X}_{.b}) - \mathbf{m}$$

Term $\mathbf{b}_b$ accounts for the discrepancy between the mean expectations for conditions having a specific level b_b and the mean expectations for all conditions $\mathbf{m}$.

Once the population mean and main terms have been found, it is easy to calculate the interaction terms by subtraction within the individual expectation equations. Simply by rearranging our expectation model (Section 4.3.3), we have

$$\mathbf{ab}_{ab} = \mathrm{E}(X_{ab}) - \mathbf{m} - \mathbf{a}_a - \mathbf{b}_b$$

A numerical example is presented in Section 4.3.5 to illustrate the application of these formulas in a specific context. Different interpretations of two-factor interaction are discussed in Section 4.3.6.

4.3.5 Numerical Example

In this example, specific values are assigned to the expectations to demonstrate the calculation of parameters. The expectations can be arranged in tabular form.

	b_1	b_2	b_3
a_1	10	5	9
a_2	6	3	9

The expectation for condition ab_{11}, that is, $\mathrm{E}(X_{11})$, equals 10. Similarly, the expectation for condition ab_{23}, that is, $\mathrm{E}(X_{23})$, equals 9. The population mean $\mathbf{m}$ is simply the mean of all six conditions:

$$\begin{aligned} \mathbf{m} &= \mathrm{E}(\bar{X}_{..}) \\ &= (10 + 5 + 9 + 6 + 3 + 9)/6 = 42/6 = 7 \end{aligned}$$

$$\mathbf{a}_a = \mathrm{E}(\bar{X}_{a.}) - \mathbf{m}$$

$$\begin{aligned} \mathbf{a}_1 &= [(10 + 5 + 9)/3] - 7 = [24/3] - 7 = 1 \\ \mathbf{a}_2 &= [(6 + 3 + 9)/3] - 7 = [18/3] - 7 = -1 \end{aligned}$$

$$\mathbf{b}_b = \mathrm{E}(\bar{X}_{.b}) - \mathbf{m}$$

$$\begin{aligned} \mathbf{b}_1 &= [(10 + 6)/2] - 7 = [16/2] - 7 = 1 \\ \mathbf{b}_2 &= [(5 + 3)/2] - 7 = [8/2] - 7 = -3 \\ \mathbf{b}_3 &= [(9 + 9)/2] - 7 = [18/2] - 7 = 2 \end{aligned}$$

$$\mathbf{ab}_{ab} = \mathrm{E}(X_{ab}) - (\mathbf{m} + \mathbf{a}_a + \mathbf{b}_b)$$

$$\begin{aligned}
\mathbf{ab}_{11} &= 10 - 7 - 1 - 1 = 1 \\
\mathbf{ab}_{12} &= 5 - 7 - 1 - (-3) = 0 \\
\mathbf{ab}_{13} &= 9 - 7 - 1 - 2 = -1 \\
\mathbf{ab}_{21} &= 6 - 7 - (-1) - 1 = -1 \\
\mathbf{ab}_{22} &= 3 - 7 - (-1) - (-3) = 0 \\
\mathbf{ab}_{23} &= 9 - 7 - (-1) - 2 = +1
\end{aligned}$$

Calculations can be simplified by using the side conditions. Once $\mathbf{a}_1$ has been found, we know from $\mathbf{a}_. = 0$ that $\mathbf{a}_2 = -\mathbf{a}_1$. Once $\mathbf{b}_1$ and $\mathbf{b}_2$ have been found, we know from $\mathbf{b}_. = 0$ that $\mathbf{b}_3 = -(\mathbf{b}_1 + \mathbf{b}_2)$. Similarly, the side conditions on the interactions can be used to simplify the calculation of interactions. It is necessarily true, for example, that $\mathbf{ab}_{21} = -\mathbf{ab}_{11}$. However, it is well to calculate the parameters in the manner illustrated and to use the side conditions to check the results. If any side condition fails to hold, there has been an error in the calculations.

4.3.6 Interpretation of Interaction

This subsection presents three interpretations of the two-factor interaction term: *algebraic*, *simple effects*, and *graphic*. The three interpretations are different but consistent ways of viewing interaction, not competing formulations or theories.

Algebraic interpretation. In a sense, the interaction term is simply a necessary quantity to enable the model to be adequate for every experimental situation. For example, we can imagine a model for $A \times B$ with only a population mean and main terms. We might call this a *main effects model*, which would be written

$$\mathrm{E}(X_{ab}) = \mathbf{m} + \mathbf{a}_a + \mathbf{b}_b$$

(A main effects model is also called an *additive model*.) The problem is that such a model fails to fit many situations because it is too simple. In a sense, the interaction term is simply the quantity required on the right-hand side of the equation to assure equality to $\mathrm{E}(X_{ab})$, regardless of what quantities the $\mathrm{E}(X_{ab})$'s may equal. Indeed, we can think of $\mathbf{ab}_{ab}$, by definition, as simply $\mathrm{E}(X_{ab}) - \mathbf{m} - \mathbf{a}_a - \mathbf{b}_b$. It is the discrepancy between the expectation and the main effects model. With the interaction thus defined, it necessarily follows from rearranging the definition that

$$\mathrm{E}(X_{ab}) = \mathbf{m} + \mathbf{a}_a + \mathbf{b}_b + \mathbf{ab}_{ab}$$

Although such a model holds regardless of the quantities assumed by the expectations, for some sets of expectations the interaction term is zero. When we say that the interaction term is zero, without specifying a particular condition ab_{ab}, we mean that $\mathbf{ab}_{ab} = 0$ for every combination of subscripts

a and b. Equivalently, we can say that there is "no interaction," that there is "no interaction effect," or that "interaction is absent." When there is no interaction, we also can say that "the main effects model holds."

Simple effects interpretation. As we have seen, the subscripted model terms other than error are often called effects—such as main effects or interaction effects. Simple effects are similar to main effects and interaction effects, except that they are interpreted within the context of part of a design only. Suppose that our design is $A \times B$, but that we ignore all conditions of the design except those having a_2. These conditions alone form a one-factor design with factor B. Any effect within this simple design is a "simple effect," compared with effects based on $A \times B$ as a whole.

For example, for the simple design B we have a main effect, symbolized by $\mathbf{b}_{b(a_2)}$, assuming that our simple design is at a_2. The simple main effect $\mathbf{b}_{b(a_2)}$ is defined in the same way as the main effect would be, if only conditions at a_2 were included.

The simple population mean $\mathbf{m}_{(a_2)} = \mathrm{E}(\bar{X}_{2.})$, and $\mathbf{b}_{b(a_2)} = \mathrm{E}(X_{2b}) - \mathbf{m}_{(a_2)}$. As previously stated,

(1) $$\mathrm{E}(X_{ab}) = \mathbf{m} + \mathbf{a}_a + \mathbf{b}_b + \mathbf{ab}_{ab}$$

Also, a simple main effect has just been defined as

(2) $$\mathbf{b}_{b(a_a)} = \mathrm{E}(X_{ab}) - \mathbf{m}_{(a_a)}$$

Alternatively, substituting for $\mathrm{E}(X_{ab})$ from (1) yields

(3) $$\mathbf{b}_{b(a_a)} = \mathbf{m} + \mathbf{a}_a + \mathbf{b}_b + \mathbf{ab}_{ab} - \mathbf{m}_{(a_a)}$$

Since

(4) $$\mathbf{m}_{(a_a)} = \mathrm{E}(\bar{X}_{a.})$$

and

(5) $$\mathbf{a}_a = \mathrm{E}(\bar{X}_{a.}) - \mathbf{m},$$

we have

$$\mathbf{a}_a = \mathbf{m}_{(a_a)} - \mathbf{m}, \text{ or}$$

(6) $$\mathbf{m}_{(a_a)} = \mathbf{a}_a + \mathbf{m}$$

Substituting this for $\mathbf{m}_{(a_a)}$ in equation (3) yields

(7) $$\mathbf{b}_{b(a_a)} = \mathbf{m} + \mathbf{a}_a + \mathbf{b}_b + \mathbf{ab}_{ab} - \mathbf{a}_a - \mathbf{m}$$
$$= \mathbf{b}_b + \mathbf{ab}_{ab}$$

Rearranging, we have

(8) $$\mathbf{ab}_{ab} = \mathbf{b}_{b(a_a)} - \mathbf{b}_b$$

Equation (8) is the definition of two-factor interaction in terms of a simple effect. The interaction effect $\mathbf{ab}_{ab}$ equals the difference between the simple main effect of b_b at a_a and the main effect for b_b for the design as a whole.

We recall that the simple effect gauges the discrepancy between the condition expectation and mean expectation for one level of factor A, and the main effect gauges the same discrepancy averaged over all levels of A. If the discrepancy is the same for a level of A as for the average across all levels, the interaction is zero for those levels of factors A and B. Only if $\mathbf{b}_{b(a_a)} - \mathbf{b}_b$ equals zero for all combinations of a_a and b_b can we make a general assertion that "there is no interaction."

Although we have analyzed the simple effects for factor B, we could just as well have examined the simple effects of factor A. We could have derived that

$$\mathbf{ab}_{ab} = \mathbf{a}_{a(b_b)} - \mathbf{a}_a \tag{9}$$

The two approaches are entirely equivalent and result in the same values for each $\mathbf{ab}_{ab}$.

Graphic interpretation. The presence or absence of interaction can be illustrated graphically. Figure 4.3-1 (top) illustrates a set of expectations for which the interaction is lacking. Graphically, the absence of interaction is illustrated by the parallel lines that represent b_1 and b_2. In words, the differ-

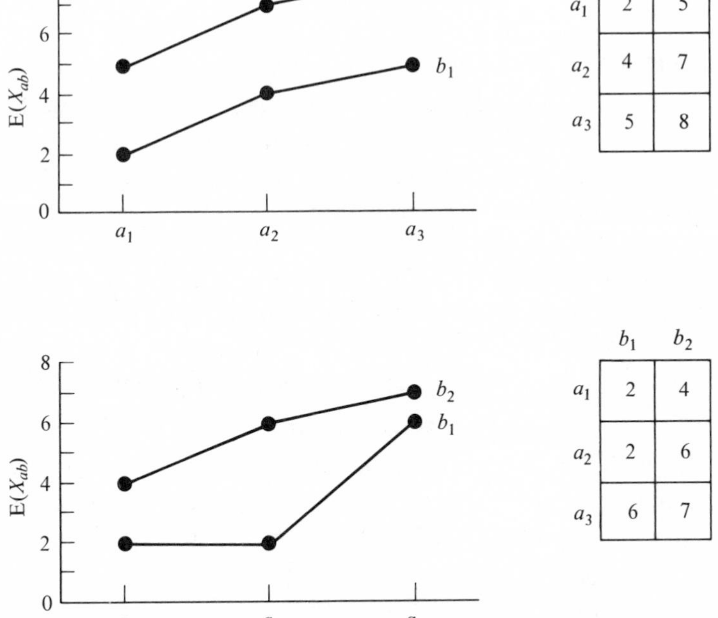

	b_1	b_2
a_1	2	5
a_2	4	7
a_3	5	8

	b_1	b_2
a_1	2	4
a_2	2	6
a_3	6	7

Figure 4.3-1. Graphs illustrating the absence of an AB interaction (*top*) and the presence of an AB interaction (*bottom*).

ences between the expectations for conditions with b_1 and with b_2 are the same at each level of a_a. If there were three or more levels of b_b, lack of interaction would require that all the lines representing these levels be parallel. We could as easily have plotted b_1 and b_2 along the abscissa; we would have three parallel lines illustrating the three a_a's. (Each line would have only two points.) It makes no difference which factor is plotted along the abscissa.

Figure 4.3-1 (bottom) illustrates a set of expectations having interaction. Interaction would be apparent if the b_b's were plotted on the abscissa.

The lack of interaction in Figure 4.3-1 (top) can be discerned directly from the data in the accompanying table of expectations. The difference between the b_2 and b_1 entries, 3, is the same for each a_a. Similarly, for any pair of a_a rows, the difference between expectations is the same for each b_b. Such differences are not the same for the expectations of Figure 4.3-1 (bottom).

In reality we do not have expectations, and even if, in truth, the interaction were zero, there might appear to be some interaction if we examined the graph of estimated expectations, since the errors in such estimates will cause the graph lines to depart, to a greater or lesser degree, from parallelism. The role of analysis of variance, discussed in Chapters 5 and 6, is in part to determine whether such discrepancies are large enough to require acceptance of the interaction term in the model. Discussion of such testing is postponed until then.

4.3.7 Estimation of Parameters

As we have seen, if expectations are known, it is possible to calculate the exact values of the model parameters. But in reality, the expectations are never known exactly; only the scores are known. The scores can be used to estimate the expectations, however, and then these estimates can be used to calculate estimates of the parameters.

To estimate parameters we follow the same procedures that are used to calculate parameters in Section 4.3.4, except that we substitute for the expectations estimates of those expectations. The estimate of the expectation for a condition is simply the mean of the scores for a condition; that is,

$$\hat{\mathrm{E}}(X_{ab}) = \bar{X}_{ab.}$$

As with design A, we can describe the resulting parameter estimates in terms of model parameters and show how the presence of error terms in the scores causes the estimates to differ from the parameters. We have

$$\begin{aligned}\hat{\mathbf{m}} &= \hat{\mathrm{E}}(\bar{X}_{..}) \\ &= \bar{X}_{...} \\ &= \frac{1}{ABE}\sum_{abe}(\mathbf{m} + \mathbf{a}_a + \mathbf{b}_b + \mathbf{ab}_{ab} + \mathbf{e}_{e(ab)}) \\ &= \mathbf{m} + \bar{\mathbf{e}}_{.(..)}\end{aligned}$$

The other terms drop out because of the side conditions. Likewise we have

$$\begin{aligned}\hat{\mathbf{a}}_a &= \hat{\mathrm{E}}(\bar{X}_{a.}) - \hat{\mathbf{m}} \\ &= \bar{X}_{a..} - \bar{X}_{...}\end{aligned}$$

Since

$$\begin{aligned}\bar{X}_{a..} &= \frac{1}{BE}\sum_{be}(\mathbf{m} + \mathbf{a}_a + \mathbf{b}_b + \mathbf{ab}_{ab} + \mathbf{e}_{e(ab)}) \\ &= \mathbf{m} + \mathbf{a}_a + \bar{\mathbf{e}}_{.(a.)},\end{aligned}$$

we have the parametric interpretation of $\hat{\mathbf{a}}_a$,

$$\hat{\mathbf{a}}_a = \mathbf{a}_a + \bar{\mathbf{e}}_{.(a)} - \bar{\mathbf{e}}_{.(..)}$$

Similarly,

$$\begin{aligned}\hat{\mathbf{b}}_b &= \hat{\mathrm{E}}(\bar{X}_{b.}) - \hat{\mathbf{m}} \\ &= \bar{X}_{.b.} - \bar{X}_{...} \\ &= \mathbf{b}_b + \bar{\mathbf{e}}_{.(.b)} - \bar{\mathbf{e}}_{.(..)}\end{aligned}$$

and

$$\begin{aligned}\widehat{\mathbf{ab}}_{ab} &= \hat{\mathrm{E}}(X_{ab}) - (\hat{\mathbf{m}} + \hat{\mathbf{a}}_a + \hat{\mathbf{b}}_b) \\ &= \bar{X}_{ab.} - \bar{X}_{.b.} - \bar{X}_{a..} + \bar{X}_{...} \\ &= \mathbf{ab}_{ab} + \bar{\mathbf{e}}_{.(ab)} - \bar{\mathbf{e}}_{.(.b)} - \bar{\mathbf{e}}_{.(a.)} + \bar{\mathbf{e}}_{.(..)}\end{aligned}$$

Also,

$$\begin{aligned}\hat{\mathbf{e}}_{e(ab)} &= X_{abe} - \mathrm{E}(X_{ab}) \\ &= X_{abe} - \bar{X}_{ab.} \\ &= \mathbf{e}_{e(ab)} - \bar{\mathbf{e}}_{.(ab)}\end{aligned}$$

In each case, an estimate of a parameter differs from the true value only as a function of error parameters, a state of affairs that obtains only when all design factors are fixed, as we shall see in the succeeding subsection.

4.3.8 Random Factors

If either or both of factors A and B are random instead of fixed, the score model remains unchanged. As with design A, interpretation of one or both the factors as random changes the score expectations because it changes the expectations for various terms. The expectation for any random main term is zero. We have already noted that the side conditions for interactions when both factors are fixed require that any sum across one subscript is equal to zero. However, the sum across a random factor subscript is not zero; only the sum across all levels in the population is zero. For example, suppose factor A is fixed and factor B is random. Then $\mathbf{ab}_{.b} = 0$ for any constant subscript b, for each replication, but $\mathbf{ab}_{a.}$ is r.v. across replications for any constant subscript a, though its expectation is zero. The change in side conditions affects the score expectations and the parameters that contribute to the estimation errors, just as it did in design A. However, the calculational formulas for estimating parameters remain the same as when all factors are fixed.

If A is fixed and B is random, let us see what theoretical components enter into various estimates.

$\hat{\mathbf{m}}$:

$$\begin{aligned}\hat{\mathbf{m}} &= \mathrm{E}(\bar{X}_{..}) \\ &= \bar{X}_{...} \\ &= \frac{1}{ABE}\sum_{abe}(\mathbf{m} + \mathbf{a}_a + \mathbf{b}_b + \mathbf{ab}_{ab} + \mathbf{e}_{e(ab)}) \\ &= \mathbf{m} + \bar{\mathbf{a}}_. + \bar{\mathbf{b}}_. + \overline{\mathbf{ab}}_{..} + \bar{\mathbf{e}}_{.(..)} \\ &= \mathbf{m} + \mathbf{b}_. + \bar{\mathbf{e}}_{.(..)}\end{aligned}$$

$\hat{\mathbf{a}}_a$:

$$\begin{aligned}\hat{\mathbf{a}}_a &= \hat{\mathrm{E}}(X_{a.}) - \hat{\mathbf{m}} \\ &= \bar{X}_{a..} - \bar{X}_{...} \\ \bar{X}_{a..} &= \frac{1}{BE}\sum_{be}(\mathbf{m} + \mathbf{a}_a + \mathbf{b}_b + \mathbf{ab}_{ab} + \mathbf{e}_{e(ab)}) \\ &= \mathbf{m} + \mathbf{a}_a + \bar{\mathbf{b}}_. + \overline{\mathbf{ab}}_{a.} + \bar{\mathbf{e}}_{.(a.)}\end{aligned}$$

Having already found $\bar{X}_{...}$, we obtain

$$\hat{\mathbf{a}}_a = \mathbf{a}_a + \overline{\mathbf{ab}}_{a.} + \bar{\mathbf{e}}_{.(a.)} - \bar{\mathbf{e}}_{.(..)}$$

$\hat{\mathbf{b}}_b$:

$$\begin{aligned}\hat{\mathbf{b}}_b &= \hat{\mathrm{E}}(\bar{X}_{.b}) - \hat{\mathbf{m}} \\ &= \bar{X}_{.b.} - \bar{X}_{...} \\ \bar{X}_{.b.} &= \frac{1}{AE}\sum_{ae}(\mathbf{m} + \mathbf{a}_a + \mathbf{b}_b + \mathbf{ab}_{ab} + \mathbf{e}_{e(ab)}) \\ &= \mathbf{m} + \bar{\mathbf{a}}_. + \mathbf{b}_b + \overline{\mathbf{ab}}_{.b} + \bar{\mathbf{e}}_{.(.b)} \\ &= \mathbf{m} + \mathbf{b}_b + \bar{\mathbf{e}}_{.(.b)} \\ \hat{\mathbf{b}}_b &= \mathbf{b}_b - \bar{\mathbf{b}}. + \bar{\mathbf{e}}_{.(.b)} - \bar{\mathbf{e}}_{.(..)}\end{aligned}$$

$\widehat{\mathbf{ab}}_{ab}$:

$$\begin{aligned}\widehat{\mathbf{ab}}_{ab} &= \hat{\mathrm{E}}(X_{ab}) - (\hat{\mathbf{m}} + \hat{\mathbf{a}}_a + \hat{\mathbf{b}}_b) \\ &= \bar{X}_{ab.} - \bar{X}_{.b.} - \bar{X}_{a..} + \bar{X}_{...} \\ X_{ab.} &= \frac{1}{E}\sum_{e}(\mathbf{m} + \mathbf{a}_a + \mathbf{b}_b + \mathbf{ab}_{ab} + \mathbf{e}_{e(ab)}) \\ \widehat{\mathbf{ab}}_{ab} &= \mathbf{ab}_{ab} - \overline{\mathbf{ab}}_{a.} + \mathbf{e}_{e(ab)} - \bar{\mathbf{e}}_{.(.b)} - \bar{\mathbf{e}}_{.(a.)} + \bar{\mathbf{e}}_{.(..)}\end{aligned}$$

$\hat{\mathbf{e}}_{e(ab)}$:

$$\hat{\mathbf{e}}_{e(ab)} = \mathbf{e}_{e(ab)} - \bar{\mathbf{e}}_{.(ab)},$$

as when both A and B are fixed.

The errors of estimation when both factors A and B are fixed (Section 4.3.7), are caused only by error parameters. In comparison, the following situation holds when factor B is random:

1. Estimate $\hat{\mathbf{m}}$ contains a random component $\bar{\mathbf{b}}_{.}$.
2. Estimate $\hat{\mathbf{a}}_a$ contains a random component $\overline{\mathbf{ab}}_{a.}$.
3. Estimate $\hat{\mathbf{b}}_b$ contains a random component $\bar{\mathbf{b}}_{.}$.
4. Estimate $\widehat{\mathbf{ab}}_{ab}$ contains a random component $\overline{\mathbf{ab}}_{a.}$.

Although we use the same estimation formulas as when both A and B are fixed, various components that sum to zero, because of the side conditions existing when factor B is fixed, sum, instead, to a random nonzero value when factor B is random. (The sums are random with respect to different replications, not for a particular experiment.)

Derivations of the type just shown lead to an understanding of why and how expected mean squares (Chapter 6) differ depending on the interpretation of design factors as fixed or random, and why different ratios of mean squares are therefore required for significance testing.

4.3.9 One Score per Condition

Sometimes design $A \times B$ has only one score per condition, that is, one subject per condition (per cell). Under such a circumstance, the error and interaction terms are confounded; a particular interaction term, such as $\mathbf{ab}_{23}$, appears but once in the score model, as does the error parameter for the same score.

When there is only one score per cell, we alter our score model for $A \times B$ slightly. We omit the subscript e from X and from $\mathbf{e}_{e(ab)}$ because it is superfluous. (Its value would remain at 1 for all scores.) We then write $\mathbf{e}_{ab}$ for the error term, omitting the parentheses. Our model, then, is

$$X_{ab} = \mathbf{m} + \mathbf{a}_a + \mathbf{b}_b + \mathbf{ab}_{ab} + \mathbf{e}_{ab}$$

In this book, whenever the error term and X are written without subscript e, error is confounded with the term having the same subscripts as $\mathbf{e}$ (and X).

If we have only one score per cell, then, assuming A and B to be fixed,

$$\begin{aligned}
\hat{\mathbf{m}} &= \bar{X}_{..} \\
&= \mathbf{m} + \bar{\mathbf{e}}_{..} \\
\hat{\mathbf{a}}_a &= \bar{X}_{a.} - \hat{\mathbf{m}} \\
&= \mathbf{a}_a + \bar{\mathbf{e}}_{a.} - \bar{\mathbf{e}}_{..} \\
\hat{\mathbf{b}}_b &= \bar{X}_{.b} - \hat{\mathbf{m}} \\
&= \mathbf{b}_b + \bar{\mathbf{e}}_{.b} - \bar{\mathbf{e}}_{..} \\
\widehat{\mathbf{ab}}_{ab} &= X_{ab} - (\hat{\mathbf{m}} + \hat{\mathbf{a}}_a + \hat{\mathbf{b}}_b) \\
&= \mathbf{ab}_{ab} + \mathbf{e}_{ab} - \bar{\mathbf{e}}_{.b} - \bar{\mathbf{e}}_{a.} + \bar{\mathbf{e}}_{..}
\end{aligned}$$

What we call $\widehat{\mathbf{ab}}_{ab}$ can actually be considered to be the estimate of $\mathbf{ab}_{ab} + \mathbf{e}_{ab}$. We cannot obtain separate estimates for these terms. If we have several scores per condition, we estimate an error parameter to be the deviation between a score and the mean score for the condition. If we have only one score per cell, there can be no such deviation.

We shall see in Chapters 5 and 6 that, in order to proceed with certain significance tests, either a separate estimate is required for the error term, such as we have with several scores per condition, or the interaction term $\mathbf{ab}_{ab}$ must be assumed to be zero. If the latter, then the formula we used to estimate $\mathbf{ab}_{ab}$, $X_{ab} - (\hat{\mathbf{m}} + \hat{\mathbf{a}}_a + \hat{\mathbf{b}}_b)$, serves to estimate $\mathbf{e}_{ab}$. The $\mathbf{ab}_{ab}$ component of this estimate would be absent by the same assumption.

4.4 DESIGN $A \times B \times C$

4.4.1 Introduction

Design $A \times B \times C$ is a single measurement design from either Class I or Class II—the class does not affect the score model, estimation, or analysis of variance. There are E subjects per condition, and hence a total of $A \cdot B \cdot C \cdot E$ subjects and a like number of scores. The distinctive feature of the score model for $A \times B \times C$ is the presence of a triple interaction term, $\mathbf{abc}_{abc}$, to which special attention is given in this section. Initially, we assume that $E > 1$ and that A, B, and C are all fixed factors. Alternative assumptions are considered in Sections 4.4.7 and 4.4.8.

4.4.2 The Score Model

Whereas the model for $A \times B$ has one two-factor interaction term, $\mathbf{ab}_{ab}$, the model for $A \times B \times C$ requires three two-factor interaction terms, one for each pair of factors. In addition, the model requires a main term for C, $\mathbf{c}_c$, and a three-factor interaction term, $\mathbf{abc}_{abc}$. The model is

$$X_{abce} = \mathbf{m} + \mathbf{a}_a + \mathbf{b}_b + \mathbf{c}_c + \mathbf{ab}_{ab} + \mathbf{ac}_{ac} + \mathbf{bc}_{bc} + \mathbf{abc}_{abc} + \mathbf{e}_{e(abc)}$$

This key equation represents $A \cdot B \cdot C \cdot E$ parameter equations.

4.4.3 The Score Expectation Model and Side Conditions

Again, we begin by assuming that the score expectations are known and then calculate the parameters. Actually, score expectations are never known exactly, so we can never compute the parameters exactly, but in going through the motions we gain some understanding of the parameters. The formulas used for parameter calculations can be used for parameter estimation when estimates of the expectations replace the expectations themselves.

Assuming A, B, and C to be fixed factors, a score expectation is:

$$\text{E}(X_{abc}) = \mathbf{m} + \mathbf{a}_a + \mathbf{b}_b + \mathbf{c}_c + \mathbf{ab}_{ab} + \mathbf{ac}_{ac} + \mathbf{bc}_{bc} + \mathbf{abc}_{abc}$$

There are $A \cdot B \cdot C$ conditions and hence $A \cdot B \cdot C$ such parameter equations. The number of unknowns in the set of linear equations is $A + B + C + A \cdot B + A \cdot C + B \cdot C + A \cdot B \cdot C$. Again, we add side conditions to equalize the number of equations and unknowns. The side conditions for main terms and two-factor interaction terms are like those for design $A \times B$, except that we now have more terms. For the triple interaction term, assuming all factors to be fixed,

$$\mathbf{abc}_{.bc} = 0 \quad \text{(for any constant values of subscripts } b \text{ and } c)$$
$$\mathbf{abc}_{a.c} = 0 \quad \text{(for any constant values of subscripts } a \text{ and } c)$$
$$\mathbf{abc}_{ab.} = 0 \quad \text{(for any constant values of subscripts } a \text{ and } b)$$

Although the number of side conditions appears to be excessive relative to the number of equations required, not all the side conditions are independent, so the expectation equations and side conditions are consistent with exactly one value for each parameter.

4.4.4 Solving for the Parameters

The formulas for calculating the population mean, main terms, and two-factor interaction terms from given score expectations are similar to the formulas used for design $A \times B$, except that the averaging proceeds over one additional factor. The population mean $\mathbf{m}$ is the mean of all $A \cdot B \cdot C$ condition expectations in the experiment:

$$\mathbf{m} = \frac{1}{ABC} \sum_{abc} \mathrm{E}(X_{abc}) = \mathrm{E}(\bar{X}_{...})$$

A main term, such as $\mathbf{a}_1$, is the deviation of the mean of all expectations having level a_1 from the population mean:

$$\begin{aligned}
\mathbf{a}_a &= \frac{1}{BC} \sum_{bc} \mathrm{E}(X_{abc}) - \mathbf{m} \\
&= \mathrm{E}(\bar{X}_{a..}) - \mathrm{E}(\bar{X}_{...}) \\
\mathbf{b}_b &= \frac{1}{AC} \sum_{ac} (X_{abc}) - \mathbf{m} \\
&= \mathrm{E}(\bar{X}_{.b.}) - \mathrm{E}(\bar{X}_{...}) \\
\mathbf{c}_c &= \frac{1}{AB} \sum_{ab} \mathrm{E}(X_{abc}) - \mathbf{m} \\
&= \mathrm{E}(\bar{X}_{..c}) - \mathrm{E}(\bar{X}_{...})
\end{aligned}$$

A two-factor interaction term, such as $\mathbf{ab}_{21}$, accounts for the difference between the mean expectation for all conditions, including both a_2 and b_1, and the main effects model based on the main terms related to the particular interaction. Specifically,

$$\begin{aligned}
\mathbf{ab}_{ab} &= \frac{1}{C} \sum_{c} \mathrm{E}(X_{abc}) - (\mathbf{m} + \mathbf{a}_a + \mathbf{b}_b) \\
&= \mathrm{E}(\bar{X}_{ab.}) - \mathrm{E}(\bar{X}_{.b.}) - \mathrm{E}(\bar{X}_{a..}) + \mathrm{E}(\bar{X}_{...}) \\
\mathbf{ac}_{ac} &= \frac{1}{B} \sum_{b} \mathrm{E}(X_{abc}) - (\mathbf{m} + \mathbf{a}_a + \mathbf{c}_c)
\end{aligned}$$

$$= \mathrm{E}(\bar{X}_{a.c}) - \mathrm{E}(\bar{X}_{..c}) - \mathrm{E}(\bar{X}_{a..}) + \mathrm{E}(\bar{X}_{...})$$

$$\mathbf{bc}_{bc} = \frac{1}{A}\sum \mathrm{E}(X_{abc}) - (\mathbf{m} + \mathbf{b}_b + \mathbf{c}_c)$$

$$= \mathrm{E}(\bar{X}_{.bc}) - \mathrm{E}(\bar{X}_{..c}) - \mathrm{E}(\bar{X}_{.b.}) + \mathrm{E}(\bar{X}_{...})$$

In $A \times B$, the two-factor interaction model with $\mathbf{ab}_{ab}$ completely accounts for a condition expectation, namely, for $\mathrm{E}(X_{ab})$ (cf. "algebraic interpretation" in Section 4.3.6). In $A \times B \times C$, however, the two-factor interaction model $\mathbf{m} + \mathbf{a}_a + \mathbf{b}_b + \mathbf{ab}_{ab}$ only accounts for the mean of a group of C expectations. Similar remarks apply for $\mathbf{bc}_{bc}$ and $\mathbf{ac}_{ac}$.

In the two-factor interaction model for $A \times B$, there is one two-factor interaction parameter for each condition. The parameter accounts for a discrepancy, which, in general, occurs between the condition expectation and a main effects model. In $A \times B \times C$, analogously, there is a triple interaction term $\mathbf{abc}_{abc}$ for each condition. This term accounts for the discrepancy that generally occurs between a condition expectation and, in this case, the following two-factor interaction model:

$$\mathrm{E}(X_{abc}) = \mathbf{m} + \mathbf{a}_a + \mathbf{b}_b + \mathbf{c}_c + \mathbf{ab}_{ab} + \mathbf{ac}_{ac} + \mathbf{bc}_{bc}$$

Such a model, in general, fails to account completely for the condition expectations. So that the model can fit any quantities whatsoever that the expectations might have, a three-factor interaction $\mathbf{abc}_{abc}$ is defined, in essence, as

$$\mathbf{abc}_{abc} = \mathrm{E}(X_{abc}) - (\mathbf{m} + \mathbf{a}_a + \mathbf{b}_b + \mathbf{c}_c + \mathbf{ab}_{ab} + \mathbf{ac}_{ac} + \mathbf{bc}_{bc})$$

4.4.5 Interpretation of Interaction

Two-factor interaction. In Section 4.3.6 we learned how to interpret the presence or absence of two-factor interaction in a two-factor design. In the three-factor design, there are three different types of two-factor interactions, $\mathbf{ab}_{ab}$, $\mathbf{ac}_{ac}$, and $\mathbf{bc}_{bc}$. If we collapse $A \times B \times C$ into the form $A \times B$ by averaging across C, the two-factor interaction AB for $A \times B \times C$ can be interpreted from this collapsed data in exactly the same way that two-factor interaction was interpreted originally for design $A \times B$. We can, for example, plot curves on the collapsed data, just as we did for the (uncollapsed) data of $A \times B$. If the curves are parallel, there is no AB interaction. Otherwise, there is an interaction. By collapsing across A, we can examine the BC interaction graphically, and by collapsing across B, we can examine the AC interaction graphically.

Three-factor interaction. Although the three-factor interaction term can be understood mathematically as a necessity to ensure that the general score model can fit any set of expectations whatsoever, it is difficult to obtain a simple, intuitive grasp of what the presence or absence of three-factor interaction implies for a set of expectations. There is no graphic explanation as simple as the type given for the two-factor interaction. However, we can give a simple effects explanation of three-factor interaction, just as we gave a simple effects interpretation of two-factor interaction.

Suppose that $A \times B \times C$ were interpreted as a set of $A \times B$ designs, C in number. In other words, for each level of factor C there is an $A \times B$ layout that might be thought of as a separate $A \times B$ design. For each such $A \times B$ design we can calculate a simple population mean, simple main effects, and simple two-factor interactions $\mathbf{ab}_{ab}$. The designations are $\mathbf{m}_{(c_c)}$, $\mathbf{a}_{a(c_c)}$, $\mathbf{b}_{b(c_c)}$, and $\mathbf{ab}_{ab(c_c)}$.

Using a derivation similar to that used in Section 4.3.6, we can show that

$$\mathbf{abc}_{abc} = \mathbf{ab}_{ab(c_c)} - \mathbf{ab}_{ab}$$

This means that a three-factor interaction term $\mathbf{abc}_{abc}$ can be described as the difference between a simple interaction $\mathbf{ab}_{ab(c_c)}$ at level c_c and the interaction of A and B for the design as a whole, $\mathbf{ab}_{ab}$. If there is no three-factor interaction, all simple two-factor interaction parameters for each level of the third factor must equal the comparable two-factor interaction parameters for the design as a whole; that is, $\mathbf{ab}_{ab(c_c)}$ must equal $\mathbf{ab}_{ab}$ for each combination of subscripts a, b, and c.

Although the preceding explanation used $\mathbf{ab}_{ab}$ at c_c levels, we could use $\mathbf{ac}_{ac}$ at b_b levels or $\mathbf{bc}_{bc}$ at a_a levels. The same three-factor interaction parameters result regardless of which of these three simple effects is used to compute them.

4.4.6 Numerical Example

Following is an example of an $A \times B \times C$ design having all three types of two-factor interaction ($\mathbf{ab}_{ab}$, $\mathbf{ac}_{ac}$, and $\mathbf{bc}_{bc}$), but lacking a three-factor interaction. Expectations are given here and parameters are calculated. However, we could interpret the numbers given as estimates of expectations, determined by averaging the scores in each cell. Then the same computations would yield parameter estimates instead of parameters. The expectations are:

	c_1			c_2	
	b_1	b_2		b_1	b_2
a_1	5	3	a_1	7	17
a_2	9	9	a_2	5	17
a_3	16	6	a_3	12	14

$$\mathbf{m} = \mathrm{E}(\bar{X}_{...}) = 120/12 = 10$$

$$\mathbf{a}_a = \mathrm{E}(\bar{X}_{a..}) - \mathbf{m}$$

$$\mathbf{a}_1 = [(5 + 3 + 7 + 17)/4] - 10 = -2$$
$$\mathbf{a}_2 = [(9 + 9 + 5 + 17)/4] - 10 = 0$$
$$\mathbf{a}_3 = [(16 + 6 + 12 + 14)/4] - 10 = 2$$

$$\mathbf{b}_b = \mathrm{E}(\bar{X}_{.b.}) - \mathbf{m}$$

$$\mathbf{b}_1 = [(5 + 9 + 16 + 7 + 5 + 12)/6] - 10 = -1$$
$$\mathbf{b}_2 = [(3 + 9 + 6 + 17 + 17 + 14)/6] - 10 = 1$$

$$\mathbf{c}_c = \mathrm{E}(\bar{X}_{..c}) - \mathbf{m}$$

$$\mathbf{c}_1 = [(5 + 9 + 16 + 3 + 9 + 6)/6] - 10 = -2$$
$$\mathbf{c}_2 = [(7 + 5 + 12 + 17 + 17 + 14)/6] - 10 = 2$$

$$\mathbf{ab}_{ab} = \mathrm{E}(\bar{X}_{ab.}) - (\mathbf{m} + \mathbf{a}_a + \mathbf{b}_b)$$

$$\mathbf{ab}_{11} = [(5 + 7)/2] - (10 - 2 - 1) = 6 - 7 = -1$$
$$\mathbf{ab}_{12} = [(3 + 17)/2] - (10 - 2 + 1) = 10 - 9 = 1$$
$$\mathbf{ab}_{21} = [(9 + 5)/2] - (10 + 0 - 1) = 7 - 9 = -2$$
$$\mathbf{ab}_{22} = [(9 + 17)/2] - (10 + 0 + 1) = 13 - 11 = 2$$
$$\mathbf{ab}_{31} = [(16 + 12)/2] - (10 + 2 - 1) = 14 - 11 = 3$$
$$\mathbf{ab}_{32} = [(6 + 14)/2] - (10 + 2 + 1) = 10 - 13 = -3$$

$$\mathbf{ac}_{ac} = \mathrm{E}(\bar{X}_{a.c}) - (\mathbf{m} + \mathbf{a}_a + \mathbf{c}_c)$$

$$\mathbf{ac}_{11} = [(5 + 3)/2] - (10 - 2 - 2) = 4 - 6 = -2$$
$$\mathbf{ac}_{12} = [(7 + 17)/2] - (10 - 2 + 2) = 12 - 10 = 2$$
$$\mathbf{ac}_{21} = [(9 + 9)/2] - (10 + 0 - 2) = 9 - 8 = 1$$
$$\mathbf{ac}_{22} = [(5 + 17)/2] - (10 - 0 + 2) = 11 - 12 = -1$$
$$\mathbf{ac}_{31} = [(16 + 6)/2] - (10 + 2 - 2) = 11 - 10 = 1$$
$$\mathbf{ac}_{32} = [(12 + 14)/2] - (10 + 2 + 2) = 13 - 14 = -1$$

$$\mathbf{bc}_{bc} = \mathrm{E}(\bar{X}_{.bc}) - (\mathbf{m} + \mathbf{b}_b + \mathbf{c}_c)$$

$$\mathbf{bc}_{11} = [(5 + 9 + 16)/3] - (10 - 1 - 2) = 10 - 7 = 3$$
$$\mathbf{bc}_{12} = [(7 + 5 + 12)/3] - (10 - 1 + 2) = 8 - 11 = -3$$
$$\mathbf{bc}_{21} = [(3 + 9 + 6)/3] - (10 + 1 - 2) = 6 - 9 = -3$$
$$\mathbf{bc}_{22} = [(17 + 17 + 14)/3] - (10 + 1 + 2) = 16 - 13 = 3$$

$$\mathbf{abc}_{abc} = \mathrm{E}(X_{abc}) - (\mathbf{m} + \mathbf{a}_a + \mathbf{b}_b + \mathbf{c}_c + \mathbf{ab}_{ab} + \mathbf{ac}_{ac} + \mathbf{bc}_{bc})$$

$$\mathbf{abc}_{111} = 5 - (10 - 2 - 1 - 2 - 1 - 2 + 3) = 5 - 5 = 0$$
$$\mathbf{abc}_{112} = 7 - (10 - 2 - 1 + 2 - 1 + 2 - 3) = 7 - 7 = 0$$

$\mathbf{abc}_{abc} = 0$ for all values of subscripts a, b, and c.

The calculation of simple effects for $A \times B$ at c_1 is as follows:

$$\mathbf{m}_{(c_1)} = \mathrm{E}(\bar{X}_{..1})$$
$$= (5 + 3 + 9 + 9 + 16 + 16)/3 \cdot 2 = 8$$

$$\mathbf{a}_{a(c_1)} = \text{E}(\bar{X}_{a.1}) - \mathbf{m}$$

$$\mathbf{a}_{1(c_1)} = [(5 + 3)/2] - 8 = -4$$
$$\mathbf{a}_{2(c_1)} = [(9 + 9)/2] - 8 = 1$$
$$\mathbf{a}_{3(c_1)} = [(16 + 6)/2] - 8 = 3$$

$$\mathbf{b}_{b(c_1)} = \text{E}(\bar{X}_{.b1}) - \mathbf{m}_{(c_1)}$$

$$\mathbf{b}_{1(c_1)} = [(5 + 9 + 16)/3] - 8 = 2$$
$$\mathbf{b}_{2(c_1)} = [(3 + 9 + 6)/3] - 8 = -2$$

$$\mathbf{ab}_{ab(c_1)} = \text{E}(X_{ab1}) - (\mathbf{m}_{(c_1)} + \mathbf{a}_{1(c_1)} + \mathbf{b}_{1(c_1)})$$

$$\mathbf{ab}_{11(c_1)} = 5 - (8 - 4 + 2) = -1$$
$$\mathbf{ab}_{12(c_1)} = 3 - (8 - 4 - 2) = 1$$
$$\mathbf{ab}_{21(c_1)} = 9 - (8 + 1 + 2) = -2$$
$$\mathbf{ab}_{22(c_1)} = 9 - (8 + 1 - 2) = 2$$
$$\mathbf{ab}_{31(c_1)} = 16 - (8 + 3 + 2) = 3$$
$$\mathbf{ab}_{32(c_2)} = 6 - (8 + 3 - 2) = -3$$

In every case the simple interaction $\mathbf{ab}_{ab(c_1)}$ equals the interaction $\mathbf{ab}_{ab}$. (The same is true for $\mathbf{ab}_{ab(c_2)}$, though demonstration of this is left to the student.) Since $\mathbf{abc}_{abc} = \mathbf{ab}_{ab(c_c)} - \mathbf{ab}_{ab}$ (Section 4.4.5), it follows that the three-factor interaction is zero, as indeed, we can also show by direct calculations.

The expectations are graphed in Figures 4.4-1(a and b). Figure 4.4-1c shows the same data collapsed (averaged) over C. In essence, it is this collapsed data from which the existence of an AB interaction is inferred. If the lines for b_1 and b_2 were parallel, the interaction would be zero. From Figure 4.4-1(a and b) it can be discerned that the simple AB interactions at c_1 and c_2 are not zero, for the b_b lines are not parallel. It is not easy to see, however, that the AB interaction parameters are exactly the same for all three diagrams.

4.4.7 Random Factors

If any or all of factors A, B, and C are random, the score model is the same as when all the factors are fixed. The score expectations differ, however, because the expectation for any term based on one or more random factors is zero. The estimates of terms, on the other hand, are the same regardless of whether the factors are interpreted as fixed or random. The theoretical components of the estimates change, however, because of the change in side conditions.

4.4.8 One Score per Condition

If $A \times B \times C$ has only one score per cell, we adjust the model equation, just as we did for design $A \times B$. In this example, however, error is not

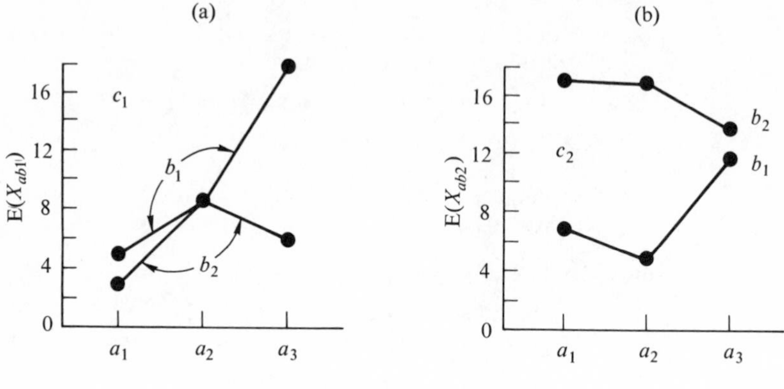

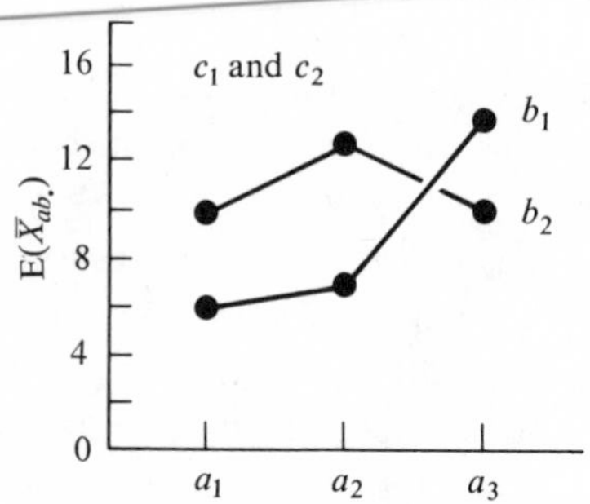

Figure 4.4-1. Graphs showing the expectations (a) for level c_1 only, (b) for level c_2 only, and (c) averaged across c_1 and c_2.

confounded with a two-factor interaction, but with the three-factor interaction $\mathbf{abc}_{abc}$. Under such circumstances, we write the score model as follows:

$$X_{abc} = \mathbf{m} + \mathbf{a}_a + \mathbf{b}_b + \mathbf{c}_c + \mathbf{ab}_{ab} + \mathbf{ac}_{ac} + \mathbf{bc}_{bc} + \mathbf{abc}_{abc} + \mathbf{e}_{abc}$$

Note that subscript e has been omitted from X and from term $\mathbf{e}$, and the parentheses have been removed from the abc subscript on term $\mathbf{e}$. The confounded terms $\mathbf{abc}_{abc}$ and $\mathbf{e}_{abc}$ have exactly the same subscripts.

4.5 DESIGN $S \times A$

4.5.1 Introduction

Sections 4.5.1 and 4.5.2 concern the completely crossed repeated measurements (Class IV) designs, $S \times A$ and $S \times A \times B$. There are two special considerations regarding the modeling and analysis of repeated measurements designs. First, all such designs include the factor S, and S is virtually always treated as a random factor. Second, in repeated measurements experiments, a condition of the experiment virtually always has only one score per condition. In contrast, the completely crossed single measurement designs that have been discussed so far usually have several scores per condition, although such designs sometimes have one score per condition as well. The modeling and analysis of $S \times A$ and $S \times A \times B$ are the same as the modeling and analysis for $A \times B$ and $A \times B \times C$, respectively, when these latter designs have one score per condition and the same number of random factors as the former designs.

Design $S \times A$ can be diagrammed as follows:

	a_1	a_2	$\cdots$	a_A
s_1	X_{11}	X_{21}		X_{A1}
s_2				
$\cdot$				
$\cdot$				
$\cdot$				
s_S	X_{1S}	X_{2S}		X_{AS}

This design can be compared with both design A and design $A \times B$. It is like design A in that the purpose of both design A and $S \times A$ is to study the effects of one factor, A, on some criterion. It is like design $A \times B$ in that, according to the convention used in this book, both are two-factor designs. The similarity is clear when we compare the models and the analyses for the two designs. Design $S \times A$ is particularly similar to $A \times B$ when the latter has only one score per condition, or $A \cdot B$ scores altogether. Design $S \times A$ virtually always has but one score for each as_{as} condition.

Typically, an experimenter has no real interest in the score differences among the subjects for design $S \times A$, even though S is a design factor. It can be taken for granted that there will be differences, since such differences among subjects are virtually always found when the experimenter looks for them. There is no possibility of studying an AS interaction by using this design, for reasons that will be discussed later.

Although the purpose of both A and $S \times A$ is to investigate the effects of factor A, each has particular advantages and disadvantages. The relative merits will not be analyzed at this time, however, because the purpose of this chapter is to present and explain the design models, not to consider the relative merits of competing designs. Design $S \times A$ is frequently classified as a one-factor design. However, in this book S is considered to be a design factor, and $S \times A$ is classified as a two-factor design.

4.5.2 The Score Model

A score is symbolized by X_{as}. The order of the subscripts on X is alphabetical, even though in the design symbolization S is placed first. Because $S \times A$ has but one score per condition, subscript e is superfluous and is omitted from X as well as from **e**.

The score model for $S \times A$ is

$$X_{as} = \mathbf{m} + \mathbf{a}_a + \mathbf{s}_s + \mathbf{as}_{as} + \mathbf{e}_{as}$$

Except for a change in symbolism from **b** to **s**, this is exactly the score model that was presented for $A \times B$ with one score per condition. Here, as with $A \times B$ with one score per condition, error is confounded with the two-factor interaction, as the equality in the subscripts indicates.

Main term $\mathbf{a}_a$, as before, represents an effect due to level $\mathbf{a}_a$. Main term $\mathbf{s}_s$ is basically like any other main term. It represents an effect on scores due to subject s_s. The interaction term $\mathbf{as}_{as}$ is fundamentally like $\mathbf{ab}_{ab}$, though in general it is described slightly differently. The existence of an $\mathbf{as}_{as}$ interaction means that one subject's tendency to score higher (or lower) than another is not uniform across all levels of factor A. For example, the true scoring advantage of s_4 over s_2 might be 5 for a_1 but only 2 for a_3. If there were no interaction, the true scoring advantage would be the same at each level a_a. This discussion of interaction is of small practical bearing, however, for in design $S \times A$ with one score per condition, we cannot even test for the presence of such an interaction.

Because S is formalized as a design factor and because, in repeated measurements designs, it is possible to have an $\mathbf{s}_s$ term that is not completely confounded with error, the error term here is what was called variable error in Section 4.2.3. It is free of the systematic scoring tendency $\mathbf{s}_s$ associated with a particular subject.

In some sources, the student may see the model for $S \times A$ written without the interaction term, which, according to the notation used in this book, would be

$$X_{as} = \mathbf{m} + \mathbf{a}_a + \mathbf{s}_s + \mathbf{e}_{as}$$

Such a model assumes that interaction $\mathbf{as}_{as}$ is zero. Under such an assumption, the discrepancy between a score and the estimated main effects model estimates error as follows:

$$\hat{\mathbf{e}}_{as} = X_{as} - (\hat{\mathbf{m}} + \hat{\mathbf{a}}_a + \hat{\mathbf{s}}_s)$$

The assumption that $\mathbf{as}_{as}$ is zero would seldom have much justification in practice, but such an assumption might be made at times for instructive purposes—to see whether it would, or would not, affect the analysis.

4.5.3 Side Conditions, Score Expectations, and Estimates

We assume that factor A is fixed, but, as we have noted, factor S is always random. The side conditions, then, are:

$$\mathbf{a}_{.} = 0$$
$$\mathbf{s}_{.} \text{ is r.v.}$$
$$\mathbf{as}_{.s} = 0 \quad \text{(for any constant value of subscript } s\text{)}$$
$$\mathbf{as}_{a.} \text{ is r.v.} \quad \text{(for any constant value of subscript } a\text{)}$$

The expectation of a constant over replications is that constant, but the expectation of any term having a random factor referent is zero. Thus

$$\text{E}(X_{as}) = \mathbf{m} + \mathbf{a}_a$$

Formulas for parameter estimation are not based on solutions to the preceding equation, however, but, as always, on the score expectation equations that assume all design factors to be fixed:

$$\text{E}(X_{as}) = \mathbf{m} + \mathbf{a}_a + \mathbf{s}_s + \mathbf{as}_{as}$$

Estimation for $S \times A$ is similar to estimation for $A \times B$ with B random (Section 4.3.8), except that here there is no subscript e:

$$\begin{aligned}
\hat{\mathbf{m}} &= \bar{X}_{..} \\
&= \mathbf{m} + \bar{\mathbf{s}}_{.} + \bar{\mathbf{e}}_{..} \\
\hat{\mathbf{a}}_a &= \bar{X}_{a.} - \bar{X}_{..} \\
&= \mathbf{a}_a + \overline{\mathbf{as}}_{a.} + \bar{\mathbf{e}}_{a.} - \bar{\mathbf{e}}_{..} \\
\hat{\mathbf{s}}_s &= \bar{X}_{.s} - \bar{X}_{..} \\
&= \mathbf{s}_s - \bar{\mathbf{s}}_{.} + \bar{\mathbf{e}}_{.s} - \bar{\mathbf{e}}_{..} \\
\widehat{\mathbf{as}}_{as} &= X_{as} - \bar{X}_{.s} - \bar{X}_{a.} + \bar{X}_{..} \\
&= \mathbf{as}_{as} - \overline{\mathbf{as}}_{a.} + \mathbf{e}_{as} - \bar{\mathbf{e}}_{.s} - \bar{\mathbf{e}}_{a.} + \bar{\mathbf{e}}_{..}
\end{aligned}$$

What we call $\widehat{\mathbf{as}}_{as}$ can be thought of as an estimate of $\mathbf{as}_{as} + \mathbf{e}_{as}$. It is not possible to make separate estimates for interaction and error with only one score per condition. If $\widehat{\mathbf{as}}_{as}$ were assumed to be zero, then the formula for $\widehat{\mathbf{as}}_{as}$ would estimate $\mathbf{e}_{as}$; that is, we would have

$$\begin{aligned}
\hat{\mathbf{e}}_{as} &= X_{as} - \bar{X}_{.s} - \bar{X}_{a.} + \bar{X}_{..} \\
&= \mathbf{e}_{as} - \bar{\mathbf{e}}_{.s} - \bar{\mathbf{e}}_{a.} + \bar{\mathbf{e}}_{..}
\end{aligned}$$

4.5.4 Numerical Example

The following table gives the hypothetical scores from an $S \times A$ experiment:

	a_1	a_2	a_3	a_4
s_1	5	4	1	8
s_2	7	4	1	10
s_3	12	9	8	16
s_4	4	9	6	9
s_5	8	9	5	13

Each of five subjects ($S = 5$) has scores for each of $A = 4$ treatments, yielding 20 scores altogether. The fact that the scores appear in the table in the same order from a_1 to a_4 for each subject does not imply that all subjects received the treatments in the order a_1 to a_4. We may assume, instead, that a different random order of presentation was used for each subject. The estimates of the parameters, calculated by the method of successive extraction, are as follows:

$$\hat{\mathbf{m}} = \bar{X}_{..} = 148/20 = 7.4$$

$$\hat{\mathbf{a}}_a = \bar{X}_{a.} - \hat{\mathbf{m}}$$

$$\begin{aligned}
\hat{\mathbf{a}}_1 &= [36/5] - 7.4 = -.2\\
\hat{\mathbf{a}}_2 &= [35/5] - 7.4 = -.4\\
\hat{\mathbf{a}}_3 &= [21/5] - 7.4 = -3.2\\
\hat{\mathbf{a}}_4 &= [56/5] - 7.4 = 3.8
\end{aligned}$$

$$\hat{\mathbf{s}}_s = \bar{X}_{.s} - \hat{\mathbf{m}}$$

$$\begin{aligned}
\hat{\mathbf{s}}_1 &= [18/4] - 7.4 = -2.9\\
\hat{\mathbf{s}}_2 &= [22/4] - 7.4 = -1.9\\
\hat{\mathbf{s}}_3 &= [45/4] - 7.4 = 3.85\\
\hat{\mathbf{s}}_4 &= [28/4] - 7.4 = -.4\\
\hat{\mathbf{s}}_5 &= [35/4] - 7.4 = 1.35
\end{aligned}$$

$$\widehat{\mathbf{as}}_{as} = X_{as} - (\hat{\mathbf{m}} + \hat{\mathbf{a}}_a + \hat{\mathbf{s}}_s)$$

$$\begin{aligned}
\widehat{\mathbf{as}}_{11} &= 5 - (7.4 - .2 - 2.9) = .7\\
\widehat{\mathbf{as}}_{12} &= 7 - (7.4 - .2 - 1.9) = 1.7\\
\widehat{\mathbf{as}}_{13} &= 12 - (7.4 - .2 + 3.85) = .95\\
\widehat{\mathbf{as}}_{14} &= 4 - (7.4 - .2 - .4) = -2.8\\
\widehat{\mathbf{as}}_{15} &= 8 - (7.4 - .2 + 1.35) = -.55
\end{aligned}$$

$$\begin{aligned}
\widehat{\mathbf{as}}_{21} &= 4 - (7.4 - .4 - 2.9) = -.1\\
\widehat{\mathbf{as}}_{22} &= 4 - (7.4 - .4 - 1.9) = -1.1\\
\widehat{\mathbf{as}}_{23} &= 9 - (7.4 - .4 + 3.85) = -1.85\\
\widehat{\mathbf{as}}_{24} &= 9 - (7.4 - .4 - .4) = 2.4\\
\widehat{\mathbf{as}}_{25} &= 9 - (7.4 - .4 + 1.35) = .65
\end{aligned}$$

$$\begin{aligned}
\widehat{\mathbf{as}}_{31} &= 1 - (7.4 - 3.2 - 2.9) = -.3\\
\widehat{\mathbf{as}}_{32} &= 1 - (7.4 - 3.2 - 1.9) = -1.3\\
\widehat{\mathbf{as}}_{33} &= 8 - (7.4 - 3.2 + 3.85) = -.05\\
\widehat{\mathbf{as}}_{34} &= 6 - (7.4 - 3.2 - .4) = 2.2\\
\widehat{\mathbf{as}}_{35} &= 5 - (7.4 - 3.2 + 1.35) = -.55
\end{aligned}$$

$$\begin{aligned}
\widehat{\mathbf{as}}_{41} &= 8 - (7.4 + 3.8 - 2.9) = -.3\\
\widehat{\mathbf{as}}_{42} &= 10 - (7.4 + 3.8 - 1.9) = .7\\
\widehat{\mathbf{as}}_{43} &= 16 - (7.4 + 3.8 + 3.85) = .95\\
\widehat{\mathbf{as}}_{44} &= 9 - (7.4 + 3.8 - .4) = -1.8\\
\widehat{\mathbf{as}}_{45} &= 13 - (7.4 + 3.8 + 1.35) = .45
\end{aligned}$$

In general, $\mathbf{s}_.$ and $\mathbf{as}_{a.}$ for an experiment do not equal zero, since S is a random factor. However, estimates for random factor terms conform to the side conditions for fixed factors, so $\hat{\mathbf{s}}_. = 0$ and $\widehat{\mathbf{as}}_{a.} = 0$, as the student may verify for the preceding estimates.

4.6 DESIGN $S \times A \times B$

4.6.1 Introduction

Design $S \times A \times B$ is a Class IV repeated measurements design. In essence, the design is an alternative to $A \times B$ in that its basic purpose is to investigate the effects of factors A and B. Factor S is included for technical reasons, but $S \times A \times B$ is still called a three-factor design, since S is always included in the count of design factors. Furthermore, the design's score model resembles the model for a three-factor design $A \times B \times C$, especially when the latter design has but one score per condition. In this section A and B are assumed to be fixed factors even though, as usual, S is random.

4.6.2 The Score Model

A score is symbolized by X_{abs}. Since there is only one score per condition, subscript e is omitted. Again, the order of the subscripts on X is alphabetical, even though S appears first in the design symbolization. The score model is

$$X_{abs} = \mathbf{m} + \mathbf{a}_a + \mathbf{b}_b + \mathbf{s}_s + \mathbf{ab}_{ab} + \mathbf{as}_{as} + \mathbf{bs}_{bs} + \mathbf{abs}_{abs} + \mathbf{e}_{abs}$$

Note that the order of the terms follows our convention: main terms, two-factor interaction terms, three-factor interaction term, then error. Main terms appear in alphabetical order, and two-factor interaction terms are ordered by the first letter insofar as possible, then by the second letter when the first letter is the same for two terms. The score model for $S \times A \times B$ is the same as the model for $A \times B \times C$ with one score per condition, except that $\mathbf{s}$ replaces $\mathbf{c}$.

For $S \times A$, the two-factor interaction term $\mathbf{as}_{as}$ is confounded with the error term. For $S \times A \times B$, however, error is confounded neither with $\mathbf{as}_{as}$ nor with any other two-factor interaction term, but with the three-factor interaction term $\mathbf{abs}_{abs}$. In general, for completely crossed (Class IV) designs with factor S, only the highest-order interaction term—the one involving all design factors including S—is confounded with the error term. Therefore, an interaction term confounded with error in one design will not be confounded with error in a design with more factors.

The error term e in the model for $S \times A \times B$, as in all repeated measurements design models, represents the "variable error" component only, not a confounding of this error with systematic scoring tendencies of individual subjects (Section 4.2.3).

4.6.3 Side Conditions

It is no longer necessary to spell out the side conditions in detail. The sum of parameters across all levels of the subscript for a fixed factor is zero. This is true even for interaction terms having one or more random factor symbols. For a random factor, the sum of parameters across all levels in the *population* is zero; the sum across all levels for a particular experiment, however, would not, in general, be zero, and would be a random variable across replications.

4.6.4 Estimation of Parameters

The procedures for estimating the parameters for $S \times A \times B$ parallel those used for $A \times B \times C$ with one score per condition, the only change being in the symbolism. The formulas required for estimating parameters for $A \times B \times C$ with one score per condition were not given, but in Section 4.4.4 formulas for were given calculating parameters assuming that the expectations were known. Estimates can be made using these same formulas, except that estimated expectations substitute for the true values, and for $S \times A \times B$ there is no subscript e to average over. When there is only one score per condition, the condition expectation equals that score. For example, if there are several scores per condition in $A \times B \times C$, $\hat{\mathrm{E}}(X_{abc}) = \bar{X}_{abc.}$. If there is only one score, $\hat{\mathrm{E}}(X_{abc}) = X_{abc}$. Also, if there is only one score per condition, the discrepancy between a score and the estimated two-factor interaction model for it is an estimate of $\mathbf{abc}_{abc}$. There is no separate estimate for $\mathbf{e}_{abc}$. The fact that S is random makes no difference for purposes of estimation, although it does affect the interpretation of the parametric components of the estimates.

4.7 DESIGN $B(A)$

4.7.1 Introduction

Thus far this chapter has considered only completely crossed designs. Henceforth the chapter considers designs with nesting. The present section concerns a single measurement design involving only nesting. The following two sections concern repeated measurements designs involving both nesting and crossing relations.

There are no repeated measurements designs involving only nesting, for unless S is crossed with at least one factor, by definition we cannot have repeated measurements. We can have single measurement designs involving both nesting and crossing, but no section in this chapter is devoted to such designs. The most common designs involving both crossing and nesting are repeated measurements designs. The models for single measurement designs are the same as those for repeated measurements designs, however (except for the particular symbols used), if adjustments are made, as previously, depending on the number of scores per condition—one or more than one. The theoretical components of the estimates and the analysis of variance are also the same, except for symbolism and for any adjustments that might be required as a result of variations in the classification of factors as random or fixed.

In design $B(A)$ there are B ($B > 1$) levels of factor B associated with each level of factor A. A level of B with the same symbol (say, "b_2") appears in combination with the various levels of A, but these various levels of b_2 are meaningfully different. Level b_2 under a_1 has no more relation to b_2 under a_2 or a_3 than it has to "b_1," "b_4," or any other "b_b" under any other a_a. The b_2 symbols are the same because of initialized labeling (Section 3.4.2). If the various levels b_1, b_2, and so on were meaningfully the same, we would have design $A \times B$, not design $B(A)$.

Factor A might be cities of a certain size and factor B might be schools within the respective cities. Suppose B is 10 schools from each city, randomly selected for inclusion in the investigation. The chosen schools in each city are arbitrarily labeled $b_1, b_2, \ldots, b_{10}$. Within each school 50 pupils from the fourth grade are randomly selected and given a reading achievement test. The investigation could determine whether there are differences among such cities in fourth grade reading achievement, and whether there are differences among the schools within a given city. Such a design would be of Class II; factors A and B are blocking factors, so a particular subject could not have been included in any other condition.

We can also have a Class I design $B(A)$. Let factor A be a treatment factor and let factor B be a group factor. (The design would then normally be symbolized $G(A)$.) Subjects chosen randomly from some common source are randomly assigned in groups of four to the treatments (say, 5 groups to each).

4.7.2 The Score Model

A score for a nested design is symbolized in the same way as a score for a crossed design with the same design factors: simply symbol X with one subscript for each design factor, listed alphabetically, with subscript e added. The score model for $B(A)$ is

$$X_{abe} = \mathbf{m} + \mathbf{a}_a + \mathbf{b}_{b(a)} + \mathbf{e}_{e(ab)}$$

The model includes a nested main term, $\mathbf{b}_{b(a)}$; except for $\mathbf{e}$, nested terms never appear in completely crossed designs. On the other hand, the model lacks an $\mathbf{ab}_{ab}$ interaction term, which is present in the model for $A \times B$. Score models for completely crossed designs always lack nested terms, and score models for designs having only nesting relations always lack interaction terms. As we shall see in subsequent sections of this chapter, however, score models for designs having both nesting and crossing relations have both nested terms and interaction terms.

If a design has only nesting relations, we always must have more than one score per condition, so the score model would never appear without subscript *e*. But could we not conceivably use but one subject per school, say, in the preceding example for $B(A)$? We conceivably could, but we would then consider the design to be simply A, not $B(A)$. With only one subject per condition, the subject-error effect, symbolized by $\mathbf{e}$, would be confounded with the school effect. We would then drop this "school effect" as a separate consideration, even though, indeed, the effect would be included within $\mathbf{e}$, which would now be a confounded school-subject-error effect, although we simply refer to it as "error."

4.7.3 Interpretation of the Nested Term

In design $A \times B$ there is a $\mathbf{b}_b$ term; there is no $\mathbf{b}_b$ term in $B(A)$, but there is a comparable nested term $\mathbf{b}_{b(a)}$. In $A \times B$, $\mathbf{b}_b$ accounts for the systematic difference between the scores at level b_b and the mean of all scores. Term $\mathbf{b}_b$ is estimated as the difference between the mean of all scores at b_b and the mean of all scores in the experiment. In $A \times B$, however, a particular level of B, such as b_2, has the same meaning for each level of A it occurs in combination with. For example, if b_2 refers to some level of a treatment factor, "b_2" means exactly that same level throughout the experiment.

In $B(A)$, however, "b_2's" at different a_a levels have no special relation to each other. If there are three levels of A, there would be no more point in averaging $\mathbf{b}_{1(1)}$, $\mathbf{b}_{1(2)}$, and $\mathbf{b}_{1(3)}$, than in averaging $\mathbf{b}_{1(1)}$, $\mathbf{b}_{4(2)}$, and $\mathbf{b}_{2(3)}$, or any other combination of b's across the levels of A. A meaningful level of B occurs only within one level of A. A particular school, for example, occurs only within one city. In $G(A)$, a particular group of four subjects appears only within a single level of A. A parameter describing the effect of a nested level thus is applicable only to scores at one level of the nest factor, that level being indicated by the nest subscript. This is indicated by use of the term $\mathbf{b}_{b(a)}$ instead of $\mathbf{b}_b$.

Term $\mathbf{b}_{b(a)}$ differs from $\mathbf{b}_b$ in another way as well. It does not represent the systematic discrepancy between scores at level $b_{b(a)}$ and the population mean. Part of that discrepancy is already accounted for by $\mathbf{a}_a$. Term $\mathbf{b}_{b(a)}$ merely accounts for systematic differences among $b_{b(a)}$ levels at a given level of A. For example, $\mathbf{b}_{2(3)}$ accounts for the systematic difference between scores at $b_{2(3)}$ and all scores at a_3. It is estimated, therefore, as

$$\hat{\mathbf{b}}_{b(a)} = \bar{X}_{ab.} - \bar{X}_{a..}$$

Term $\mathbf{b}_{b(a)}$ is analogous to a simple effect of design $A \times B$. Were the design actually $A \times B$, that is, were the b_b's the same for each a_a, the simple effect of b_b at a particular level of A, $\mathbf{b}_{b(a_a)}$, would be estimated as $\bar{X}_{ab.} - \bar{X}_{a..}$, in the same way that we estimated $\mathbf{b}_{b(a)}$. Nested effects (including nested interaction effects) are, in general, analogous to simple effects.

4.7.4 The Missing Interaction Term

The interaction term $\mathbf{ab}_{ab}$ was a necessary component of the score model for $A \times B$. Why is no such term required for $B(A)$? Term $\mathbf{ab}_{ab}$ is necessary because the effect of level b_b on scores is very often not the same at each level of a_a (hence the failure of the lines to be parallel in Figure 4.3-1b). However, in $B(A)$, a particular (meaningful) level of B occurs only with a single level of A. If this level of B had occurred with a different level of A, the magnitude of its effect might have differed. In other words, the theoretical possibility of an $\mathbf{ab}_{ab}$ interaction is not denied; however, with $B(A)$, there is no need for a supplementary term $\mathbf{ab}_{ab}$. Whatever special effect b_b's own level of A has on the scores is included within $\mathbf{b}_{b(a)}$ already, by the way that $\mathbf{b}_{b(a)}$ is defined and estimated.

If the same levels of B and A were used in a crossed design, then $\mathbf{b}_{b(a)}$ of the nested design would equal $\mathbf{b}_b + \mathbf{ab}_{ab}$ of the crossed design. We can think of $\mathbf{b}_{b(a)}$ as representing a confounded sum of $\mathbf{b}_b + \mathbf{ab}_{ab}$, terms that can be picked apart in $A \times B$, but not in $B(A)$. As we shall see, in general we can interpret any nested term as being equivalent to a confounded sum of non-nested terms from a crossed design involving the same factors.

4.7.5 Side Conditions and Estimation

Side conditions for non-nest subscripts are formed as before, regardless of whether the nest subscript (or subscripts) is interpreted as fixed or random. If B is fixed, $\mathbf{b}_{.(a)} = 0$; if B is random, $\mathbf{b}_{.(a)}$ is r.v. The nature of A is irrelevant. As a rule, nested factors are random.

What about the summation over a in $\mathbf{b}_{b(a)}$ for some specific b? Such a summation never occurs in the formulas that interest us; however, we would not expect $\mathbf{b}_{b(.)}$ to be zero, even if A were fixed. We do encounter summations such as $\mathbf{b}_{.(.)}$. This summation is zero if $\mathbf{b}_{.(a)}$ is zero, for it is simply the sum of A quantities, each zero. If $\mathbf{b}_{.(a)}$ is r.v., then $\mathbf{b}_{.(.)}$ is also r.v., for it is the sum of A independent random variables.

We could derive the formulas for estimating the parameters by imagining A and B to be fixed and solving the expectation model, as previously. Let us forego that process now, since it is familiar, and simply write

$$\begin{aligned} \hat{\mathbf{m}} &= \bar{X}_{...} \\ \hat{\mathbf{a}}_a &= \bar{X}_{a..} - \bar{X}_{...} \\ \hat{\mathbf{b}}_{b(a)} &= \bar{X}_{ab.} - \bar{X}_{a..} \\ \hat{\mathbf{e}}_{e(ab)} &= \bar{X}_{abe} - \bar{X}_{ab.} \end{aligned}$$

Estimation for nested designs proceeds in much the same manner as for crossed designs, except that, as we have noted, nested terms are estimated as if they were simple effects. The theoretical components of the estimates depend on whether A and B are interpreted as random or fixed. The student should have no trouble deriving the formulas.

4.8 DESIGN $S(A) \times B$

4.8.1 Introduction

Up to now this chapter has discussed designs involving only crossing or only nesting. We shall now consider designs involving both crossing and nesting. Such designs may be either single measurement or repeated measurements designs. Most often they are repeated measurements designs, and the discussion here concerns only this type. The repeated measurements designs we now consider fall in Class III and, as a rule, have only one score per condition, just as do Class IV designs. However, if the symbolization of the Class III design is the same as for a single measurement design (except for the specific factor symbols involved), the model has the same form also, except for the usual adjustments required when there is but one score per condition.

Let us first consider the simplest possible Class III design, $S(A) \times B$. Factor S is nested within A, but both S and A are crossed with B. Recall (Section 3.4.4) that $S(A) \times B$ is the simplest design involving a nested factor S, there being no design that is symbolized $S(A)$.

A particular subject in $S(A) \times B$ is scored under every one of the levels of factor B, but only under one level of A. Each subject, then, has B scores. In the following design, $A = 2$, $B = 3$, and $S = 4$. Initialized labeling is used for the levels of S so that s_1 within a_1 is a different subject than s_1 within a_2, the two having no special relation to each other.

This design can be compared with $A \times B$ and $S \times A \times B$ in that the primary purpose of all three is to investigate the effects of factors A and B.

		b_1	b_2	b_3
a_1	s_1			
	s_2			
	s_3			
	s_4			
a_2	s_1			
	s_2			
	s_3			
	s_4			

In other words, we can investigate these factors with designs requiring that a single subject have one treatment, all treatments, or an intermediate number of treatments.

4.8.2 The Score Model

A score for $S(A) \times B$ is symbolized by X_{abs}. Here, as with $S \times A \times B$, there is but one score per condition, so error is confounded with an interaction term, and we have no error subscript e. The subscripts are in alphabetical order, and there is no subscript nest on X, even for nested designs.

The score model is

$$X_{abs} = \mathbf{m} + \mathbf{a}_a + \mathbf{b}_b + \mathbf{s}_{s(a)} + \mathbf{ab}_{ab} + \mathbf{bs}_{bs(a)} + \mathbf{e}_{abs}$$

The model has not only a nested main term, $\mathbf{s}_{s(a)}$, but a nested interaction term, $\mathbf{bs}_{bs(a)}$. Note that the nested terms lack a bold-faced letter corresponding to the nest subscript—a general rule for nested terms. Although, in completely crossed designs with one score per condition, error is confounded with the highest-order interaction term, in the present model there appears to be no "highest-order" interaction term, since there are two two-factor interaction terms. It is the nested two-factor term that is confounded. More generally, it is the interaction term having subscripts for all design factors, whether the subscripts are in a nest or not, that is confounded with error.

We adhere to the following rules for ordering model terms when there are nested factors:

1. Terms with fewer subscripts are written before terms with a greater number, regardless of whether the subscripts are nested or nest.
2. When terms have the same total number of subscripts, those having fewer non-nest subscripts precede those having a greater number.
3. Among terms having the same total number of subscripts and the same number of nested subscripts, ordering is alphabetical.

4.8.3 Interpretation of Nested Terms

In $S \times A \times B$ there is an $\mathbf{s}_s$ term; there is no $\mathbf{s}_s$ term in $S(A) \times B$, but there is a comparable nested term $\mathbf{s}_{s(a)}$. In $S \times A \times B$, $\mathbf{s}_s$ accounts for the systematic difference between the score of a particular subject and the mean score of all subjects. It is estimated as the difference between the mean of all scores for a subject and the mean of all scores in the experiment.

In $S(A) \times B$ such an estimate would reflect not only an effect due to a particular subject, but an effect due to his level of factor A as well. Subjects differ systematically in scoring in $S(A) \times B$, not only because of inherent differences with respect to the scoring instrument, but because they are differentially affected by factor A. In $S \times A \times B$, by way of contrast, all subjects are exposed equally often to all levels of A and B, so systematic differences in their scoring cannot be attributed even in part to the treatment factors.

If, with $S(A) \times B$, we consider only a particular level of A, however, all subjects at that level are exposed to exactly the same levels of treatment factors, so systematic differences between those subjects are due only to inherent subject differences. Term $\mathbf{s}_{s(a)}$ accounts for such differences. It is estimated as the difference between the mean of all scores for a subject and the mean of all scores at the same level of A.

Recall that in Section 4.7, the nested effect $\mathbf{b}_{b(a)}$ of $B(A)$ was compared with the simple effect $\mathbf{b}_{b(a_a)}$ of $A \times B$. Likewise, the nested effects for $S(A) \times B$ can be compared with simple effects for $S \times A \times B$. The nested effect $\mathbf{s}_{s(a)}$ of $S(A) \times B$ is analogous to the simple effect $\mathbf{s}_{s(a_a)}$ of $S \times A \times B$, and they are estimated by the same computations. In both cases we can imagine that the design exists only at one level of A, thus making the design, effectively, $S \times B$. Then either $\mathbf{s}_{s(a)}$ or $\mathbf{s}_{s(a_a)}$ is equivalent to the main effect $\mathbf{s}_s$ for the reduced design $S \times B$. The term accounts for the systematic discrepancy between the mean scores for one subject and the mean scores for all subjects (at that level a_a). In a similar way, the nested interaction term $\mathbf{bs}_{bs(a)}$ can be thought of as the simple interaction $\mathbf{bs}_{bs(a_a)}$ of the reduced design $S \times B$ for some specific level $\mathbf{a}_a$.

In comparing the model for $S(A) \times B$ with the model for $S \times A \times B$ (Section 4.6.2) we note that the former model lacks the $\mathbf{as}_{as}$ interaction term that is present in the latter. This term is used in $S \times A \times B$ to account for systematically different patterns of scores across the a_a levels for different subjects. In other words, the combination of a particular a_a level with a particular subject results in a mean scoring tendency that is different from the one that can be predicted from the main terms $\mathbf{s}_s$ and $\mathbf{a}_a$. The effect of the combination of subject and a_a level differs among subjects.

Would not the same situation exist for $S(A) \times B$, and would not an interaction term likewise be required? Indeed, we do not presume that the idiosyncratic manner in which particular subjects respond to particular a_a levels is absent because a subject is exposed to only one a_a level, as he is in $S(A) \times B$. But because a subject is exposed to only one a_a level in $S(A) \times B$, any special response tendency of the subject due to that a_a is included as part of $\mathbf{s}_{s(a)}$. In other words, we can view $\mathbf{s}_{s(a)}$ as a confounded sum of the terms $\mathbf{s}_s$ and $\mathbf{as}_{as}$ that would apply had the design been $S \times A \times B$ with the same levels of the same factors.

The other nested term of $S(A) \times B$ is $\mathbf{bs}_{bs(a)}$. This is a nested interaction term, and it also represents a confounded sum of terms from $S \times A \times B$, specifically of $\mathbf{bs}_{bs}$ and $\mathbf{abs}_{abs}$.

4.8.4 Side Conditions and Estimation Formulas

The general rules for side conditions for nested terms were given in Section 4.7.5, and those rules apply here as well. For example, $\sum_s^\infty \mathbf{s}_{s(a)} = 0$, and $\mathbf{s}_{.(a)}$ is r.v., since S is a random factor. In general, we are not concerned with sums on nest subscripts alone, for example, $\mathbf{s}_{s(.)}$. However, even if A were fixed, such a sum would not be zero, since it is the sum of S random parameters, each of which takes on a new random value for each replication. The result of summations over both nest and non-nest subscripts is governed by the summations on the non-nest subscript(s). For example, $\mathbf{s}_{.(.)}$ is a random variable, since it is the sum of A different independent random variables $\mathbf{s}_{.(a)}$.

The calculational formulas for the estimates are as follows:

$$\hat{\mathbf{m}} = \bar{X}_{...}$$

$$\begin{aligned}\hat{\mathbf{a}}_a &= \bar{X}_{a..} - \hat{\mathbf{m}} \\ &= \bar{X}_{a..} - \bar{X}_{...}\end{aligned}$$

$$\begin{aligned}\hat{\mathbf{b}}_b &= \bar{X}_{.b.} - \hat{\mathbf{m}} \\ &= \bar{X}_{.b.} - \bar{X}_{...}\end{aligned}$$

$$\begin{aligned}\hat{\mathbf{s}}_{s(a)} &= \bar{X}_{a.s} - (\hat{\mathbf{m}} + \hat{\mathbf{a}}_a) \\ &= \bar{X}_{a.s} - \bar{X}_{a..}\end{aligned}$$

$$\begin{aligned}\widehat{\mathbf{ab}}_{ab} &= \bar{X}_{ab.} - (\hat{\mathbf{m}} + \hat{\mathbf{a}}_a + \hat{\mathbf{b}}_b) \\ &= \bar{X}_{ab.} - \bar{X}_{.b.} - \bar{X}_{a..} + \bar{X}_{..}\end{aligned}$$

$$\begin{aligned}\widehat{\mathbf{bs}}_{bs(a)} &= X_{abs} - (\hat{\mathbf{m}} + \hat{\mathbf{a}}_a + \hat{\mathbf{b}}_b + \hat{\mathbf{s}}_{s(a)} + \widehat{\mathbf{ab}}_{ab}) \\ &= X_{abs} - \bar{X}_{ab.} - \bar{X}_{a.s} + \bar{X}_{a..}\end{aligned}$$

4.8.5 Numerical Example

Suppose the scores from an $S(A) \times B$ design are as follows:

		b_1	b_2
a_1	s_1	5	3
	s_2	−1	7
	s_3	−7	−1
a_2	s_1	6	12
	s_2	4	12
	s_3	−1	3
a_3	s_1	4	12
	s_2	−3	11
	s_3	−7	13

In this design $A = 3$, $B = 2$, and $S = 3$.

We calculate parameter estimates using the $\bar{X}$ formulas. We use a more methodical and efficient approach than previously. This new approach is of small importance for the simple designs and few scores of the examples given thus far, but is worthwhile for more realistic situations. First we calculate all the $\bar{X}$ quantities we shall need. There is such an $\bar{X}$ term for each score model term (except those having all the subscripts that X has), and the corresponding terms have the same subscripts (other subscripts on $\bar{X}$ being dotted). We begin toward the end of the score model and work leftward.

$\mathbf{ab}_{ab}$: $\bar{X}_{ab.}$

$$\bar{X}_{11.} = -3/3 = -1$$
$$\bar{X}_{12.} = 9/3 = 3$$
$$\bar{X}_{21.} = 9/3 = 3$$
$$\bar{X}_{22.} = 27/3 = 9$$
$$\bar{X}_{31.} = -6/3 = -2$$
$$\bar{X}_{32.} = 36/3 = 12$$

$\mathbf{s}_{s(a)}$: $\bar{X}_{a.s}$

$$\bar{X}_{1.1} = 8/2 = 4$$
$$\bar{X}_{1.2} = 6/2 = 3$$
$$\bar{X}_{1.3} = -8/2 = -4$$
$$\bar{X}_{2.1} = 18/2 = 9$$
$$\bar{X}_{2.2} = 16/2 = 8$$
$$\bar{X}_{2.3} = 2/2 = 1$$
$$\bar{X}_{3.1} = 16/2 = 8$$
$$\bar{X}_{3.2} = 8/2 = 4$$
$$\bar{X}_{3.3} = 6/2 = 3$$

$\mathbf{b}_b$: $\bar{X}_{.b.}$

$$\bar{X}_{.1.} = 0/9 = 0$$
$$\bar{X}_{.2.} = 72/9 = 8$$

$\mathbf{a}_a$: $\bar{X}_{a..}$

$$\bar{X}_{1..} = 6/6 = 1$$
$$\bar{X}_{2..} = 36/6 = 6$$
$$\bar{X}_{3..} = 30/6 = 5$$

$\mathbf{m}$:

$$\bar{X}_{...} = 72/18 = 4$$

Where the subscript(s) on $\bar{X}$ constitutes a subset of the subscripts on a previously calculated $\bar{X}$, it is simplest to calculate the new $\bar{X}$'s from these values rather than from the original scores. For example,

$$\bar{X}_{a..} = \sum_b \bar{X}_{ab.}/B$$

$$\bar{X}_{.b.} = \sum_a \bar{X}_{ab.}/A$$

$$\bar{X}_{...} = \sum_a \bar{X}_{a..}/A = \sum_b \bar{X}_{.b.}/B$$

It is simplest to calculate one $\bar{X}$ from another having only one more subscript, rather than two more or several more. For example, we find $\bar{X}_{...}$ as

$$\bar{X}_{...} = \sum_a \bar{X}_{a..}/A$$

rather than

$$\bar{X}_{...} = \sum_{ab} \bar{X}_{ab.}/AB,$$

even though both give the same result.

The mean of all $\bar{X}$'s in a set should equal $\bar{X}_{...}$. For example,

$$\sum_{as} \bar{X}_{a.s}/AS = \bar{X}_{...},$$

in addition to the other averages already mentioned that equal $\bar{X}_{...}$. Such means for each set of X's should be calculated as a check on computational error.

Having found, by the procedures given, all $\bar{X}$ quantities, we proceed to estimate the parameters by the method of $\bar{X}$'s.

$$\hat{\mathbf{m}} = \bar{X}_{...} = 4$$

$$\hat{\mathbf{a}}_a = \bar{X}_{a..} - \bar{X}_{...}$$

$$\begin{aligned}
\hat{\mathbf{a}}_1 &= \bar{X}_{1..} - \bar{X}_{...} = 1 - 4 = -3\\
\hat{\mathbf{a}}_2 &= \bar{X}_{2..} - \bar{X}_{...} = 6 - 4 = 2\\
\hat{\mathbf{a}}_3 &= \bar{X}_{3..} - \bar{X}_{...} = 5 - 4 = 1
\end{aligned}$$

$$\hat{\mathbf{b}}_b = \bar{X}_{.b.} - \bar{X}_{...}$$

$$\begin{aligned}
\hat{\mathbf{b}}_1 &= \bar{X}_{.1.} - \bar{X}_{...} = 0 - 4 = -4\\
\hat{\mathbf{b}}_2 &= \bar{X}_{.2.} - \bar{X}_{...} = 8 - 4 = 4
\end{aligned}$$

$$\hat{\mathbf{s}}_{s(a)} = \bar{X}_{a.s} - \bar{X}_{a..}$$

$$\begin{aligned}
\hat{\mathbf{s}}_{1(1)} &= \bar{X}_{1.1} - \bar{X}_{1..} = 4 - 1 = 3\\
\hat{\mathbf{s}}_{2(1)} &= \bar{X}_{1.2} - \bar{X}_{1..} = 3 - 1 = 2\\
\hat{\mathbf{s}}_{3(1)} &= \bar{X}_{1.3} - \bar{X}_{1..} = -4 - 1 = -5\\
\hat{\mathbf{s}}_{1(2)} &= \bar{X}_{2.1} - \bar{X}_{2..} = 9 - 6 = 3\\
\hat{\mathbf{s}}_{2(2)} &= \bar{X}_{2.2} - \bar{X}_{2..} = 8 - 6 = 2\\
\hat{\mathbf{s}}_{3(2)} &= \bar{X}_{2.3} - \bar{X}_{2..} = 1 - 6 = -5\\
\hat{\mathbf{s}}_{1(3)} &= \bar{X}_{3.1} - \bar{X}_{3..} = 8 - 5 = 3\\
\hat{\mathbf{s}}_{2(3)} &= \bar{X}_{3.2} - \bar{X}_{3..} = 4 - 5 = -1\\
\hat{\mathbf{s}}_{3(3)} &= \bar{X}_{3.3} - \bar{X}_{3..} = 3 - 5 = -2
\end{aligned}$$

$$\widehat{\mathbf{ab}}_{ab} = \bar{X}_{ab.} - \bar{X}_{.b.} - \bar{X}_{a..} + \bar{X}_{...}$$

$$\begin{aligned}
\widehat{\mathbf{ab}}_{11} &= \bar{X}_{11.} - \bar{X}_{.1.} - \bar{X}_{1..} + \bar{X}_{...}\\
&= -1 - 0 - 1 + 4 = 2\\
\widehat{\mathbf{ab}}_{12} &= \bar{X}_{12.} - \bar{X}_{.2.} - \bar{X}_{1..} + \bar{X}_{...}\\
&= 3 - 8 - 1 + 4 = -2
\end{aligned}$$

$$\begin{aligned}
\widehat{\mathbf{ab}}_{21} &= \bar{X}_{21.} - \bar{X}_{.1.} - \bar{X}_{2..} + \bar{X}_{...} \\
&= 3 - 0 - 6 + 4 = 1 \\
\widehat{\mathbf{ab}}_{22} &= \bar{X}_{22.} - \bar{X}_{.2.} - \bar{X}_{2..} + \bar{X}_{...} \\
&= 9 - 8 - 6 + 4 = -1 \\
\widehat{\mathbf{ab}}_{31} &= \bar{X}_{31.} - \bar{X}_{.1.} - \bar{X}_{3..} + \bar{X}_{...} \\
&= 2 - 0 - 5 + 4 = -3 \\
\widehat{\mathbf{ab}}_{32} &= \bar{X}_{32.} - \bar{X}_{.2.} - \bar{X}_{3..} + \bar{X}_{...} \\
&= 12 - 8 - 5 + 4 = 3
\end{aligned}$$

$$\widehat{\mathbf{bs}}_{bs(a)} = X_{abs} - \bar{X}_{ab.} - \bar{X}_{a.s} + \bar{X}_{a..}$$

$$\begin{aligned}
\widehat{\mathbf{bs}}_{11(1)} &= X_{111} - \bar{X}_{11.} - \bar{X}_{1.1} + \bar{X}_{1..} \\
&= 5 + 1 - 4 + 1 = 3 \\
\widehat{\mathbf{bs}}_{12(1)} &= X_{112} - \bar{X}_{11.} - \bar{X}_{1.2} + \bar{X}_{1..} \\
&= -1 + 1 - 3 + 1 = -2 \\
\widehat{\mathbf{bs}}_{13(1)} &= X_{113} - \bar{X}_{11.} - \bar{X}_{1.3} + \bar{X}_{1..} \\
&= -7 + 1 + 4 + 1 = -1 \\
\widehat{\mathbf{bs}}_{21(1)} &= -3 \\
\widehat{\mathbf{bs}}_{22(1)} &= 2 \\
\widehat{\mathbf{bs}}_{23(1)} &= 1 \\
\widehat{\mathbf{bs}}_{11(2)} &= 0 \\
\widehat{\mathbf{bs}}_{12(2)} &= -1 \\
\widehat{\mathbf{bs}}_{13(2)} &= 1 \\
\widehat{\mathbf{bs}}_{21(2)} &= 0 \\
\widehat{\mathbf{bs}}_{22(2)} &= 1 \\
\widehat{\mathbf{bs}}_{23(2)} &= -1 \\
\widehat{\mathbf{bs}}_{11(3)} &= 3 \\
\widehat{\mathbf{bs}}_{12(3)} &= 0 \\
\widehat{\mathbf{bs}}_{13(3)} &= -3 \\
\widehat{\mathbf{bs}}_{21(3)} &= -3 \\
\widehat{\mathbf{bs}}_{22(3)} &= 0 \\
\widehat{\mathbf{bs}}_{23(3)} &= 3
\end{aligned}$$

The confinement of all means and estimates to integral numbers in this example is for simplicity, and virtually never occurs with real data. As always, the sum of the *estimates* across any *non-nest* subscript is zero (when other subscripts are held constant), even though the sum of the parameters generally is not zero for a random factor. Whenever making such estimates, the student should perform all such summations as a check on his calculations.

4.8.6 Parametric Components of Estimates

As before, by replacing the $\bar{X}$ terms with the parametric equivalents, we can determine the parametric components of an estimate. Matters are simplified if we construct a table of parameter equivalents for each $\bar{X}$ term. There is an $\bar{X}$ term for each model term (except error, when error is confounded with an interaction term). Each $\bar{X}$ term has the same subscripts as the corresponding model term; the other subscript positions are dotted. For **m**, all subscript positions of the corresponding $\bar{X}$ are dotted. Such a table for $S(A) \times B$ follows. The entries are based on the assumption that A and B are fixed and that S is random.

	$\mathbf{m}$	$\mathbf{a}_a$	$\mathbf{b}_b$	$\mathbf{s}_{s(a)}$	$\mathbf{ab}_{ab}$	$\mathbf{bs}_{bs(a)}$	$\mathbf{e}_{abs}$
X_{abs}	$\mathbf{m}$	$\mathbf{a}_a$	$\mathbf{b}_b$	$\mathbf{s}_{s(a)}$	$\mathbf{ab}_{ab}$	$\mathbf{bs}_{bs(a)}$	$\mathbf{e}_{abs}$
$\bar{X}_{ab.}$	$\mathbf{m}$	$\mathbf{a}_a$	$\mathbf{b}_b$	$\bar{\mathbf{s}}_{.(a)}$	$\mathbf{ab}_{ab}$	$\overline{\mathbf{bs}}_{b.(a)}$	$\bar{\mathbf{e}}_{ab.}$
$\bar{X}_{a.s}$	$\mathbf{m}$	$\mathbf{a}_a$	0	$\bar{\mathbf{s}}_{s(a)}$	0	0	$\bar{\mathbf{e}}_{a.s}$
$\bar{X}_{.b.}$	$\mathbf{m}$	0	$\mathbf{b}_b$	$\bar{\mathbf{s}}_{.(.)}$	0	$\overline{\mathbf{bs}}_{b.(.)}$	$\bar{\mathbf{e}}_{.b.}$
$\bar{X}_{a..}$	$\mathbf{m}$	$\mathbf{a}_a$	0	$\bar{\mathbf{s}}_{.(a)}$	0	0	$\bar{\mathbf{e}}_{a..}$
$\bar{X}_{...}$	$\mathbf{m}$	0	0	$\bar{\mathbf{s}}_{.(.)}$	0	0	$\bar{\mathbf{e}}_{...}$

There is one column for each key term of the model. An $\bar{X}$ row heading is equal to the sum of the entries in the row. The differences within a column arise because, for different rows, the averaging is across different subscripts.

Having already derived the formulas for the parameter estimates using $\bar{X}$'s, we can use our table to find the parametric components of our estimates. For example, we stated that

$$\hat{\mathbf{a}}_a = \bar{X}_{a..} - \bar{X}_{...}$$

Then to find $\hat{\mathbf{a}}_a$ in terms of parameters, we simply subtract the respective table entries for $\bar{X}_{...}$ from those for $\bar{X}_{a..}$, yielding

$$\begin{aligned} \hat{\mathbf{a}}_a &= \bar{X}_{a..} - \bar{X}_{...} \\ &= (\mathbf{m} - \mathbf{m}) + (\mathbf{a}_a - 0) + \bar{\mathbf{s}}_{.(a)} - \bar{\mathbf{s}}_{.(.)} + \bar{\mathbf{e}}_{a..} - \bar{\mathbf{e}}_{...} \\ &= \mathbf{a}_a + \bar{\mathbf{s}}_{.(a)} - \bar{\mathbf{s}}_{.(.)} + \bar{\mathbf{e}}_{a..} - \bar{\mathbf{e}}_{...} \end{aligned}$$

Similarly,

$$\begin{aligned} \hat{\mathbf{b}}_b &= \bar{X}_{.b.} - \bar{X}_{...} \\ &= \mathbf{b}_b + \overline{\mathbf{bs}}_{b.(.)} + \bar{\mathbf{e}}_{.b.} - \bar{\mathbf{e}}_{...} \end{aligned}$$

$$\begin{aligned} \hat{\mathbf{s}}_{s(a)} &= \bar{X}_{a.s} - \bar{X}_{a..} \\ \mathbf{s} &= {}_{s(a)} - \bar{\mathbf{s}}_{.(a)} + \bar{\mathbf{e}}_{a.s} - \bar{\mathbf{e}}_{a..} \end{aligned}$$

$$\begin{aligned}
\widehat{\mathbf{ab}}_{ab} &= \bar{X}_{ab.} - \bar{X}_{.b.} - \bar{X}_{a..} + \bar{X}_{...} \\
&= \mathbf{ab}_{ab} + \overline{\mathbf{bs}}_{b.(a)} - \overline{\mathbf{bs}}_{b.(.)} + \bar{\mathbf{e}}_{ab.} - \bar{\mathbf{e}}_{.b.} - \bar{\mathbf{e}}_{a..} + \bar{\mathbf{e}}_{...} \\
\widehat{\mathbf{bs}}_{bs(a)} &= X_{abs} - \bar{X}_{a.s} - \bar{X}_{ab.} + \bar{X}_{a..} \\
&= \mathbf{bs}_{bs(a)} - \overline{\mathbf{bs}}_{b.(a)} + \mathbf{e}_{abs} - \bar{\mathbf{e}}_{a.s} - \bar{\mathbf{e}}_{ab.} + \bar{\mathbf{e}}_{a..}
\end{aligned}$$

4.9 OTHER NESTED DESIGNS

4.9.1 Design $S(A) \times B \times C$

The layout for design $S(A) \times B \times C$ is shown in the following table.

		b_1			$\cdots$	b_B		
		c_1	$\cdots$	c_C	$\cdots$	c_1	$\cdots$	c_C
a_1	s_1 s_2 $\cdot$ $\cdot$ $\cdot$ s_S							
$\cdot$ $\cdot$ $\cdot$	$\cdot$ $\cdot$ $\cdot$							
a_A	s_1 s_2 $\cdot$ $\cdot$ $\cdot$ s_S							

The labeling is initialized for factor S but literal for factors A, B, and C. A particular subject appears at only a single level of A, but it appears at every combination of levels bc_{bc}. Each subject, then, yields $B \cdot C$ scores. The layout does not imply that all subjects are exposed to the various bc_{bc} levels in the same order. We may assume that, unless B or C is a trial factor, each of the $A \cdot S$ subjects has an independently randomized order. If C were a trial factor and B a treatment factor, only the order of exposure to levels of B could be randomized.

The score model is

$$X_{abcs} = \mathbf{m} + \mathbf{a}_a + \mathbf{b}_b + \mathbf{c}_c + \mathbf{s}_{s(a)} + \mathbf{ab}_{ab} + \mathbf{ac}_{ac} + \mathbf{bc}_{bc} + \mathbf{bs}_{bs(a)} + \mathbf{cs}_{cs(a)} + \mathbf{abc}_{abc} + \mathbf{bcs}_{bcs(a)} + \mathbf{e}_{abcs}$$

Note that all key terms with **s**, and only key terms with **s**, have a nest subscript *a*.

4.9.2 Design $S(A \times B) \times C$

In the two previous nested designs discussed in this chapter, S was nested in only one other factor. We now consider $S(A \times B) \times C$, in which S is nested in two crossed factors. The design layout is as follows:

			c_1	c_2	$\cdots$	c_C
b_1	a_1	s_1 ⋮ s_S				
	⋮					
	a_A	s_1 ⋮ s_S				
⋮	⋮					
b_B	a_1	s_1 ⋮ s_S				
	⋮					
	a_A	s_1 ⋮ s_S				

The labeling is initialized for the levels of S but literal for the levels of A and B. A particular subject appears at only a single combination ab_{ab}, but it appears at each of the C levels of factor C. If both A and B were treatment factors, subjects would be assigned randomly, but in equal numbers (we assume for now), to the various ab_{ab} levels. If A or B were a blocking factor, however, a subject could, of course, be randomly assigned only within his own block. Unless C is a trial factor, each subject will have an independently randomized order of exposure to the levels of C.

The score model is:

$$X_{abcs} = \mathbf{m} + \mathbf{a}_a + \mathbf{b}_b + \mathbf{c}_c + \mathbf{ab}_{ab} + \mathbf{ac}_{ac} + \mathbf{bc}_{bc} + \mathbf{s}_{s(ab)} + \mathbf{abc}_{abc} + \mathbf{cs}_{cs(ab)} + \mathbf{e}_{abcs}$$

Note that all key terms with **s**, and only key terms with **s**, have nest subscripts a and b.

The relations between A, B, and C are the same for designs $S(A) \times B \times C$ and $S(A \times B) \times C$. The only difference between the two designs is the relation between S and B. In the former design, S is crossed with B, and in the latter design, S is nested within B. How is this difference reflected in the two score models? The terms involving only **a**, **b**, and **c** are exactly the same for the two models. Both have **s** and **cs** terms; in the former design, these terms occur with the nest subscript a, and in the latter design they occur with nest subscripts a and b. Whereas the model for $S(A) \times B \times C$ includes a **bs** and a **bcs** term, the model for $S(A \times B) \times C$ has no interaction terms that combine **b** and **s**. A score model always lacks interaction terms that include both a nested factor and any of its nest factors. In effect, as we noted in relation to designs $B(A)$ and $S(A) \times B$, such interaction terms are confounded parts of the nested terms.

For the present two designs, we have the following equivalences:

$S(A \times B) \times C$	$S(A) \times B \times C$
$\mathbf{s}_{s(ab)}$	$\mathbf{s}_{s(a)} + \mathbf{bs}_{bs(a)}$
$\mathbf{cs}_{cs(ab)}$	$\mathbf{cs}_{cs(a)} + \mathbf{bcs}_{bcs(a)}$

4.9.3 Design $S(A(B)) \times C$

This design has the same four factors as $S(A) \times B \times C$ and $S(A \times B) \times C$, which have just been discussed. Factor S is nested within both A and B, as it was in design $S(A \times B) \times C$, but now we have $A(B)$ instead of $A \times B$. The layout used for $S(A \times B) \times C$ will serve equally well for $S(A(B)) \times C$, except that now the labeling for both S and A is initialized.

The score model is:

$$X_{abcs} = \mathbf{m} + \mathbf{b}_b + \mathbf{c}_c + \mathbf{a}_{a(b)} + \mathbf{bc}_{bc} + \mathbf{s}_{s(ab)} + \mathbf{ac}_{ac(b)} + \mathbf{cs}_{cs(ab)} + \mathbf{e}_{abcs}$$

We have fewer terms than we did for $S(A \times B) \times C$, which always happens when a factor formerly crossed is changed to a nest factor. The model for $S(A(B)) \times C$ has two fewer terms than the model for $S(A \times B) \times C$, since the former model lacks **ab** and **abc** interaction terms. Since A is nested within B, no interaction terms combining **a** and **b** occur in the model.

Although in symbolizing the design, parentheses within parentheses have been used, more than one set of parentheses is never used for the subscripts of key terms. For example, the nested main effect for S is written $\mathbf{s}_{s(ab)}$, not $\mathbf{s}_{s(a(b))}$.

4.9.4 Comparison of Model Terms

Table 4.9-1 compares the model terms for the three nested designs discussed in this section and the completely crossed design $S \times A \times B \times C$, not previously considered. Each of the four designs is a column heading, below which appear the terms of the model. The order of the model terms deviates here from the convention previously followed in order to show more clearly which terms become confounded when nesting occurs.

Proceeding from the left to the adjacent design, two factors crossed in one design are nested in the next. As previously explained, when one factor is nested within another, the score model contains no interaction term combining both factors. Table 4.9-1 shows which terms are confounded by the nesting and which term of the new model equals which confounded sum.

For each design with nesting, we can find from the table what the equivalent terms would be in the completely crossed design by reading the terms to the left of the braces. For example, term $\mathbf{cs}_{cs(ab)}$ of $S(A \times B) \times C$ is equivalent to $\mathbf{cs}_{cs} + \mathbf{acs}_{acs} + \mathbf{bcs}_{bcs} + \mathbf{abcs}_{abcs}$ of $S \times A \times B \times C$. Actually, we can find such equivalents rather easily without a table by using a simple rule. The nested term is equivalent to the sum of all terms of the completely crossed design having all the nested subscripts of the nested term, either alone or in

Table 4.9-1 Comparison of model terms in four four-factor designs

$S \times A \times B \times C$	$S(A) \times B \times C$	$S(A \times B) \times C$	$S(A(B)) \times C$
$\mathbf{m}$	$\mathbf{m}$	$\mathbf{m}$	$\mathbf{m}$
$\mathbf{a}_a$	$\mathbf{a}_a$	$\mathbf{a}_a$ }	$\mathbf{a}_{a(b)}$
$\mathbf{ab}_{ab}$	$\mathbf{ab}_{ab}$	$\mathbf{ab}_{ab}$ }	
$\mathbf{b}_b$	$\mathbf{b}_b$	$\mathbf{b}_b$	$\mathbf{b}_b$
$\mathbf{c}_c$	$\mathbf{c}_c$	$\mathbf{c}_c$	$\mathbf{c}_c$
$\mathbf{s}_s$ }	$\mathbf{s}_{s(a)}$ }	$\mathbf{s}_{s(ab)}$	$\mathbf{s}_{s(ab)}$
$\mathbf{as}_{as}$ }			
$\mathbf{bs}_{bs}$ }	$\mathbf{bs}_{bs(a)}$ }		
$\mathbf{abs}_{abs}$ }			
$\mathbf{ac}_{ac}$	$\mathbf{ac}_{ac}$	$\mathbf{ac}_{ac}$ }	$\mathbf{ac}_{ac(b)}$
$\mathbf{abc}_{abc}$	$\mathbf{abc}_{abc}$	$\mathbf{abc}_{abc}$ }	
$\mathbf{bc}_{bc}$	$\mathbf{bc}_{bc}$	$\mathbf{bc}_{bc}$	$\mathbf{bc}_{bc}$
$\mathbf{cs}_{cs}$ }	$\mathbf{cs}_{cs(a)}$ }	$\mathbf{cs}_{cs(ab)}$	$\mathbf{cs}_{cs(ab)}$
$\mathbf{acs}_{acs}$ }			
$\mathbf{bcs}_{bcs}$ }	$\mathbf{bcs}_{bcs(a)}$ }		
$\mathbf{abcs}_{abcs}$ }			
$\mathbf{e}_{abcs}$	$\mathbf{e}_{abcs}$	$\mathbf{e}_{abcs}$	$\mathbf{e}_{abcs}$

combination with subscripts from the nest. If we apply this rule to the preceding example, we see that the equivalent of $\mathbf{cs}_{cs(ab)}$ in $S \times A \times B \times C$ is $\mathbf{cs}_{cs}$ (the term having all the nested subscripts by themselves) plus terms that combine **c** and **s** with **a** and **b**, either singly or together. As we shall see in Chapter 5, the understanding of such equivalents can be used to save an experimenter considerable work in his analytical calculations.

4.10 DERIVING MODELS IN GENERAL

Thus far this chapter has presented and explained the models for a variety of simple designs. In addition, the estimation of model terms has been described for a number of the designs. This section explains how the student can derive the score model for any design that he can symbolize by following the procedures described in Chapter 3. The following section discusses a method of estimating model terms for any such design.

It is convenient to describe model derivation in the language of the tree diagram (Section 3.6), but it is also possible to derive the score model directly from the linear design symbolization. Going up a branch in the tree diagram is comparable to going from a nest factor or factors to the factors nested within them. Going down a branch means going from a nested factor to a nest factor one level deeper.

First we shall consider the model derivation for design $S(A \times B(C)) \times D(G)$, whose tree diagram is:

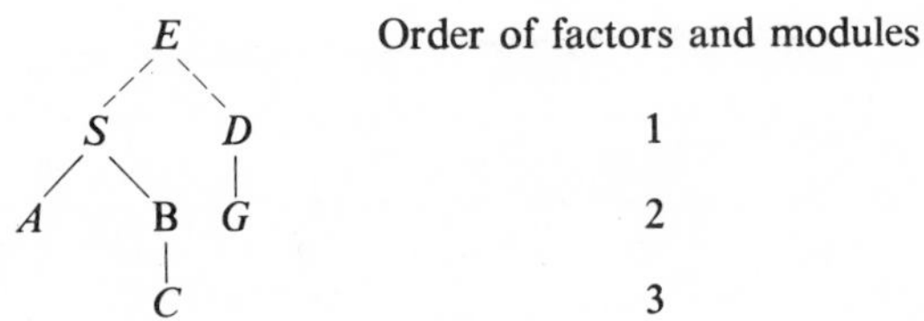

A *partial model* (PM) is a summation of model terms from the score model, plus 1. Model derivation consists of aggregating terms to the PM for a terminal factor (Section 3.6), creating PM's that are even more inclusive, until the score model is obtained.

A PM for a terminal factor is simply the main term for that factor plus 1. For example, the PM for C of the present example, symbolized by PM_C, is $\mathbf{c}_c + 1$. Beginning with the PM for any terminal factor, aggregation proceeds by one of two processes:

1. Crossing (combining) the PM at hand with the PM for another module in the same set. (Section 3.5.1 defines a set of modules or factors.) This crossing process, which will be described momentarily, must always be completed to the fullest extent possible before proceeding to process (2).

In this example, C is the only factor (or module) in its set, so we may proceed with process (2).

2. Adding a nested main term for the factor directly up the branch or branches from the set whose PM we have obtained. In the example, there is only one factor, C, in the set, so we proceed up the single branch to factor B, and add a nested main effect for B to the PM at hand, which yields

$$\mathrm{PM}_B = \mathbf{b}_{b(c)} + \mathrm{PM}_C = \mathbf{c}_c + \mathbf{b}_{b(c)} + 1$$

Within the subscript nest we include the subscripts for all factors of the partial model to which we add the nested main effect. In the example, there is only one such factor, C, so the nest contains only one subscript, c.

Before proceeding further up the branch, we must incorporate all other modules that are in the same set as B. Here, the only other such module is the simple module A. Aggregation of modules in the same set proceeds by the crossing of PM's, so we need PM_A. Since A is a terminal factor, $\mathrm{PM}_A = \mathbf{a}_a + 1$.

The crossing of two PM's of the same set proceeds by a process analogous to multiplication, called *quasi-multiplication*. The two PM's, and the new PM obtained from crossing them, are as follows:

$$\begin{array}{ll} \mathrm{PM}_B: & \mathbf{b}_{b(c)} + \mathbf{c}_c + 1 \\ \mathrm{PM}_A: & \qquad\qquad \mathbf{a}_a + 1 \\ \hline \mathrm{PM}_{AB}: & \mathbf{a}_a + \mathbf{c}_c + \mathbf{b}_{b(c)} + \mathbf{ac}_{ac} + \mathbf{ab}_{ab(c)} + 1 \end{array}$$

The quasi-multiplication of PM_1 and PM_2 consists of combining ("multiplying") each term in PM_1 by each term in PM_2. The result of combining 1 with a term is simply the term itself. The result of combining 1 with 1 is 1. The result of combining two model terms is a new interaction term consisting of all bold-faced letters and subscripts of the two terms combined. Nest subscripts remain so after combining, and non-nest subscripts likewise remain non-nest.

Having crossed PM_A and PM_B, we have to cross PM_{AB} with the PM for any other module in the same set, continuing in this manner until all modules in the set have been included. Since the set consists only of A and $B(C)$, we climb to the next higher level, which is always a single factor branching down to all factors in the set just considered. This factor is S, so we now have to include S in the PM. Factor S is nested in all the factors involved in constructing PM_{AB}. Therefore, we proceed as we did when proceeding to B from its nested factor C. We simply add a nested main effect for S, $\mathbf{s}_{s(abc)}$, with nest subscripts for all factors of PM_{AB}.

The result is a new PM, symbolized by PM_S:

$$\mathrm{PM}_S = \mathbf{a}_a + \mathbf{c}_c + \mathbf{b}_{b(c)} + \mathbf{ac}_{ac} + \mathbf{ab}_{ab(c)} + \mathbf{s}_{s(abc)} + 1$$

There is another module in the same set as S, namely, $D(G)$, and we now must cross PM_D with PM_S. At present, however, PM_D is lacking. Furthermore, D is not a terminal factor, so we cannot write its PM simply as $\mathbf{d}_d + 1$. Instead, we have to descend down the branch from D until we reach a terminal factor, however many levels are necessary, and start building a PM upward from there.

In this example, we need descend only one level to terminal factor G, whose PM is $\mathbf{g}_g + 1$. We derive PM_D by process (2):

$$PM_D = \mathbf{d}_{d(g)} + PM_G = \mathbf{g}_g + \mathbf{d}_{d(g)} + 1$$

We are now able to cross PM_S with PM_D, which yields

$$\begin{aligned} PM_{SD} = {} & \mathbf{a}_a + \mathbf{c}_c + \mathbf{g}_g + \mathbf{b}_{b(c)} + \mathbf{d}_{d(g)} + \mathbf{ac}_{ac} + \mathbf{ag}_{ag} + \mathbf{cg}_{cg} + \mathbf{ab}_{ab(c)} \\ & + \mathbf{ad}_{ad(g)} + \mathbf{bg}_{bg(c)} + \mathbf{cd}_{cd(g)} + \mathbf{acg}_{acg} + \mathbf{s}_{s(abc)} + \mathbf{bd}_{bd(cg)} \\ & + \mathbf{abg}_{abg(c)} + \mathbf{acd}_{acd(g)} + \mathbf{gs}_{gs(abc)} + \mathbf{ds}_{ds(abcg)} + \mathbf{abd}_{abd(cg)}) + 1 \end{aligned}$$

We now have the PM for all design factors. To obtain the score model from this, we simply remove (subtract) the 1 and add the population mean $\mathbf{m}$ and an error term.

If we realize at this point that there is only one score per condition, as would normally be true when factor S is present, we can omit the e subscript for error. Then we would have

$$X_{abcdgs} = \mathbf{m} + PM_{SD} - 1 + \mathbf{e}_{abcdgs}$$

If the design has more than one score per condition, the error is $\mathbf{e}_{e(abcdgs)}$; all design factor subscripts appear in the subscript nest, and subscript e should be added to the other subscripts on X.

If we were to accept error as a factor nested within all others, we could have brought the error into the model by process (2), that is, $PM_E = \mathbf{e}_{e(abcdgs)} + PM_{SD}$, and then we would simply replace 1 with $\mathbf{m}$ to get the score model. However, it is just as well to stop at order 1, since error requires some special consideration when there is but one score per condition.

The procedure for deriving the score model can be summarized by means of an algorithm.

1. Begin at any terminal factor.
2. Write the PM for the terminal factor.
   Go to step 3.
3. Are there one or more other factors in the same set besides the one most recently considered?
   If yes, go to step 4.
   If no, go to step 8.
4. Has a PM been constructed thus far for one or more of these other factors in the same set?

If yes, go to step 5.
If no, go to step 6.

5. Using process (1), cross the PM last constructed with the partial model already existing for the same set. Is there another factor in the same set not yet incorporated into a PM?
If yes, go to step 6.
If no, go to step 8.
6. Is the factor under consideration terminal?
If yes, go to step 2.
If no, go to step 7.
7. Descend from this factor along any path until a terminal factor is reached.
Go to step 2.
8. Is the set last considered of order 1?
If yes, go to step 10.
If no, go to step 9.
9. Ascend up one level from the set of modules last considered and write the PM of the factor reached by process (2).
Go to step 3.
10. Finally, write the score model from the PM accumulated by removing the 1 and adding **m** and an error term.

4.11 ESTIMATING MODEL TERMS IN GENERAL

This chapter has considered not only which key terms appear in a design's score model, but also how to estimate them. In the preceding section, rules were given for deriving the score model for any compact design. Here rules are presented for estimating the terms of any compact design. First rules are given for deriving the formulas for estimation by the method of successive extraction. Then rules are presented for deriving the estimation formulas by the more direct method of $\bar{X}$'s. These latter formulas would normally be more practical for computation. We assume, as usual, that there are an equal number of scores per cell.

By now the student should have some understanding of the estimation of nested and non-nested terms via successive extraction. The population mean is estimated as the mean of all scores. A main parameter is estimated as the mean of all scores at a given level of the factor minus the mean of all scores. A two-factor interaction parameter is estimated as the mean of all scores at the two specific levels of two factors, minus the estimated main effects model for that combination of levels. Nested main and interaction parameters are estimated in the same way, except that we imagine that the design exists only at the specific level of the nest factor under consideration. The rules for deriving the formula for estimating a parameter can be summarized as follows:

1. Take the mean of all scores having the same subscript values as the parameter being estimated.
2. Subtract from this an estimated model composed of $\hat{\mathbf{m}}$ plus estimates of all terms having one or more (but less than all) of the subscripts of the term being estimated (and having no subscripts that are absent from that term).
3. Subscripts should be set at the same level for the parameter being estimated as for the parameters used in making the estimation.

For example, to estimate $\mathbf{ab}_{ab}$ of $A \times B \times C$, we calculate

$$\widehat{\mathbf{ab}}_{ab} = \bar{X}_{ab..} - (\hat{\mathbf{m}} + \hat{\mathbf{a}}_a + \hat{\mathbf{b}}_b),$$

or, for a specific parameter, say, $\mathbf{ab}_{23}$,

$$\widehat{\mathbf{ab}}_{23} = \bar{X}_{23..} - (\hat{\mathbf{m}} + \hat{\mathbf{a}}_2 + \hat{\mathbf{b}}_3)$$

We do not add to $\hat{\mathbf{m}}$ any terms having the c subscript, since c is absent from the term being estimated.

Now we apply the rules to the nested terms of $S(A) \times B$. The score model is $X_{abs} = \mathbf{m} + \mathbf{a}_a + \mathbf{b}_b + \mathbf{s}_{s(a)} + \mathbf{ab}_{ab} + \mathbf{bs}_{bs(a)}$. The only term with one or more of the same subscripts as $\mathbf{s}_{s(a)}$, but no others, is $\mathbf{a}_a$. Therefore,

$$\hat{\mathbf{s}}_{s(a)} = \bar{X}_{a.s} - (\hat{\mathbf{m}} + \hat{\mathbf{a}}_a)$$

Similarly,

$$\widehat{\mathbf{bs}}_{bs(a)} = X_{abs} - (\hat{\mathbf{m}} + \hat{\mathbf{a}}_a + \hat{\mathbf{b}}_b + \hat{\mathbf{s}}_{s(a)} + \widehat{\mathbf{ab}}_{ab})$$

Since there is only one score having the same subscripts as $\mathbf{bs}_{bs(a)}$, the mean of such "scores" is simply X_{abs}. Since $\mathbf{bs}_{bs(a)}$ includes all model subscripts, estimates of all other terms are among the subtractors.

It is possible, as we have seen, to derive the formulas for estimation using only $\bar{X}$'s from the formulas for estimation by successive extraction. It is much simpler, however, by using the rules that follow, to derive the $\bar{X}$ formulas directly. We first have to form a *symbolic product* (SP) from the subscripts of the key term. To find the SP, we replace each non-nest subscript with that subscript minus 1, and then we multiply out across all subscripts. The result is an *expanded symbolic product* (ESP). Each term in the ESP is replaced by an $\bar{X}$ term. Each $\bar{X}$ has exactly the same subscripts as the corresponding ESP. We average over all remaining subscript positions, and indicate this with dots. The term 1 of the ESP is replaced by the mean of all scores. The plus or minus sign of the ESP term is retained in the $\bar{X}$ term replacing it. The formula thus derived can be used to estimate the parameters of the key term under consideration.

Table 4.11-1 Estimation formulas for the terms of $S(A) \times B$

Key term	SP	ESP	Estimation formula
$\mathbf{a}_a$	$a - 1$	$a - 1$	$\bar{X}_{a..} - \bar{X}_{...}$
$\mathbf{b}_b$	$b - 1$	$b - 1$	$\bar{X}_{.b.} - \bar{X}_{...}$
$\mathbf{s}_{s(a)}$	$a(s - 1)$	$as - a$	$\bar{X}_{a.s} - \bar{X}_{a..}$
$\mathbf{ab}_{ab}$	$(a - 1)(b - 1)$	$ab - b - a + 1$	$\bar{X}_{ab.} - \bar{X}_{.b.} - \bar{X}_{a..} + \bar{X}_{...}$
$\mathbf{bs}_{bs(a)}$	$a(b - 1)(s - 1)$	$abs - ab - as + a$	$X_{abs} - \bar{X}_{ab.} - \bar{X}_{a.s} + \bar{X}_{a..}$

The method can be illustrated by deriving estimation formulas for design $S(A) \times B$, which has the following score model:

$$X_{abs} = \mathbf{m} + \mathbf{a}_a + \mathbf{b}_b + \mathbf{s}_{s(a)} + \mathbf{ab}_{ab} + \mathbf{bs}_{bs(a)} + \mathbf{e}_{abs}$$

The estimation formulas for the key terms are shown in Table 4.11-1. In the formulas for estimating $\mathbf{bs}_{bs(a)}$, we have X_{abs}, which is, of course, an individual score, so no averaging is required and the X has no bar over it. When term **e** is confounded with an interaction term, as here, we have no separate estimate for error, and there is no entry **e** in the table. When term **e** is not confounded, the rules given provide the formula for estimating **e**. For example, if we have $\mathbf{e}_{e(ab)}$, the SP is $(e - 1)(ab)$; the ESP is $abe - ab$; $\hat{\mathbf{e}}_{e(ab)} = X_{abe} - \bar{X}_{ab.}$.

CHAPTER FIVE

Sums of Squares

5.1 ANALYSIS OF VARIANCE

This chapter concerns the analysis of variance (ANOVA) of the scores obtained from the experiment. The sources that the author has seen that discuss ANOVA either neglect to define the term, or are rather vague and inconsistent with each other concerning exactly what the term includes. As the term is used here, ANOVA refers to the theory and procedures for analyzing the scores of an experiment, as well as the process of performing such an analysis. The general purpose of ANOVA is to determine which factors of an experiment have noteworthy effects on the scores, and to provide quantitative information about the relative importance of different factors and their levels.

As the term is used here, ANOVA includes the theory and procedures for calculating sums of squares, expected mean squares, F-ratios, and variance estimates, all topics that are covered in the present chapter. The estimation of model parameters, a topic already considered in Chapter 4, is also included in ANOVA. The usual procedure in doing an ANOVA is first to calculate the sums of squares, then, from these, to calculate expected mean squares, and then to calculate F-ratios. There is an F-ratio for most of the key terms in the score model. Each F-ratio is compared with some critical tabled value. If the ratio exceeds the tabled value, that term is said to be *significant*. If a term is significant, at least some of its parameters are considered to deviate sufficiently from zero so that the term must be considered in accounting for the experimental results. If a term is not significant, all its parameters are considered to be zero, or virtually zero for any theoretical or practical purpose. In essence, when a key term is considered to be nonsignificant, it can be eliminated from the score model.

The main purpose of most ANOVA's is to perform such F-tests, and thereby to obtain conclusions concerning the significance of the various key terms in the model. Sums of squares and expected mean squares are usually calculated for this purpose alone—they are necessary steps in finding the F-ratios. Indeed, in practice, ANOVA is nearly synonymous with "significance testing."

Although the preceding statement is, in fact, true, ANOVA also has a more general purpose: ". . . to summarize the data effectively and to help in understanding what 'goes on' in the experimental situation." [1] In regard to this more general purpose, calculation of the variance of the parameters of the different key terms might be useful. The magnitude of the variance for a key term can help describe the importance of that term in producing experimental variations.

The expected mean squares can be described in terms of such variances, and, in turn, such variances can be estimated from the expected mean squares. In truth, however, such variance calculations are seldom performed, and the experimenter who ignores them will seldom, if ever, be criticized for his omission. Nonetheless, an experimenter should be familiar with such variance calculations and should understand how they are obtained. He can then decide for himself whether such calculations will help him understand and describe his experimental results. The calculation of variances for key terms is explained in Chapter 6.

Calculation of various sums of squares (SS's) for the experimental scores is typically the first step in ANOVA. There is a total sum of squares (SS_{tot}) for the experiment that can be broken down into components in various ways. For example, there is an SS for each key term of the score model. The sum of all such SS's equals SS_{tot}. As previously noted, the SS's are needed for calculation of expected means squares, from which are derived both the F-statistics and the variance estimates for the key terms.

5.2 TOTAL SUM OF SQUARES (SS_{tot})

Two formulas are presented for SS_{tot}. The first conveys a clearer conception of SS_{tot}. The second formula is more convenient for calculation.

By definition,

(1) $$SS_{tot} = \sum(X - \bar{X})^2$$

No subscripts are placed on X, so that this discussion can remain general.

[1] B. F. Green, Jr., and J. W. Tukey, "Complex Analyses of Variance: General Problems," *Psychometrika* 25: 127–152 (1960).

The mean of all scores in the experiment is denoted by $\bar{X}$. For any specific design, X would be given the appropriate subscripts, and equation (1) would apply. The summation $\sum$ is over all subscripts of the experiment.

To calculate SS_{tot} from equation (1), we proceed as follows:

1. Calculate the mean, $\bar{X}$, of all of the N scores in the experiment.
2. Subtract $\bar{X}$ from each score, X, yielding N differences.
3. Square each difference individually.
4. Sum together all N such squared differences.

Following is a simple example. Suppose an experiment has four scores: 2, 4, 6, and 8. Then $\bar{X} = (2 + 4 + 6 + 8)/4 = 20/4 = 5$. The differences between the scores and the means are -3, -1, $+1$, and $+3$, respectively. The four squares of the differences are 9, 1, 1, and 9, respectively, and the sum of these, 20, is SS_{tot}.

Although the preceding formula and method provide a good understanding of the meaning of SS_{tot}, another formula, leading to exactly the same quantity for SS_{tot}, is computationally simpler when there are many scores:

$$SS_{tot} = \sum X^2 - \frac{(\sum X)^2}{N} \tag{2}$$

When this formula is used, the differences between the scores and the mean do not have to be computed. If we are concerned with a specific design, we can write the SS_{tot} formula with the appropriate subscripts. For example, for design $A \times B$, equation (2) would be

$$SS_{tot} = \sum X_{abe}^2 - \frac{(\sum X_{abe})^2}{ABE}$$

Two components of the formula, $\sum X^2$ and $(\sum X)^2$, are easy to confuse, but they are quite different quantities and must be clearly distinguished. The quantity $\sum X^2$ is calculated in two stages. First, each score in the experiment is squared. Second, all quantities thus obtained are added together. For the simple example with four scores, $2^2 = 4$, $4^2 = 16$, $6^2 = 36$, and $8^2 = 64$, so $\sum X^2 = 4 + 16 + 36 + 64 = 120$.

The quantity $(\sum X)^2$ is also calculated in two stages, but they are different. First we sum the scores directly, obtaining $\sum X = 2 + 4 + 6 + 8 = 20$. Then we square the sum, obtaining $(\sum X)^2 = 20^2 = 400$. Thus, although $\sum X^2$ and $(\sum X)^2$ differ in notation only by the inclusion of parentheses in the latter, these parentheses drastically alter the meaning of the term.

Now substituting these components in equation (2), we obtain

$$SS_{tot} = 120 - \frac{400}{4} = 20,$$

the same result as before. With so few scores, the two methods hardly differ in computational ease, but with many scores, the second method is much simpler.

Of course, if a computer is used to calculate SS's, the experimenter may not be too concerned about different methods of computation (though the programmer should be). But this book is addressed basically to those using a desk calculator. Even persons who expect to use computers routinely should understand how such calculations are done "by hand," on a desk calculator, and should have some experience in doing them.

Preferably, the desk calculator should require that each score be entered only once and that only one operation button be pressed after each score; the calculator should also accumulate and display both $\sum X^2$ and $\sum X$. Then SS_{tot} can be readily obtained by using these quantities and equation (2). Merely to calculate the mean, $\bar{X}$, for use in equation (1), would require approximately the same time and effort as required to obtain $\sum X^2$ and $\sum X$, since each score would have to be entered on the calculator also to compute $\bar{X}$.

On simpler desk (or hand) calculators, it may be necessary to square each number by entering the same X twice to multiply "X times X." Also, accumulation may not be possible, so that the X^2 for each score must be written down on paper for later summation. Although computation is more laborious with these simpler calculators, such devices will adequately serve the student using this textbook and the accompanying workbook, since the number of scores in the examples and exercises is kept artificially small.

From equation (1) it is apparent that the more the scores vary above and below the mean, $\bar{X}$, the greater is the sum of squares. If all the scores are very close to $\bar{X}$, SS_{tot} will be small, but if the scores deviate greatly above and below $\bar{X}$ (that is to say, if the scores deviate greatly from each other), SS_{tot} will be large. In this respect, SS_{tot} is similar to the variance measure for a group of scores. In fact, $\text{Var}(X) = SS_{tot}/N$.

If we increase the number of scores in an experiment by increasing the number of subjects per treatment, SS_{tot} increases approximately in proportion. In contrast, because of the division by N, which increases the number of subjects per treatment, the variance remains approximately constant. Another important difference is that SS_{tot} for an experiment can be analyzed into component SS's associated with the key terms, and these components sum to SS_{tot}. The total variance, however, is not the sum of the variances for the separate key terms.

Thus far we have learned how to calculate SS_{tot} from the scores and to relate it to the variance. We can also relate SS_{tot} to the parameter estimates. The difference $X - \bar{X}$ equals the sum of the estimated parameters for all key terms of the design (except **e**, when there is only one score per condition). For example, for design $A \times B$, $X_{abe} - \bar{X}_{...} = \hat{\mathbf{a}}_a + \hat{\mathbf{b}}_b + \widehat{\mathbf{ab}}_{ab} + \hat{\mathbf{e}}_e$. For design $S(A) \times B$, $X_{abs} - \bar{X}_{...} = \hat{\mathbf{a}}_a + \hat{\mathbf{b}}_b + \hat{\mathbf{s}}_{s(a)} + \widehat{\mathbf{ab}}_{ab} + \widehat{\mathbf{bs}}_{bs(a)}$. Then, for design $A \times B$, $SS_{tot} = \sum_{abe}(\hat{\mathbf{a}}_a + \hat{\mathbf{b}}_b + \widehat{\mathbf{ab}}_{ab} + \hat{\mathbf{e}}_{e(ab)})^2$, and for design $S(A) \times B$, $SS_{tot} = \sum_{abs}(\hat{\mathbf{a}}_a + \hat{\mathbf{b}}_b + \hat{\mathbf{s}}_{s(a)} + \widehat{\mathbf{ab}}_{ab} + \widehat{\mathbf{bs}}_{bs(a)})^2$. It is well known that $(a + b)^2 = a^2 + 2ab + b^2$; that is, in general, when a sum of terms is squared, we can rewrite the quantity showing individual terms squared plus two times the cross products of the terms. We can apply this rule to our sums of estimates squared, but after applying $\sum$, the cross products drop out, so we have, for example, for $A \times B$,

$$\text{SS}_{\text{tot}} = \sum_{abe} \hat{\mathbf{a}}_a^2 + \sum_{abe} \hat{\mathbf{b}}_b^2 + \sum_{abe} \widehat{\mathbf{ab}}_{ab}^2 + \sum_{abe} \hat{\mathbf{e}}_{e(ab)}^2$$

Whenever the term summed is constant with respect to a summation subscript, the summation over that subscript amounts to multiplication by the upper limit of the subscript. For example, $\sum_{abe}\hat{\mathbf{a}}_a^2 = \sum_b\sum_e\sum_a\hat{\mathbf{a}}_b^2 = BE\sum_a\hat{\mathbf{a}}_a^2$. Therefore, for $A \times B$,

$$\text{SS}_{\text{tot}} = BE\sum\hat{\mathbf{a}}_a^2 + AE\sum\hat{\mathbf{b}}_b^2 + E\sum\widehat{\mathbf{ab}}_{ab}^2 + \sum\hat{\mathbf{e}}_{e(ab)}^2$$

Likewise, for $S(A) \times B$, we have

$$\text{SS}_{\text{tot}} = BS\sum\hat{\mathbf{a}}_a^2 + AS\sum\hat{\mathbf{b}}_b^2 + B\sum\hat{\mathbf{s}}_{s(a)}^2 + S\sum\widehat{\mathbf{ab}}_{ab}^2 + \sum\widehat{\mathbf{bs}}_{bs(a)}^2$$

(When not subscripted, a $\sum$ is assumed to be over only subscripts of the quantity being summed.) Any SS_{tot} for any design in our system can be thus analyzed; the components derived are SS's for the key terms, to which we now turn.

5.3 SS's FOR KEY TERMS: INTERPRETATION VIA PARAMETERS

Each key term of the score model has an associated SS, and, as we previously noted, calculation of such SS's is an important step in ANOVA. These key term SS's sum to SS_{tot}, as stated in the preceding section. The key term being considered is indicated by capital letter subscripts corresponding to the boldfaced letters of the key term. For example, SS_A indicates the SS for $\mathbf{a}_a$, and SS_{AB} indicates the SS for $\mathbf{ab}_{ab}$.

As with SS_{tot}, the SS for a key term can be described either in terms of a formula that is conceptually simple, but computationally complex, or in terms of a formula that is computationally more simple, but conceptually obscure. This section considers the description that is conceptually simple; the following section considers the simpler computational formulas.

The conceptually simple approach is based on the parameter estimates for a key term. The procedure for calculating a key term SS is as follows:

1. Square each estimate for the key term parameters.
2. Sum these squares.
3. Multiply the sum by the product of the upper limits of all design subscripts absent from the key term (including E, if subscript e appears in the score model). The nest subscripts of a key term are considered to be

present rather than absent. If all design subscripts are present for the key term, the multiplier is one.

The product of upper limits referred to in step (3) equals the number of scores to which a particular parameter for that key term contributes. In other words, the product equals the number of equations in the score model in which a particular parameter appears.

For design A, there are two key terms in the score model, $\mathbf{a}_a$ and $\mathbf{e}_{e(a)}$. The corresponding SS's are:

$$SS_A = E\sum\hat{\mathbf{a}}_a^2$$
$$SS_E = \sum\hat{\mathbf{e}}_{e(a)}^2$$

In SS_A only one design subscript, e, is absent from key term $\mathbf{a}_a$, so the multiplier of $\sum\hat{\mathbf{a}}_a^2$ is E. Each $\mathbf{a}_a$ parameter (for example, $\mathbf{a}_2$) appears in E different equations in the score model, namely, those E equations for X_{ae} in which subscript $a = 2$. No design subscripts are absent from key term $\mathbf{e}_{e(a)}$, so the multiplier of $\sum\hat{\mathbf{e}}_{e(a)}^2$ is one (though it is not written explicitly). A specific parameter such as $\mathbf{e}_{2(1)}$ applies only to one score.

Suppose that $A = 2$ and that the data are as follows:

a_1	a_2
10	9
7	10

Then

$$\hat{\mathbf{m}} = \bar{X}_{..} = 9$$
$$\hat{\mathbf{a}}_1 = -.5$$
$$\hat{\mathbf{a}}_2 = .5$$
$$\hat{\mathbf{e}}_{1(1)} = 1.5$$
$$\hat{\mathbf{e}}_{2(1)} = -1.5$$
$$\hat{\mathbf{e}}_{1(2)} = -.5$$
$$\hat{\mathbf{e}}_{2(2)} = .5$$

Hence,

$$SS_A = E\sum\hat{\mathbf{a}}_a^2 = 2(.25 + .25) = 1$$
$$SS_E = \sum\hat{\mathbf{e}}_{e(a)}^2 = 2.25 + 2.25 + .25 + .25 = 5$$

For $A \times B$, the SS's are:

$$SS_A = BE\sum\hat{\mathbf{a}}_a^2$$
$$SS_B = AE\sum\hat{\mathbf{b}}_b^2$$
$$SS_{AB} = E\sum\widehat{\mathbf{ab}}_a^2$$
$$SS_E = \sum\mathbf{e}_{e(ab)}^2$$

Here $\mathbf{a}_1$ appears in $B \cdot E$ equations of the score model, $\mathbf{b}_2$ appears in $A \cdot E$ equations, and $\mathbf{ab}_{21}$ appears in E equations. An error parameter such as $\mathbf{e}_{1(21)}$ appears within but one equation, so its multiplier is 1 (though it is not written explicitly).

For design $S(A) \times B$, the SS's are:

$$\begin{aligned} SS_A &= BS\sum \hat{\mathbf{a}}_a^2 \\ SS_B &= AS\sum \hat{\mathbf{b}}_b^2 \\ SS_S &= B\sum \hat{\mathbf{s}}_{s(a)}^2 \\ SS_{AB} &= S\sum \widehat{\mathbf{ab}}_{ab}^2 \\ SS_{BS} &= \sum \widehat{\mathbf{bs}}_{bs(a)}^2 \end{aligned}$$

Since nest subscripts as well as nested ones are considered to be present rather than absent in applying step (3), A is not a multiplier for SS_S or SS_{BS}. In addition, E is not among the multipliers, since subscript e is lacking in the score model for $S(A) \times B$.

The required estimates for a numerical example were calculated in Section 4.8.5, so let us now use these estimates to compute the SS's. We omit the minus signs on the parameter estimates, since the squared quantities are positive regardless of sign.

$$\begin{aligned} SS_A &= 6(3^2 + 2^2 + 1^2) = 6 \cdot 14 = 84 \\ SS_B &= 9(4^2 + 4^2) = 9 \cdot 32 = 288 \\ SS_S &= 2(3^2 + 2^2 + 5^2 + 3^2 + 2^2 + 5^2 + 3^2 + 1^2 + 2^2) \\ &= 2 \cdot 90 = 180 \\ SS_{AB} &= 3(2^2 + 2^2 + 1^2 + 1^2 + 3^2 + 3^2) \\ &= 3 \cdot 28 = 84 \\ SS_{BS} &= 3^2 + 2^2 + 1^2 + 3^2 + 2^2 + 1^2 + 0^2 + 1^2 + 1^2 + 0^2 + 1^2 \\ &\quad + 1^2 + 3^2 + 0^2 + 3^2 + 3^2 + 0^2 + 3^2 \\ &= 68 \end{aligned}$$

5.4 COMPUTATIONAL FORMULAS FOR KEY TERM SS's

In practice, SS's are not calculated from parameter estimates. This section demonstrates how to derive the simpler computational formulas used to obtain the same quantities. Recall from Section 4.11 how the symbolic product (SP) and expanded symbolic product (ESP) are derived from the subscripts of a key term. We simply substitute the subscript -1 for non-nest subscripts, keep nest subscripts as they are, and interpret the result as a formula, the SP. Multiplied out, it becomes the ESP, which is the key to deriving the formula for estimating parameters. The ESP is also the key to deriving the computational formula for an SS. In deriving the computational formula for calculating the SS for a key term, then, we begin with the ESP for that key term.

For each term of the ESP there is a *sum of squares element*, which is denoted by SS subscripted with the term of the ESP. Hence, each subscript consists of one or more lower-case letters or one. We must clearly distinguish between an SS, which has an upper-case subscript, and an SS element. An SS consists of a summation of SS elements, as indicated by the ESP formula for a key term. For example, in Section 4.11, we found that the ESP for $\mathbf{s}_{s(a)}$ of $S(A) \times B$ is $as - a$. Therefore, $SS_S = SS_{as} - SS_a$, where SS_{as} and SS_a are SS elements. The signs, plus or minus, of the ESP terms are maintained for the SS elements. For $\mathbf{ab}_{ab}$ of $S(A) \times B$, the ESP is $ab - b - a + 1$, so $SS_{AB} = SS_{ab} - SS_b - SS_a + SS_1$.

In order to compute an SS, we first compute the SS elements and then sum these elements. The following steps describe how to derive the computational formula for an SS element.

1. Write X with all subscripts present in the SS element. Replace by dots (for summation) all subscripts absent from the SS element. (If the element subscript is one, replace all X subscripts with dots.)
2. Square the sum so indicated.
3. Indicate by $\sum$ a summing of such squared quantities across all subscripts present on X.
4. Divide the result from step (3) by the product of the upper limits of the absent (dotted) subscripts.

For example, for design A, $SS_a = \sum X^2_{a.}/E$. For design $S(A) \times B$, $SS_b = \sum X^2_{.b.}/AS$.

The same SS element typically appears in the formulas for a number of different SS's. It is not necessary, however, to search through the formulas to determine how many distinct SS elements there are before writing the computational formulas for these elements. For any compact design, the number of distinct SS elements equals the number of terms in the score model, including **m** but excluding **e** when there is only one score per condition (that is, when e is confounded). For each term of the score model, there is a corresponding SS element having exactly the same subscripts (including the nest subscripts of the key term—but nests are never used in this book to denote SS elements). The SS element corresponding to **m** is SS_1.

The steps in the derivation of SS formulas and preparation for computation of SS's are:

1. Write the score model.
2. Write the computational formulas for the SS elements, using the four rules just given.
3. Derive the ESP's.
4. Using the ESP's, write the SS formulas in terms of the SS elements.
5. Prepare an SS elements table.

The SS elements table lists the SS elements in the left-hand column and the SS's in a row at the top. Proceeding one column (one SS) at a time, and using the SS formula found in step (2), we write a plus sign in each cell of the table where the SS element enters positively in the SS formula, and a minus sign where it enters negatively. The table also has an SS_{tot} column with a plus sign in the top row and a minus sign in the bottom (SS_1) row. Examples of such tables are given in the forthcoming sections.

Once derivation and preparation are complete, we compute the SS's as follows:

1. Compute the required summations on X.
2. Using the summations and the computational formulas for the SS elements, compute the SS elements.
3. Enter each element thus computed in its row of the SS elements table, but only in spaces having a plus or minus sign. Sum the entries in each column. Each sum is the SS indicated by the column heading.

The procedures described in this section will now be illustrated for designs A, $A \times B$, $S \times A$, $S(A) \times B$, and $S \times A \times B$.

5.5 DESIGN A

First we prepare for computation by deriving the formulas and the SS elements tables. No data are required at this point. The steps are as follows:

1. Write the score model:

$$X_{ae} = \mathbf{m} + \mathbf{a}_a + \mathbf{e}_{e(a)}$$

2. Write the computational formulas for the SS elements. There is a distinct SS element corresponding to each key term and to **m**. It is helpful to list the SS elements in an order opposite from that for the corresponding model terms:

$$\begin{aligned} \mathbf{e}_{e(a)}: &\quad SS_{ae} = \textstyle\sum X_{ae}^2 \\ \mathbf{a}_a: &\quad SS_a = \textstyle\sum X_{a.}^2/E \\ \mathbf{m}: &\quad SS_1 = X_{..}^2/AE \end{aligned}$$

Normally, it is not necessary to write the model terms corresponding to each element, but they are included here to help the student see the correspondence. The student should verify that the formula in each case was derived according to the rules given in Section 5.4.

3. Derive the ESP's as follows:

Key term	SP	ESP
$\mathbf{a}_a$	$a - 1$	$a - 1$
$\mathbf{e}_{e(a)}$	$(e - 1)a$	$ae - a$

4. Write the SS formulas in terms of SS elements. These formulas follow directly from the ESP's. There is an SS for each key term:

$$\begin{aligned} SS_A &= SS_a - SS_1 \\ SS_E &= SS_{ae} - SS_a \end{aligned}$$

5. Prepare an SS elements table as follows:

	SS_A	SS_E	SS_{tot}
SS_{ae}		+	+
SS_a	+	−	
SS_1	−		−
Sum			

The plus and minus signs were filled in column by column after the previously given formulas for the SS's were examined. The SS_{tot} column should have a plus sign in the top row and a minus sign in the bottom SS_1 row. The top row should otherwise have only one plus sign; the bottom row should otherwise have one more minus sign than plus signs. All rows except those at the top and bottom of the table should have an equal number of plus and minus signs. Otherwise, an error was made in the derivation. In the present example, the only other row is for SS_a, which has one plus and one minus sign.

We are now prepared to consider the scores for a specific experiment and to compute the SS's. We use the same data for design *A* as were given in Section 5.3:

a_1	a_2
10	9
7	10

1. Compute the required summations on *X*. Examination of the computational formulas for the SS elements reveals that the $X_{a.}$'s and $X_{..}$ are required:

$$\begin{aligned} X_{1.} &= 17 \\ X_{2.} &= 19 \\ X_{..} &= 36 \end{aligned}$$

2. Compute the SS elements:

$$SS_{ae} = 10^2 + 7^2 + 9^2 + 10^2 = 330$$

$$SS_a = \frac{17^2 + 19^2}{2} = 325$$

$$SS_1 = \frac{36^2}{4} = 324$$

3. Enter the SS elements in the SS elements table, and sum the columns:

	SS_A	SS_E	SS_{tot}
SS_{ae}		+330	+330
SS_a	+325	−325	
SS_1	−324		−324
Sum	1	5	6

The column sums show that $SS_A = 1$, $SS_E = 5$, and $SS_{tot} = 6$, which agrees with the results obtained in Section 5.3 based on parameter estimates. As a check on the accuracy in summing the columns, SS_{tot} should be computed not only as a column sum, but as a row sum of the SS's for the key terms. Both summations should give the same result. This check will not detect errors in the computation of the SS elements, however.

5.6 DESIGN $A \times B$

These procedures are now illustrated for design $A \times B$, which was discussed in Section 4.3. Again, derivation precedes computation.

1. Write the score model:

$$X_{abe} = \mathbf{m} + \mathbf{a}_a + \mathbf{b}_b + \mathbf{ab}_{ab} + \mathbf{e}_{e(ab)}$$

2. Write the computational formulas for the SS elements:

$$\begin{aligned}
\mathbf{e}_{e(ab)}&: & SS_{abe} &= \sum X^2_{abe} \\
\mathbf{ab}_{ab}&: & SS_{ab} &= \sum X^2_{ab.}/E \\
\mathbf{b}_b&: & SS_b &= \sum X^2_{.b.}/AE \\
\mathbf{a}_a&: & SS_a &= \sum X^2_{a..}/BE \\
\mathbf{m}&: & SS_1 &= X^2_{...}/ABE
\end{aligned}$$

3. Derive the ESP's:

Key term	SP	ESP
$\mathbf{a}_a$	$a - 1$	$a - 1$
$\mathbf{b}_b$	$b - 1$	$b - 1$
$\mathbf{ab}_{ab}$	$(a - 1)(b - 1)$	$ab - b - a + 1$
$\mathbf{e}_{e(ab)}$	$(e - 1)ab$	$abe - ab$

	SS_A	SS_B	SS_{AB}	SS_E	SS_{tot}
SS_{abe}				+	+
SS_{ab}			+	−	
SS_b		+	−		
SS_a	+		−		
SS_1	−	−	+		−
Sum					

4. Write the SS formulas in terms of SS elements:

$$SS_A = SS_a - SS_1$$
$$SS_B = SS_b - SS_1$$
$$SS_{AB} = SS_{ab} - SS_b - SS_a + SS_1$$
$$SS_E = SS_{abe} - SS_{ab}$$

5. Prepare an SS elements table (shown in Table 5.6-1). Again, except for the SS_{tot} in Table 5.6-1, the row for the element with all design subscripts has a single plus sign, and the row for SS_1 has one more minus sign than plus signs. All other rows correctly have an equal number of plus and minus signs.

Following is a simple numerical example showing computation of the SS's for $A \times B$. The scores are:

	b_1	b_2
a_1	10 7	2 6
a_2	9 10	7 5

1. Compute the required summations on X:

$X_{ab.}$: $X_{11.} = 17$
$X_{12.} = 8$
$X_{21.} = 19$
$X_{22.} = 12$

$X_{.b.}$: $X_{.1.} = 36$
$X_{.2.} = 20$

$$X_{a..}: \qquad X_{1..} = 25$$
$$X_{2..} = 31$$
$$X_{...}: \qquad X_{...} = 56$$

Only the $X_{ab.}$'s need be calculated directly from the scores. Other summations can be computed from a previous set having one more subscript. For example, the $X_{a..}$'s and $X_{.b.}$'s can be computed from the $X_{ab.}$'s:

$$X_{.b.} = \sum_a X_{ab.} = X_{1b.} + X_{2b.}$$

Specifically,

$$X_{.2.} = X_{12.} + X_{22.} = 8 + 12 = 20$$

Quantity $X_{...}$ can be computed from either the $X_{a..}$'s or the $X_{.b.}$'s.

As a check on the accuracy of computation, $X_{...}$ should also be computed directly from the scores. Also, a check should be made to ensure that the quantities within each set ($X_{ab.}$'s, $X_{.b.}$'s, $X_{a..}$'s) sum to $X_{...}$, as they should.

2. Compute the SS elements:

$$\begin{aligned} SS_{abe} &= \sum X_{abe}^2 \\ &= 10^2 + 7^2 + 9^2 + 10^2 + 2^2 + 6^2 + 7^2 + 5^2 \\ &= 100 + 49 + 81 + 100 + 4 + 36 + 49 + 25 \\ &= 444 \end{aligned}$$

$$\begin{aligned} SS_{ab} &= \sum X_{ab.}^2/E \\ &= (17^2 + 19^2 + 8^2 + 12^2)/2 \\ &= (289 + 361 + 64 + 144)/2 \\ &= 858/2 = 429 \end{aligned}$$

$$\begin{aligned} SS_{b} &= \sum X_{.b.}^2/AE \\ &= (36^2 + 20^2)/4 = (1296 + 400)/4 \\ &= 1696/4 = 424 \end{aligned}$$

$$\begin{aligned} SS_{a} &= \sum X_{a..}^2/BE \\ &= (25^2 + 31^2)/4 \\ &= (625 + 961)/4 = 1586/4 = 396.5 \end{aligned}$$

$$\begin{aligned} SS_{1} &= X_{...}^2/ABE \\ &= 56^2/8 = 3136/8 \\ &= 392 \end{aligned}$$

Some calculating machines accumulate not only the sum of squares but the sum of the entries per se. This latter sum should equal $X_{...}$ for each SS element computed. Otherwise, an error has occurred, such as the entry of a wrong number. If the entries accumulate to $X_{...}$, the SS element is also probably correct. Nonetheless, it is wise to compute each SS element twice as a check. If the entries are not accumulated, SS elements should, without question, be computed twice.

Table 5.6-2 Filled-in SS elements table for $A \times B$

	SS_A	SS_B	SS_{AB}	SS_E	SS_{tot}
SS_{abe}				+444	+444
SS_{ab}			+429	−429	
SS_b		+424	−424		
SS_a	+396.5		−396.5		
SS_1	−392	−392	+392		−392
Sum	4.5	32	.5	15	52

3. Enter the SS elements in the SS elements table, and sum the columns. The filled-in SS elements table for this exercise is Table 5.6-2. The table shows that SS_{tot} sums to 52 by both row and column summations, as it should.

5.7 CALCULATING SS's WITH REDUCED SCORES

Because of squaring, calculation of the SS's can involve very large numbers. This did not happen in the preceding example because the artificial scores were made so low. If such large numbers are awkward to deal with, smaller numbers can be used if a constant is subtracted from each score in the experiment. The SS's derived from the new reduced scores will exactly equal the SS's derived from the original scores.

Adding or subtracting a constant to all scores in the experiment changes the estimate of **m**, but does not change the estimates for other parameters, since the other parameters are used to explain *deviations* of scores from the mean, not the mean itself. For example, $\mathbf{a}_a$ is used to explain the mean deviation of the scores derived under a_a from those in the experiment as a whole. If we alter all scores by adding or subtracting a constant, we change the overall mean and the mean for scores under a_a equally, so the discrepancy between them remains constant. A similar situation applies for all parameters except **m**. Since, as was explained in Section 5.3, the SS's are sums of squares of parameter estimates, the SS's are not affected by the constant subtraction.

One particularly interesting constant we could subtract is the mean of all scores. If we subtract $\bar{X}_{...}$ from all scores for $A \times B$, the new mean is zero. This simplifies the computations of SS's because if the mean is zero, then SS_1 is zero. Let us see how the computations would be affected for the example of $A \times B$ just considered.

At the moment, the mean of all scores is $56/8 = 7$. By subtracting 7 from each score, we obtain a new set of scores as follows:

	b_1	b_2
a_1	3 0	−5 −1
a_2	2 3	0 −2

It is easy to determine that the total of all scores, $X_{...}$, is now zero.

The SS elements on these reduced scores are as follows:

$$
\begin{aligned}
SS_{abe} &= \sum X_{abe}^2 \\
&= 3^2 + 0^2 + 2^2 + 3^2 + (-5)^2 + (-1)^2 + 0^2 + (-2)^2 \\
&= 9 + 4 + 9 + 25 + 1 + 4 \\
&= 52
\end{aligned}
$$

$$
\begin{aligned}
SS_{ab} &= \sum X_{ab.}^2/E \\
&= [3^2 + 5^2 + (-6)^2 + (-2)^2]/2 \\
&= (9 + 25 + 36 + 4)/2 \\
&= 74/2 = 37
\end{aligned}
$$

$$
\begin{aligned}
SS_b &= \sum X_{.b.}^2/AE \\
&= [8^2 + (-8)^2]/4 \\
&= 128/4 = 32
\end{aligned}
$$

$$
\begin{aligned}
SS_a &= \sum X_{a..}^2/BE \\
&= [(-3)^2 + 3^2]/4 = 18/4 = 4.5
\end{aligned}
$$

$$
\begin{aligned}
SS_1 &= X_{...}^2/ABE \\
&= 0
\end{aligned}
$$

The SS elements and SS's are shown in Table 5.7-1. The SS's are the same as those shown in Table 5.6-2, even though the elements differ. Because SS_1 is zero for the adjusted scores, the SS's for the main effects and SS_{tot} are particularly simple; they are equal to single SS elements.

In the example, the mean was an integral number and the subtractions were easy. If, however, we only wanted to simplify the calculations without

Table 5.7-1 SS elements table, with elements calculated on reduced scores

	SS_A	SS_B	SS_{AB}	SS_E	SS_{tot}
SS_{abe}				+52	+52
SS_{ab}			+37	−37	
SS_b		+32	−32		
SS_a	+4.5		−4.5		
SS_1	−0	−0	+0		−0
Sum	4.5	32	.5	15	52

first bothering to find the mean, which might turn out to have decimal places and be a nuisance in subtraction, we could merely choose a simple integral number that seemed approximately close to the mean, and subtract it from all scores. Of course, SS_1 then would probably not be zero. (It would not be unless serendipity assisted us in choosing the mean to subtract with.) Although it is perfectly all right to have negative scores after subtraction (or before), nothing is gained by using a very large subtractor, for then very large negative numbers result; squaring of these numbers produces large and unwieldy quantities, which we were trying to avoid.

5.8 DESIGN $S \times A$

Design $S \times A$ was described in Section 4.5. The score model for this design is

$$X_{as} = \mathbf{m} + \mathbf{a}_a + \mathbf{s}_s + \mathbf{as}_{as} + \mathbf{e}_{as}$$

For each model term except $\mathbf{e}_{as}$ (because it is confounded with $\mathbf{as}_{as}$), there is a corresponding SS element, as follows:

$$\begin{aligned} SS_{as} &= \sum X_{as}^2 \\ SS_s &= \sum X_{.s}^2/A \\ SS_a &= \sum X_{a.}^2/S \\ SS_1 &= X_{..}^2/AS \end{aligned}$$

Using the ESP's for the model terms, we derive the following formulas for the SS's:

$$\begin{aligned} SS_A &= SS_a - SS_1 \\ SS_S &= SS_s - SS_1 \\ SS_{AS} &= SS_{as} - SS_s - SS_a + SS_1 \end{aligned}$$

As an example of the use of these formulas, the same scores can be used that were used to illustrate $S \times A$ in Section 4.5.4. In that section, the required summations on the X's were computed, except that each sum was divided to get an $\bar{X}$. Nonetheless, Section 4.5.4 shows these sums before division, and, by using these data, we can compute the SS elements:

$$\begin{aligned} SS_{as} &= 5^2 + 7^2 + 12^2 + 4^2 + 8^2 + 4^2 + 4^2 + 9^2 + 9^2 + 9^2 + 1^2 \\ &\quad + 1^2 + 8^2 + 6^2 + 5^2 + 8^2 + 10^2 + 16^2 + 9^2 + 13^2 \\ &= 25 + 49 + 144 + 16 + 64 + 16 + 16 + 81 + 81 + 81 + 1 \\ &\quad + 1 + 64 + 36 + 25 + 64 + 100 + 256 + 81 + 169 \\ &= 1370 \end{aligned}$$

$$\begin{aligned} SS_s &= (18^2 + 22^2 + 45^2 + 28^2 + 35^2)/4 \\ &= (324 + 484 + 2025 + 784 + 1225)/4 \\ &= 4842/4 = 1210.5 \end{aligned}$$

$$\begin{aligned} SS_a &= (36^2 + 35^2 + 21^2 + 56^2)/5 \\ &= (1296 + 1225 + 441 + 3136)/5 \\ &= 6098/5 = 1219.6 \end{aligned}$$

Table 5.8-1 SS elements table for $S \times A$

	SS_A	SS_S	SS_{AS}	SS_{tot}
SS_{as}			+1370	+1370
SS_s		+1210.5	−1210.5	
SS_a	+1219.6		−1219.6	
SS_1	−1095.2	−1095.2	+1095.2	−1095.2
Sum	124.4	115.3	35.1	274.8

$$SS_1 = 148^2/20$$
$$= 21904/20 = 1095.2$$

These elements have been entered in Table 5.8-1 (which was constructed from the previously given formulas for the SS's), and summed to provide the SS's.

5.9 DESIGN $S(A) \times B$

Design $S(A) \times B$ was described in Section 4.8. The score model for the design is

$$X_{abs} = \mathbf{m} + \mathbf{a}_a + \mathbf{b}_b + \mathbf{s}_{s(a)} + \mathbf{ab}_{ab} + \mathbf{bs}_{bs(a)} + \mathbf{e}_{abs}$$

For each model term except $\mathbf{e}_{abs}$ (because it is confounded with $\mathbf{bs}_{bs(a)}$), there is a corresponding SS element, as follows:

$$SS_{abs} = \sum X_{abs}^2$$
$$SS_{ab} = \sum X_{ab.}^2/S$$
$$SS_{as} = \sum X_{a.s}^2/B$$
$$SS_b = \sum X_{.b.}^2/AS$$
$$SS_a = \sum X_{a..}^2/BS$$
$$SS_1 = X_{...}^2/ABS$$

The ESP's for $S(A) \times B$ are shown in Table 4.11-1, and from them we derive the formulas for the SS's:

$$SS_A = SS_a - SS_1$$
$$SS_B = SS_b - SS_1$$
$$SS_S = SS_{as} - SS_a$$
$$SS_{AB} = SS_{ab} - SS_b - SS_a + SS_1$$
$$SS_{BS} = SS_{abs} - SS_{ab} - SS_{as} + SS_a$$

To exemplify the use of these formulas, the same scores are used as were used to illustrate $S(A) \times B$ in Section 4.8.5. In that section the required summations on the X's were computed, except that each sum was then divided in order to get an $\bar{X}$. Nonetheless, Section 4.8.5 shows these sums before division, and using these data we can compute the SS elements. We omit the minus signs on quantities being squared since the result is positive regardless of the sign:

$$\begin{aligned}
SS_{abs} &= 5^2 + 1^2 + 7^2 + 3^2 + 7^2 + 1^2 + 6^2 + 4^2 + 1^2 + 12^2 \\
&\quad + 12^2 + 3^2 + 4^2 + 3^2 + 7^2 + 12^2 + 11^2 + 13^2 \\
&= 25 + 1 + 49 + 9 + 49 + 1 + 36 + 16 + 1 + 144 \\
&\quad + 144 + 9 + 16 + 9 + 49 + 144 + 121 + 169 \\
&= +992 \\
SS_{ab} &= (3^2 + 9^2 + 9^2 + 27^2 + 6^2 + 36^2)/3 \\
&= (9 + 81 + 81 + 729 + 36 + 1296)/3 \\
&= 2232/3 = 744 \\
SS_{as} &= (8^2 + 6^2 + 8^2 + 18^2 + 16^2 + 2^2 + 16^2 + 8^2 + 6^2)/2 \\
&= (64 + 36 + 64 + 324 + 256 + 4 + 256 + 64 + 36)/2 \\
&= 1104/2 = 552 \\
SS_b &= (0^2 + 72^2)/9 \\
&= (0 + 5184)/9 \\
&= 576 \\
SS_a &= (6^2 + 36^2 + 30^2)/6 \\
&= (36 + 1296 + 900)/6 \\
&= 2232/6 \\
&= 372 \\
SS_1 &= 72^2/18 = 5184/18 = 288
\end{aligned}$$

These elements are entered in Table 5.9-1, which was constructed from the formulas for the SS's given previously; we sum the columns to get the SS's.

5.10 DESIGN $S \times A \times B$

We shall now derive the SS formulas for $S \times A \times B$. Their use can be demonstrated with a numerical example. We use the same data that were used for $S(A) \times B$, in Section 4.8.5, except that we interpret the labeling

Table 5.9-1 SS elements table for $S(A) \times B$

	SS_A	SS_B	SS_S	SS_{AB}	SS_{BS}	SS_{tot}
SS_{abs}					+992	+992
SS_{ab}				+744	−744	
SS_{as}			+552		−552	
SS_b		+576		−576		
SS_a	+372		−372	−372	+372	
SS_1	−288	−288		+288		−288
Sum	84	288	180	84	68	704

Table 5.10-1 SP's and ESP's for $S \times A \times B$

Key term	SP	ESP
$\mathbf{a}_a$	$a - 1$	$a - 1$
$\mathbf{b}_b$	$b - 1$	$b - 1$
$\mathbf{s}_s$	$s - 1$	$s - 1$
$\mathbf{ab}_{ab}$	$(a - 1)(b - 1)$	$ab - b - a + 1$
$\mathbf{as}_{as}$	$(a - 1)(s - 1)$	$as - s - a + 1$
$\mathbf{bs}_{bs}$	$(b - 1)(s - 1)$	$bs - s - b + 1$
$\mathbf{abs}_{abs}$	$(a - 1)(b - 1)(s - 1)$	$abs - bs - as - ab + s + b + a - 1$

as literal so that s_1, for example, refers to the same subject whenever it appears, and likewise for s_2 and s_3.

The score model is:

$$X_{abs} = \mathbf{m} + \mathbf{a}_a + \mathbf{b}_b + \mathbf{s}_s + \mathbf{ab}_{ab} + \mathbf{as}_{as} + \mathbf{bs}_{bs} + \mathbf{abs}_{abs} + \mathbf{e}_{abs}$$

The SP's and ESP's are shown in Table 5.10-1.

For each model term (except the confounded $\mathbf{e}_{abs}$), there is an SS element as follows:

$$\begin{aligned} SS_{abs} &= \textstyle\sum X_{abs}^2 \\ SS_{bs} &= \textstyle\sum X_{.bs}^2/A \\ SS_{as} &= \textstyle\sum X_{a.s}^2/B \\ SS_{ab} &= \textstyle\sum X_{ab.}^2/S \\ SS_s &= \textstyle\sum X_{..s}^2/AB \\ SS_b &= \textstyle\sum X_{.b.}^2/AS \\ SS_a &= \textstyle\sum X_{a..}^2/BS \\ SS_1 &= X_{...}^2/ABS \end{aligned}$$

The SS elements are shown in Table 5.10-2. Most of them are precisely the same as those for design $S(A) \times B$, since the formulas and the data are the

Table 5.10-2 SS elements table for $S \times A \times B$

	SS_A	SS_B	SS_S	SS_{AB}	SS_{AS}	SS_{BS}	SS_{ABS}	SS_{tot}
SS_{abs}							+992	+992
SS_{bs}						+768	−768	
SS_{as}					+552		−552	
SS_{ab}				+744			−744	
SS_s			+444		−444	−444	+444	
SS_b		+576		−576		−576	+576	
SS_a	+372			−372	−372		+372	
SS_1	−288	−288	−288	+288	+288	+288	−288	−288
Sum	84	288	156	84	24	36	32	704

same. Specifically, the equivalence holds for SS_{abs}, SS_{as}, SS_{ab}, SS_b, SS_a, and SS_1.

Each element of $S(A) \times B$ appears also for $S \times A \times B$, but the latter has additional elements as well, namely, SS_{bs} and SS_s:

$$
\begin{aligned}
SS_{bs} &= X^2_{.bs}/A \\
&= (15^2 + 0^2 + 15^2 + 27^2 + 30^2 + 15^2)/3 \\
&= (225 + 0 + 225 + 729 + 900 + 225)/3 \\
&= 2304/3 = 768
\end{aligned}
$$

$$
\begin{aligned}
SS_s &= X^2_{..s}/AB \\
&= (42^2 + 30^2 + 0^2)/6 \\
&= (1764 + 900 + 0)/6 \\
&= 2664/6 = 444
\end{aligned}
$$

Many of the SS's are exactly the same for designs $S \times A \times B$ and $S(A) \times B$. Where there is a difference, as for SS_S and SS_{BS}, it is because the terms of $S(A) \times B$ to which these SS's relate are nested terms, $\mathbf{s}_{s(a)}$ and $\mathbf{bs}_{bs(a)}$.

It was stated in Section 4.8.3 that $\mathbf{s}_{s(a)}$ of $S(A) \times B$ equals $\mathbf{s}_s + \mathbf{as}_{as}$ of $S \times A \times B$. Also, $\mathbf{bs}_{bs(a)}$ of the former design equals $\mathbf{bs}_{bs} + \mathbf{abs}_{abs}$ of the latter. These equivalences also occur for the SS's. If we add $SS_S = 156$ and $SS_{AS} = 24$ of $S \times A \times B$, we get the $SS_S = 180$ of $S(A) \times B$. Likewise, if we add the $SS_{BS} = 36$ and $SS_{ABS} = 32$ of $S \times A \times B$, we get the $SS_{BS} = 68$ of $S(A) \times B$.

It is always true that if we compare a design of the type $S(Z') \times Z$ with $S \times Z' \times Z$, each nested key term in the former equals a sum of key terms in the latter. If we calculate the SS's for these latter key terms and sum them, they total the SS for the equivalent key term of the $S(Z') \times Z$ design. This fact can serve a practical result, as will be seen in the next section.

5.11 CALCULATING NESTED SS's WITH A PROGRAM FOR CROSSED DESIGNS

5.11.1 General Principles

Oftentimes an experimenter has access to computer programs for ANOVA that work only for completely crossed designs, but lacks a program for the analysis of nested designs, though the latter designs are frequently used. Another experimenter may have a more general program that can analyze nested design data, but cannot determine how to make it run correctly with his data and his computer installation. In such circumstances experimenters sometimes struggle along, trying to get the general program to work, or they give up and proceed to do the analysis with a desk calculator. A third approach is to "trick" the program for completely crossed designs into doing the required analysis, or at least most of it, and thereby avoid a good deal of annoyance.

This last possibility, which we consider now, is not widely appreciated by experimenters, but it can be a valuable approach. As was explained in the previous section, data from a nested design such as $S(A) \times B$ can be interpreted as coming from a completely crossed design, and analyzed as such. The SS's from the crossed analysis frequently directly equal their counterparts in the nested design. If not, the SS's of the nested design can be obtained by adding together certain SS's of the crossed design.

Adding together the required SS's from the computer analysis of the "crossed" design requires relatively little work, compared with computing all required SS's on a calculator by hand, and the possibilities for errors are fewer. The problem for most experimenters is to recognize that the program for completely crossed designs can be used and to know which terms of the output must be summed together. We have already seen which terms must be combined from the $S \times A \times B$ analysis of $S(A) \times B$. We now consider, in more general terms, a method for finding the SS's for an $S(Z) \times Z'$ design by analyzing the data as if it were from an $S \times Z \times Z'$ design. The analysis of the data as a crossed design, of course, requires that we be able to interpret the data in this way. We can, so long as for each case of nesting, the nested factor has an equal number of levels within each level of the nest factor. For example, in $S(A) \times B$, there would have to be an equal number of subjects (s_s levels) within each level of A. Then to interpret the data as crossed, we simply consider s_3, for example, to represent the same subject in all its appearances, though in actuality "s_3" represents a different subject for each a_a level.

In determining which SS's to combine, we first require the score models for both $S(Z) \times Z'$ and for $S \times Z \times Z'$. It is assumed that by now the student can derive such models, if necessary. We begin by considering the key terms for $S(Z) \times Z'$ and following these rules:

1. For terms without subscript nests, no combining of SS's is required. The SS's can be taken directly from the SS's for the same terms of the $S \times Z \times Z'$ analysis.
2. For a term with a subscript nest, the SS is obtained by combining the SS for the following key terms of the crossed design:
   a. the term having exactly the same subscripts as the nested subscripts in the latter.
   b. the terms having all the subscripts as in (a), plus one or more of the nest subscripts of the nested key term (but no other subscripts).

5.11.2 SS Equivalents for $S \times A \times B$ and $S(A) \times B$

Let us see how these rules would apply for an $S \times A \times B$ analysis of $S(A) \times B$. The combining of terms is shown in Table 5.11-1. The key terms $\mathbf{a}_a$, $\mathbf{b}_b$, and $\mathbf{ab}_{ab}$ of the nested design have no subscript nests, and according to

Table 5.11-1 Key terms and SS equivalents for $S(A) \times B$ and $S \times A \times B$

$S(A) \times B$		$S \times A \times B$	
Key term	SS	Key term	SS
$\mathbf{a}_a$	SS_A	$\mathbf{a}_a$	SS_A
$\mathbf{b}_b$	SS_B	$\mathbf{b}_b$	SS_B
$\mathbf{ab}_{ab}$	SS_{AB}	$\mathbf{ab}_{ab}$	SS_{AB}
$\mathbf{s}_{s(a)}$	SS_S	$\left\{\begin{array}{l}\mathbf{s}_s \\ \mathbf{as}_{as}\end{array}\right.$	SS_S SS_{AS}
$\mathbf{bs}_{bs(a)}$	SS_{BS}	$\left\{\begin{array}{l}\mathbf{bs}_{bs} \\ \mathbf{abs}_{abs}\end{array}\right.$	SS_{BS} SS_{ABS}

rule (1) the SS's can be taken directly, without summing, from the SS's of the crossed analysis. Since the terms $\mathbf{s}_{s(a)}$ and $\mathbf{bs}_{bs(a)}$ have subscript nests, some combining of SS's is necessary. For $\mathbf{s}_{s(a)}$ we include, first, from the crossed design, $\mathbf{s}_s$, since by rule (2a) we need the term having exactly the same subscripts as the nested subscripts in the nested key term. The common subscript here is s. According to rule (2b), we need in addition the terms from the crossed design having not only s, but a (and no others), so we include $\mathbf{as}_{as}$. The SS's for these two terms of the crossed design must be summed to get the SS_S of the nested design. By similarly following the rules, we find terms whose SS's must be summed to get SS_{BS} of the nested design.

5.11.3 SS Equivalents for $S \times A \times B \times C \times D$ and $S(A \times B) \times C \times D$

Now let us see how the rules apply to a more complicated example. The SS's for $S(A \times B) \times C \times D$ are found by combining SS's of $S \times A \times B \times C \times D$. We begin by determining the equivalences between key terms of the two models. The model for $S \times A \times B \times C \times D$ is:

$$\begin{aligned} X_{abcds} = {} & \mathbf{m} + \mathbf{a}_a + \mathbf{b}_b + \mathbf{c}_c + \mathbf{d}_d + \mathbf{s}_s + \mathbf{ab}_{ab} + \mathbf{ac}_{ac} + \mathbf{ad}_{ad} + \mathbf{as}_{as} \\ & + \mathbf{bc}_{bc} + \mathbf{bd}_{bd} + \mathbf{bs}_{bs} + \mathbf{cd}_{cd} + \mathbf{cs}_{cs} + \mathbf{ds}_{ds} + \mathbf{abc}_{abc} \\ & + \mathbf{abd}_{abd} + \mathbf{abs}_{abs} + \mathbf{acd}_{acd} + \mathbf{acs}_{acs} + \mathbf{ads}_{ads} + \mathbf{bcd}_{bcd} \\ & + \mathbf{bcs}_{bcs} + \mathbf{bds}_{bds} + \mathbf{cds}_{cds} + \mathbf{abcd}_{abcd} + \mathbf{abcs}_{abcs} \\ & + \mathbf{abds}_{abds} + \mathbf{acds}_{acds} + \mathbf{bcds}_{bcds} + \mathbf{abcds}_{abcds} + \mathbf{e}_{abcds} \end{aligned}$$

The model for $S(A \times B) \times C \times D$ is:

$$\begin{aligned} X_{abcds} = {} & \mathbf{m} + \mathbf{a}_a + \mathbf{b}_b + \mathbf{c}_c + \mathbf{d}_d + \mathbf{ab}_{ab} + \mathbf{ac}_{ac} + \mathbf{ad}_{ad} + \mathbf{bc}_{bc} + \mathbf{bd}_{bd} \\ & + \mathbf{cd}_{cd} + \mathbf{s}_{s(ab)} + \mathbf{abc}_{abc} + \mathbf{abd}_{abd} + \mathbf{acd}_{acd} + \mathbf{bcd}_{bcd} \\ & + \mathbf{cs}_{cs(ab)} + \mathbf{ds}_{ds(ab)} + \mathbf{abcd}_{abcd} + \mathbf{cds}_{cds(ab)} + \mathbf{e}_{abcds} \end{aligned}$$

We require summing only for the nested terms of $S(A \times B) \times C \times D$. The SS's for the non-nested terms have direct equivalences in the crossed analysis. The only nested terms of $S(A \times B) \times C \times D$ are those involving

Table 5.11-2 Key term and SS equivalents for $S(A \times B) \times C \times D$ and $S \times A \times B \times C \times D$

$S(A \times B) \times C \times D$	$S \times A \times B \times C \times D$
$\mathbf{s}_{s(ab)}$	$\mathbf{s}_s + \mathbf{as}_{as} + \mathbf{bs}_{bs} + \mathbf{abs}_{abs}$
SS_S	$SS_S + SS_{AS} + SS_{BS} + SS_{ABS}$
$\mathbf{cs}_{cs(ab)}$	$\mathbf{cs}_{cs} + \mathbf{acs}_{acs} + \mathbf{bcs}_{bcs} + \mathbf{abcs}_{abcs}$
SS_{CS}	$SS_{CS} + SS_{ACS} + SS_{BCS} + SS_{ABCS}$
$\mathbf{ds}_{ds(ab)}$	$\mathbf{ds}_{ds} + \mathbf{ads}_{ads} + \mathbf{bds}_{bds} + \mathbf{abds}_{abds}$
SS_{DS}	$SS_{DS} + SS_{ADS} + SS_{BDS} + SS_{ABDS}$
$\mathbf{cds}_{cds(ab)}$	$\mathbf{cds}_{cds} + \mathbf{acds}_{acds} + \mathbf{bcds}_{bcds} + \mathbf{abcds}_{abcds}$
SS_{CDS}	$SS_{CDS} + SS_{ACDS} + SS_{BCDS} + SS_{ABCDS}$

s, so we attend here only to these terms. To find the SS for each term involving **s**, we must sum two or more SS's of the crossed analysis. Table 5.11-2 shows the nested terms of $S(A \times B) \times C \times D$ and the equivalent terms of $S \times A \times B \times C \times D$. All terms from both models containing **s** are included in this table, and all terms that are entered in the table include **s**.

5.12 WITHIN- AND BETWEEN-SS's

Discussions of ANOVA frequently refer to a breakdown of SS_{tot} into a within- and a between-SS. There are two common breakdowns: within-cell versus between-cell, and within-subject versus between-subject.

According to the first breakdown,

$$SS_{tot} = SS_{w\ cell} + SS_{b\ cell}$$

To be specific, consider design $A \times B$. From equation (1) of Section 5.2, we write

$$SS_{tot} = \sum(X_{abe} - X_{...})^2$$

According to this equation, SS_{tot} is based on deviations of the individual scores X_{abe} from the grand mean $\bar{X}_{...}$. We can divide this deviation into two components, using the cell mean $\bar{X}_{ab.}$:

$$X_{abe} - X_{...} = (X_{abe} - \bar{X}_{ab.}) + (\bar{X}_{ab.} - \bar{X}_{...})$$

The equation holds because we have merely added and subtracted the same quantity, $\bar{X}_{ab.}$. If we expand the equation according to the algebraic rule $(a + b)^2 = a^2 + 2ab + b^2$, we can prove that the cross product $2ab$ equals zero, leaving us with

$$SS_{tot} = \underset{\text{(I)}}{\sum(X_{abe} - \bar{X}_{ab.})^2} + \underset{\text{(II)}}{E\sum(\bar{X}_{ab.} - \bar{X}_{...})^2}$$

The multiplier E occurs in term II because $\sum_e$ for term II is over a quantity that is constant relative to e. Hence, the rule given in Section 2.1.4 concerning summation over a constant applies.

Term I is $SS_{\text{w cell}}$. It is based on the deviations of the scores from the cell means. If each score equaled the mean for its cell, then $SS_{\text{w cell}}$ would equal zero. According to Section 4.3.7, $X_{abe} - \bar{X}_{ab.} = \hat{\mathbf{e}}_{e(ab)}$, so $SS_{\text{w cell}} = \sum \hat{\mathbf{e}}^2_{e(ab)}$. According to Section 5.3, $\sum \hat{\mathbf{e}}^2_{e(ab)} = SS_E$, so $SS_{\text{w cell}} = SS_E$, which is always true for single measurement designs.

Term II is $SS_{\text{b cell}}$. The quantity $\bar{X}_{ab.} - \bar{X}_{...}$ is the discrepancy between a cell mean and the grand mean. If all cell means were the same, and hence equal to $\bar{X}_{...}$, $SS_{\text{b cell}}$ would equal zero. Quantity $\bar{X}_{ab.} - \bar{X}_{...} = \hat{\mathbf{a}}_a + \hat{\mathbf{b}}_b + \widehat{\mathbf{ab}}_{ab}$ (Section 4.3.7), and if these latter estimates are substituted in term II we obtain

$$\begin{aligned} SS_{\text{b cell}} &= BE\sum \hat{\mathbf{a}}^2_a + AE\sum \hat{\mathbf{b}}^2_b + E\sum \widehat{\mathbf{ab}}^2_{ab} \\ &= SS_A + SS_B + SS_{AB} \end{aligned}$$

In general, it is true that $SS_{\text{w cell}}$ equals SS_E for single measurement designs, and $SS_{\text{b cell}}$ equals the sum of the SS's for the other key terms. With repeated measurements designs the situation is somewhat different. Consider design $S(A) \times B$:

$$SS_{\text{tot}} = \sum (X_{abs} - \bar{X}_{...})^2$$

Using the cell mean $\bar{X}_{ab.}$, we can write this as

$$\begin{aligned} SS_{\text{tot}} &= \sum [(X_{abs} - \bar{X}_{ab.}) + (\bar{X}_{ab.} - \bar{X}_{...})]^2 \\ &= \underset{\text{(I)}}{\sum (X_{abs} - \bar{X}_{ab.})^2} + \underset{\text{(II)}}{S\sum (X_{ab.} - X_{...})^2} \end{aligned}$$

Term I is $SS_{\text{w cell}}$. Referring back to Section 4.8.4, we can ascertain that $X_{abs} - \bar{X}_{ab.} = \hat{\mathbf{s}}_{s(a)} + \widehat{\mathbf{bs}}_{bs(a)}$, and hence that $SS_{\text{w cell}} = B\sum \hat{\mathbf{s}}^2_{s(a)} + \sum \widehat{\mathbf{bs}}^2_{bs(a)}$ $= SS_S + SS_{BS}$. Term II is $SS_{\text{b cell}}$, and we can determine that

$$SS_{\text{b cell}} = SS_A + SS_B + SS_{AB}$$

In general, in repeated measurements designs, SS's involving (subscripted with) factor S compose $SS_{\text{w cell}}$, whereas those SS's not involving factor S compose $SS_{\text{b cell}}$.

A second kind of within- and between- breakdown that is frequently encountered applies only to repeated measurements designs: The total SS equals the within-subject SS plus the between-subject SS. Let us exemplify with $S(A) \times B$:

$$\begin{aligned} SS_{\text{tot}} &= \sum_{abs} (X_{abs} - X^2_{...}) \\ SS_{\text{w subj}} &= \sum_{abs} (X_{abs} - \bar{X}_{a.s})^2 \\ SS_{\text{b subj}} &= \sum_{abs} (\bar{X}_{a.s} - \bar{X}_{...})^2 \end{aligned}$$

To get a breakdown of SS_{tot} into within- and between-cell components, we inserted the cell mean into the formula for SS_{tot}; similarly, to get a breakdown into within- and between-subject components, we insert the mean score for a subject, $\bar{X}_{a.s}$, into the formula for SS_{tot}. Since S is nested within A, a particular subject's scores occur only at a specific level of a. Hence, in computing the mean score for a subject, we average over subscript b but not over a.

The $SS_{w\ subj}$ is based on the deviation between a score, X_{abs}, and the mean score for the subject, $\bar{X}_{a.s}$. It is large if the different scores for individual subjects vary greatly, and it is small if they tend to be approximately the same. Since $X_{abs} - \bar{X}_{a.s} = \hat{\mathbf{b}}_b + \widehat{\mathbf{ab}}_{ab} + \widehat{\mathbf{bs}}_{bs(a)}$ (confirm by referring back to Section 4.8.4), $SS_{w\ subj} = SS_B + SS_{AB} + SS_{BS}$.

The $SS_{b\ subj}$ is based on the deviation between the mean score for a subject and the mean of all scores. It is large if the mean scores of subjects differ greatly from each other, and it is small if all subjects have approximately the same mean scores. Since $\bar{X}_{a.s} - \bar{X}_{...} = \hat{\mathbf{a}}_a + \hat{\mathbf{s}}_{s(a)}$, $SS_{b\ subj} = SS_A + SS_S$.

In general, we can say the following about repeated measurements designs: SS's for key terms involving factors with which S is crossed are part of $SS_{w\ subj}$. Other SS's are part of $SS_{b\ subj}$. For Class IV designs, S is crossed with all other design factors. Hence only key term $\mathbf{s}_s$ contributes to $SS_{b\ subj}$, so $SS_{b\ subj} = SS_S$.

No formulas are presented for computing within- and between-SS's directly, since an experimenter rarely has to have these quantities. They can be readily obtained, however, by summing the appropriate SS's for the key terms, which are routinely computed for any experiment.

5.13 PROPORTIONAL REPRESENTATION OF SUBJECTS

For the most part, this book concerns the structure and analysis of *balanced designs*—those having an equal number of scores in each cell, a property that implies that there are an equal number of scores at each level of any given design factor. The validity of our formulas and procedures depends on this restriction. Generally speaking, it is awkward to analyze and interpret experimental results when such a restriction is lacking. Some leniency is possible, however. In particular, the analysis is not greatly complicated if designs departing from our restriction have the property known as *proportionality*.

We shall consider here only a simplified kind of proportionality—the kind an experimenter is most likely to utilize. A design with (simple) proportionality has one factor, the *factor of proportionality*, whose levels include a different number of scores. However, any simple design defined by any level of this factor has an equal number of scores per cell.

The following design, $A \times B$, exhibits proportionality. The factor of proportionality is A. Each cell at a_1 has three scores, and each cell at a_2 has two scores.

	b_1	b_2
a_1	3 2 4	7 5 6
a_2	2 5	4 3

The following design lacks proportionality:

	b_1	b_2
a_1	3 2 4	7 5 6
a_2	2 5 –	4 3 8

The same number of scores per cell does not exist either at each level of B or at each level of A.

It is rather easy to adjust our formulas for finding the SS's if proportionality holds. The solution presented here, however, requires that the differing proportions be representative of underlying populations and not be simply the result of happenstance. The most common situation for which we would want to avail ourselves of these formulas would be a Class II or Class III design having a blocking factor. A different number of subjects might be used in each block, in proportion to the total number in the respective populations represented by the blocks. For example, if there were a blocking factor "school," we might wish to include some particular proportion of the freshman students from each school. Since the schools may differ greatly in student population, the number of scores per cell would differ considerably, but proportionality could still hold.

If proportionality holds, we find our SS's from our SS elements, as before, except that some adjustments are required in the formulas for the SS elements. For example, in $A \times B$ we cannot have a symbol E, if E differs among cells, for there is no such constant E. We may have an E_b, however, if B is the factor of proportionality, E_b being the number of scores per cell at level b_b.

Likewise, in a repeated measurements design having $S(A)$, A would be the factor of proportionality if the number of subjects differed among the levels of A. Then the number of subjects per level would not be symbolized as a constant S, but as S_a. The formulas for the SS elements must be changed whenever S appears, but not otherwise. Two kinds of changes can be made:

Table 5.13-1 Adjustments of the SS elements of $S(A) \times B$ for proportionality

SS element	S constant	S_a varies with a_a
SS_{abs}	$\sum X^2_{abs}$	No change
SS_{ab}	$\sum X^2_{ab.}/S$	$\sum X^2_{ab.}/S_a$
SS_{as}	$\sum X^2_{a.s}/B$	No change
SS_b	$\sum X^2_{.b.}/AS$	$\sum X^2_{.b.}/S_{tot}$
SS_a	$\sum X^2_{a..}/BS$	$\sum(X^2_{a..}/S_a)/B$
SS_1	$X^2_{...}/ABS$	$X^2_{...}/BS_{tot}$

1. If S appears in a formula requiring summation over subscript a, we can substitute S_a for S, but the division by S_a must now be done separately for each term of the summation; for example, for $\sum X^2_{a..}/BS$, we write $\sum(X^2_{a..}/S_a)/B$.
2. In some formulas S appears, but there is no summation over subscript a. We cannot substitute S_a, as no values of a are specified that would give S_a meaning. In such formulas we have not only S, but A multiplying S. The product $A \cdot S$ is the total number of different subjects in the experiment when there are an equal number of subjects at each level of A. The total number of subjects when proportionality holds is $\sum_{a=1}^{A} S_a$, which we also symbolize as S_{tot} ("tot" for total). We substitute S_{tot} for $A \cdot S$ whenever $A \cdot S$ appears.

Following these rules, we alter our formulas for the SS elements of $S(A) \times B$ as shown in Table 5.13-1. We use these formulas to compute the SS's for the following scores:

		b_1	b_2
a_1	s_1	5	3
	s_2	−1	7
	s_3	−7	−1
a_2	s_1	6	12
	s_2	4	12
	s_3	−1	3
	s_4	4	12
a_3	s_1	−3	11
	s_2	−7	13

These are the very same scores for which we computed parameter estimates in Section 4.8.5 and SS's in Section 5.9, except that one subject formerly within a_3 is now within a_2. With the new arrangement, there are three subjects within a_1, four within a_2, and two within a_3. According to the S_a symbolism, then, $S_1 = 3$, $S_2 = 4$, and $S_3 = 2$.

In order to compute the SS elements, we must first find the sums required by the SS element formulas:

$X_{ab.}$:

$$X_{11.} = -3$$
$$X_{12.} = 9$$
$$X_{21.} = 13$$
$$X_{22.} = 39$$
$$X_{31.} = -10$$
$$X_{32.} = 24$$

$X_{a.s}$:

$$X_{1.1} = 8$$
$$X_{1.2} = 6$$
$$X_{1.3} = -8$$
$$X_{2.1} = 18$$
$$X_{2.2} = 16$$
$$X_{2.3} = 2$$
$$X_{2.4} = 16$$
$$X_{3.1} = 8$$
$$X_{3.2} = 6$$

$X_{.b.}$:

$$X_{.1.} = 0$$
$$X_{.2.} = 72$$

$X_{a..}$:

$$X_{1..} = 6$$
$$X_{2..} = 52$$
$$X_{3..} = 14$$

$X_{...}$:

$$X_{...} = 72$$

Next we compute the SS elements, using the formulas from Table 5.13-1:

$$\begin{aligned}
\mathrm{SS}_{abs} &= 992 \\
\mathrm{SS}_{ab} &= (3^2 + 9^2)/3 + (13^2 + 39^2)/4 + (10^2 + 24^2)/2 \\
&= 90/3 + 1690/4 + 676/2 \\
&= 30 + 422.5 + 338 = 790.5 \\
\mathrm{SS}_{as} &= 552 \\
\mathrm{SS}_{b} &= (0^2 + 72^2)/9 \\
&= 5184/9 = 576
\end{aligned}$$

$$\begin{aligned}\text{SS}_a &= (6^2/3 + 52^2/4 + 14^2/2)/2\\ &= (36/3 + 2704/4 + 196/2)/2\\ &= (12 + 676 + 98)/2\\ &= 786/2 = 393\\ \text{SS}_1 &= 72^2/18 = 288\end{aligned}$$

Since the set of scores is the same as in Section 4.8, though arranged somewhat differently, SS_{abs} must be exactly the same as in Section 5.9, so no new computation is needed. Likewise, the set of $X_{a.s}$'s is the same as in Section 4.8, so SS_{as} is the same as in Section 5.9. Although the formula for SS_b changes slightly, the $X_{.b.}$'s are exactly the same in the two sections, and since the divisor of Section 5.9 was $A \cdot S = 9$, the same as S_{tot} in the present example, SS_b is unchanged from Section 5.9. Since the same set of scores is used as in Section 4.8, $X_{...}$ is unchanged, and since the divisor in both sections is the total number of scores, SS_1 is also unchanged from Section 5.9. Our rearrangement, then, only affects SS_a and SS_{ab}.

The elements thus evaluated are entered in SS elements Table 5.13-2. The table has the same pattern of plus and minus entries as Table 5.9-1 for $S(A) \times B$; the same elements are summed to get the SS's. The SS's for all key terms except $\mathbf{b}_b$ have changed, since SS_a, which has changed, enters into all these SS's. The SS_{tot} for a set of scores depends in no way on the arrangement of the scores into a design; hence, SS_{tot} is the same as in Table 5.9-1.

No adjustment in the estimation formulas is required for the case of simple proportionality. For example, estimation formulas for $S(A) \times B$ given in Table 4.11-1 apply as shown therein. The method of successive extraction may also be used. Population mean $\mathbf{m}$ is conceived to be the mean across all potential scores in the population, with each score receiving equal weight. Hence $\mathbf{m}$ would not, in general, be the mean of the cell expectations, as was true with equal-sized cells. Also, we estimate $\mathbf{m}$ as $\bar{X}_{...}$, as before, not as a weighted average of cell means. Of course, $\bar{X}_{...}$ is influenced more by cells

Table 5.13-2 SS elements table

	SS_A	SS_B	SS_S	SS_{AB}	SS_{BS}	SS_{tot}
SS_{abs}					+992	+992
SS_{ab}				+790.5	−790.5	
SS_{as}			+552		−552	
SS_b		+576		−576		
SS_a	+393		−393	−393	+393	
SS_1	−288	−288		+288		−288
Sum	105	288	159	109.5	42.5	704

with a larger number of scores, just as **m** is influenced to a greater extent by the more numerous subpopulations.

The formulas given in Section 5.3 relating the parameter estimates to the SS's require some adjustment for the case of proportionality, in much the same way as did the calculational formulas. There is now no such constant as S. Instead, we have S_a, which is not constant as subscript a changes, and thus cannot be placed to the left of $\sum_a$; it must be placed to the right. For example, instead of $\text{SS}_A = BS\sum \hat{\mathbf{a}}_a^2$, we have $\text{SS}_A = B\sum S_a\hat{\mathbf{a}}_a^2$. When multiplier S occurred previously without a summation over a, it appeared with A as $A \cdot S$. We replace $A \cdot S$, which was simply the total number of subjects, with S_{tot}. For example, instead of $\text{SS}_B = AS\sum \hat{\mathbf{b}}_b^2$, we have $\text{SS}_B = S_{\text{tot}}\sum \hat{\mathbf{b}}_b^2$.

For the numerical example given previously, the student may verify that $\hat{\mathbf{a}}_1 = -3.0$, $\hat{\mathbf{a}}_2 = 2.5$, and $\hat{\mathbf{a}}_3 = -.5$. Hence,

$$\begin{aligned} \text{SS}_A &= B\sum S_a\hat{\mathbf{a}}_a^2 \\ &= 2(3\cdot 9 + 4\cdot 6.25 + 2\cdot .25) \\ &= 2(52.5) = 105 \end{aligned}$$

which agrees with the result shown in Table 5.13-2.

When A is the factor of proportionality, the side conditions for summations over fixed factor A require some adjustment. It is no longer true that $\sum \mathbf{a}_a = 0$, nor that $\sum_a \mathbf{ab}_{ab} = 0$. Each parameter must be weighted according to the proportion of the total population that its subpopulation represents. Let these weights be symbolized by w_a. For our numerical example, $w_1 = 6/18 = 1/3$, $w_2 = 8/18 = 4/9$, and $w_3 = 4/18 = 2/9$. Instead of $\sum \mathbf{a}_a = 0$, we have $\sum w_a\mathbf{a}_a = 0$, and instead of $\sum_a \mathbf{ab}_{ab} = 0$, we have $\sum_a w_a\mathbf{ab}_{ab} = 0$. Likewise, for the estimates, $\sum w_a\hat{\mathbf{a}}_a = 0$; that is, $(1/3)(-3.0) + (4/9)2.5 + (2/9)(-.5) = (-9 + 10 - 1)/9 = 0$.

Inequality in the number of scores per cell may come about not because of planned proportionality but because of unforeseen difficulties or complications that are encountered in running the experiment. The latter situation is considered in Section 9.1.

5.14 SS's FOR SIMPLE EFFECTS

The term "simple effect" was defined in Section 4.3.6. In design $A \times B$, a simple effect $\mathbf{b}_{b(a_1)}$ is the value that the main effect for factor B would have if we concentrated on level a_1 alone and ignored all other levels of factor A. In such a situation, we have, effectively, a one-factor design B. After doing significance tests on the design as a whole, some experimenters prefer additionally to test simple effects. Suppose, for example, that in testing the design as a whole, we find that interaction $\mathbf{ab}_{ab}$ is significant, and that main effect $\mathbf{b}_b$ is significant also. It is possible, with such conclusions, that the simple effects of factor B at some level of factor A do not differ significantly; that is, it is possible that $E(X_{a1}) = E(X_{a2})$, assuming that factor B has two levels. In other words, factor B may be completely without influence at some particular level $\mathbf{a}_a$, even though B's main effect is significant, but this is possible only if the $\mathbf{ab}_{ab}$ interaction is not zero. If the $\mathbf{ab}_{ab}$ interaction is not significant, then tests of the simple effects of factors A or B would have little point. Actually, some authorities criticize the testing of simple effects regardless of the status of the interaction, but the student of experimental design must understand such tests whether he uses them or not.

To test a simple effect, we first require an SS for that simple effect. In design $A \times B$, we let SS_B be the sum of squares for factor B. There is but one value for SS_B for an experiment. However, we can compute SS's separately for factor B at each level of A. Such a simple SS is symbolized by $SS_{B(a_a)}$. There are A such simple SS's for our example. We must compute, then, A different simple SS's to test the simple effects of factor B at the various levels of A.

To compute a simple SS in design $A \times B$ is equivalent to computing an SS for a one-factor design (Section 5.5). If we had a one-factor design B, then

$$SS_B = SS_b - SS_1,$$

where

$$SS_b = \sum X_{\cdot b\cdot}^2/E$$
$$SS_1 = X_{\cdot\cdot\cdot}^2/BE$$

To compute a simple SS, we proceed in the same way, except that we require slightly more elaborate symbolism to keep track of the level of A. If factor A has two levels, we have

$$SS_{B(a_1)} = SS_{b(a_1)} - SS_{1(a_1)},$$

where

$$SS_{b(a_1)} = \sum X_{1b\cdot}^2/E$$
$$SS_{1(a_1)} = X_{1\cdot\cdot}^2/BE$$

Similar formulas apply for a_2.

We compute such simple effects for the data of Section 5.6 with $A = B = 2$:

	b_1	b_2
a_1	10 7	2 6
a_2	9 10	7 5

For level a_1,

$$\begin{aligned} SS_{b(a_1)} &= (17^2 + 8^2)/2 = (289 + 64)/2 \\ &= 353/2 = 176.5 \\ SS_{1(a_1)} &= 25^2/4 = 625/4 = 156.25 \end{aligned}$$

Hence,

$$SS_{B(a_1)} = 176.5 - 156.25 = 20.25$$

For level a_2,

$$\begin{aligned} SS_{b(a_2)} &= (19^2 + 12^2)/2 = (361 + 144)/2 = 505/2 \\ &= 252.5 \\ SS_{1(a_2)} &= 31^2/4 = 961/4 = 240.25 \end{aligned}$$

Hence,

$$SS_{B(a_2)} = 12.25$$

Summing the simple SS's, $\sum_a SS_{B(a_a)}$, yields

$$SS_{B(a_1)} + SS_{B(a_2)} = 20.25 + 12.25 = 32.5,$$

which equals $SS_B + SS_{AB}$ (see Table 5.6-2). This result has general validity: $\sum_a SS_{B(a_a)} = SS_B + SS_{AB}$, because $\mathbf{b}_{b(a_a)} = \mathbf{b}_b + \mathbf{ab}_{ab}$ (equation (7), p. 96). Likewise, $\sum_b SS_{A(b_b)} = SS_A + SS_{AB}$.

Although the preceding discussion has been limited to design $A \times B$, similar procedures apply to other designs. A simple SS is computed from SS elements computed at the simple level of concern. For $A \times B \times C$, we can speak of a simple two-factor interaction $\mathbf{ab}_{ab(c_c)}$, and compute a simple SS for $\mathbf{ab}_{ab}$ at each level of factor C, $SS_{AB(c_c)}$. Then,

$$\sum_{ab} SS_{AB(c_c)} = SS_{AB} + SS_{ABC}$$

Suppose we want to test the simple effects of B separately at each a_a in design $B(A)$. We may do so, but there is no $\mathbf{ab}_{ab}$ in the score model and no SS_{AB}. Hence, $\sum_a SS_{B(a_a)} = SS_B$, not $SS_B + SS_{AB}$. The reason for this difference is that what we call $\mathbf{b}_{b(a)}$ already includes, in essence, an $\mathbf{ab}_{ab}$ interaction, and what we call SS_B likewise includes the effect of this interaction (see Section 4.7.4). For $B(A)$ we might be interested in the simple effects of factor B at the a_a levels, but we would have no interest in a "simple effect" of A at some level b_b, since the different levels of, say, b_1 have no meaningful relation to each other. If schools are nested within cities, $B(A)$, we might be interested in testing the simple effects of schools within a given city, but it would be pointless to ask whether the cities differ at level b_1, since the schools were labeled b_1 in different cities on an arbitrary basis. Although this discussion has concerned only $B(A)$, similar principles apply when nesting occurs in more complicated designs.

CHAPTER SIX

Expected Mean Squares and Significance Tests

6.1 INTRODUCTION

Sums of squares are seldom of direct interest for the interpretation of experimental results. They are a way station only. The SS's are used to obtain mean squares (MS's), and these, in turn, are used to calculate F-ratios for most of the key terms. Depending on the size of the F-ratio, we conclude either that all parameters of the associated key term are zero (and therefore, that the term makes no contribution to the scores), or that the parameters for a key term differ from each other. The procedures followed in reaching such conclusions constitute what is called *significance testing*.

It is very simple to calculate an MS, given the associated SS. We simply divide the SS by its degrees of freedom (df). (This term is explained in Section 6.2.) An F-ratio is the ratio of two MS's. The experimenter must know which MS's to use as a numerator and which to use as a denominator for each key term whose significance he wishes to test. This is determined by the expectations of the MS's.

Over replications an MS is a random variable, since the corresponding SS is a random variable, though the df is a constant. Its expectation, that is, its mean value over an infinity of replications of the experiment, can be expressed as the sum of components related to the variances of the parameters for different key terms of the design. It is by examining such formulas for the expected mean squares (EMS's) that the experimenter determines the proper MS's for each F-ratio. By examining these formulas, he can also determine how to use the MS's to estimate the variance of the parameters of a key term.

Section 6.3 discusses the composition of the EMS formulas in general terms, but does not explain their derivation. Section 6.4 discusses the *F*-distribution and how the *F*-ratio leads to conclusions about the significance of various key terms. Section 6.5 presents some numerical examples of significance testing. Because there may be no proper MS for the denominator (MS_2) according to our basic rule, or because df_2 may be too small, alternative procedures may be used to determine an MS_2, and these are explained in Sections 6.6 and 6.7. The derivation of the EMS formulas is explained in Sections 6.9 and 6.10.

6.2 DEGREES OF FREEDOM

6.2.1 Interpretation

Each key term in the score model for a design (except *e* when confounded) has an associated *degrees of freedom* (*df*). The "degrees of freedom" for a particular key term is considered to be grammatically singular and is symbolized df. The term "degrees of freedom" for several key terms is grammatically plural and is symbolized df's. The symbol df is frequently subscripted to identify the specific key term to which it applies. The subscripting is identical to the subscripting for SS's: Capital letters correspond to the boldfaced letters of the key term.

A convenient way to interpret degrees of freedom is in terms of the estimation of parameters by successive extraction. As we calculate estimates for each key term, we use up the df's associated with that key term. Before estimation begins, the total degrees of freedom for the experiment equals the number of scores, N. We begin by estimating $\mathbf{m}$, a single parameter having 1 df. In estimating $\mathbf{m}$ we use up 1 df, leaving $N - 1$. We say that there are $N - 1$ df left, in the sense that given $\hat{\mathbf{m}}$, there are only $N - 1$ scores whose values are unconstrained. Once the $N - 1$ scores are determined, the last score, the Nth score, is constrained to have a specific value. Suppose, for example, that $N = 4$. If $\hat{\mathbf{m}} = 3$ and three scores are 1, 2, and 4, there is no "freedom" left for the last score; it must be 5. This is true because

$$\hat{\mathbf{m}} = (X_1 + X_2 + X_3 + X_4)/4,$$

and if $\hat{\mathbf{m}}$, X_1, X_2, and X_3 are determined values, then X_4 can have only one possible value:

$$X_4 = 4\hat{\mathbf{m}} - X_1 - X_2 - X_3,$$

or, for our example,

$$X_4 = 4 \cdot 3 - 1 - 2 - 4 = 12 - 7 = 5$$

It does not matter which three X's are specified. Whenever three are specified, the fourth no longer has "freedom."

Initially, then, we begin with N possible df in an experiment with N scores. There is 1 df for the estimated mean, $\hat{\mathbf{m}}$, since, viewed abstractly, $\hat{\mathbf{m}}$ might take on any single value. Once that value has been set, we have used up 1 df of the original N, leaving $N - 1$.

Each key term has at least two parameters. For example, term $\mathbf{a}_a$ has A parameters, $\mathbf{a}_1, \mathbf{a}_2, \mathbf{a}_3, \ldots, \mathbf{a}_A$. We might think, then, that there would be A degrees of freedom for key term $\mathbf{a}_a$. However, we must adjust A according to the df already used up by parameters used in estimating the $\mathbf{a}_a$'s by the method of successive extraction. Recall that to estimate $\mathbf{a}_a$ we use $\hat{\mathbf{m}}$. With design A, for example, $\hat{\mathbf{a}}_a = \bar{X}_{a.} - \hat{\mathbf{m}}$. The parameter estimate $\hat{\mathbf{m}}$ already used up 1 df. Given $\hat{\mathbf{m}}$, only $A - 1$ of the A different $\bar{X}_{a.}$ averages can be freely specified. After that, the Ath average is determined by the relation $\hat{\mathbf{m}} = (1/A)\sum_a \bar{X}_{a.}$. (We could equivalently draw on the fact that $\sum \hat{\mathbf{a}}_a = 0$.) Thus, for $\mathbf{a}_a$ the df is $A - 1$.

Likewise in $A \times B$, there are $A - 1$ df for factor A and $B - 1$ for factor B. Estimate $\hat{\mathbf{m}}$ is used to find both the $\mathbf{a}_a$'s and the $\mathbf{b}_b$'s. There are $A \cdot B$ different $\mathbf{ab}_{ab}$ parameters and estimates in $A \times B$. However, to estimate the $\mathbf{ab}_{ab}$'s we used $\hat{\mathbf{m}}$, the $\hat{\mathbf{a}}_a$'s, and the $\hat{\mathbf{b}}_b$'s. These represented a total of $1 + (A - 1) + (B - 1) = A + B - 1$ df, so the df for key term $\mathbf{ab}_{ab}$ is $A \cdot B - A - B + 1 = (A - 1)(B - 1)$.

When the main term is nested, as is $\mathbf{s}_{s(a)}$ of $S(A) \times B$, estimation requires more than $\hat{\mathbf{m}}$. Still, the df is the number of parameters for the key term minus the number of df for parameters used in making the estimates of the key term. For $\mathbf{s}_{s(a)}$ there are $A \cdot S$ parameters, whose estimation requires $\hat{\mathbf{m}}$ and the $\hat{\mathbf{a}}_a$'s. Since $\hat{\mathbf{m}}$ has 1 df and the $\mathbf{a}_a$'s have $A - 1$ df, $\mathbf{s}_{s(a)}$ has $AS - (1 + A - 1) = A(S - 1)$ df, in agreement with the value shown in Table 6.2.2.

6.2.2 Rules for Finding the df's

It is not necessary to consider estimation procedures in order to determine the df's. The df's may be found very readily once we have the symbolic product (SP) for a key term. Simply replace each subscript letter in the SP (a, b, and so on) with the numerical value of its upper limit (A, B, and so on) and evaluate the numerical value of the formula. The number found is the df for that key term.

For example, the df's for $A \times B$ are as shown in Table 6.2-1. If $A = 2$, $B = 2$, and $E = 2$, as in the example of Section 5.5, then the df's for $\mathbf{a}_a$, $\mathbf{b}_b$, $\mathbf{ab}_{ab}$, and $\mathbf{e}_{e(ab)}$ are 1, 1, 1, and 4, respectively. The sum of the df's is 7, one less than the number of scores in the experiment. It is always true that the sum of the df's for the key terms of an experiment is one less than the total number of scores in the experiment. If we counted the mean $\mathbf{m}$ as a key term, then its SP would be 1, and its df would be 1. The sum of the df's would then equal the total number of scores.

Table 6.2-1 SP's and df's for $A \times B$

Key term	SP	df
$\mathbf{a}_a$	$a - 1$	$A - 1$
$\mathbf{b}_b$	$b - 1$	$B - 1$
$\mathbf{ab}_{ab}$	$(a - 1)(b - 1)$	$(A - 1)(B - 1)$
$\mathbf{e}_{e(ab)}$	$(e - 1)ab$	$(E - 1)AB$

Table 6.2-2 gives another example, that of $S(A) \times B$ with $A = 3$, $B = 2$, and $S = 3$, as in the example of Sections 4.8.5 and 5.9. Again, the sum of the df's equals the total number of scores, 18, minus 1.

If there is a factor of proportionality, the procedures just given require some modification. For example, if $A \times B$ has a factor of proportionality A, then there is no such constant as E, the number of scores per cell. Instead there are A different constants E_a. The formula previously given for df_E, then, has no meaning. We can conceive of a df_E for each level of A, $(E_a - 1)B$, and df_E for the experiment is the sum of these terms over the levels of A, $\text{df}_E = \sum_a(E_a - 1)B$. We can further develop the formula as $\text{df}_E = B\sum_a(E_a - 1) = B\sum_a E_a - A \cdot B = N - A \cdot B$, where N is the total number of subjects (or scores). In general, for single measurement (Class I and II) designs proportionality only requires adjustment of the formula for df_E, and df_E always equals the total number of scores minus the number of cells.

For Class III designs with a factor of proportionality, there is no such constant as S; hence df formulas requiring S must be adjusted. If A is the factor of proportionality, there is a different constant S_a for each level a. The adjustment of a formula containing S is similar to the adjustment for single measurement designs: Replace S with S_a, and replace A with $\sum_a$ over the formula. For the previous example of $S(A) \times B$, then, we obtain $\text{df}_S = \sum_a(S_a - 1)$, which also equals $S_{\text{tot}} - A$. Also, $\text{df}_{BS} = (B - 1)\sum_a(S_a - 1)$, which equals $(B - 1)(S_{\text{tot}} - A)$. In general, for Class III designs, if A is the

Table 6.2-2 SP's and df's for $S(A) \times B$

Key term	SP	df	
$\mathbf{a}_a$	$a - 1$	$A - 1$	$= 2$
$\mathbf{b}_b$	$b - 1$	$B - 1$	$= 1$
$\mathbf{s}_{s(a)}$	$a(s - 1)$	$A(S - 1)$	$= 6$
$\mathbf{ab}_{ab}$	$(a - 1)(b - 1)$	$(A - 1)(B - 1)$	$= 2$
$\mathbf{bs}_{bs(a)}$	$a(b - 1)(s - 1)$	$A(B - 1)(S - 1)$	$= 6$
			17

factor of proportionality, we replace $A(S - 1)$ in any df formula by $(S_{tot} - A)$.

The df associated with a simple effect is the same as the df for the effect itself. For example, if, in $A \times B$, $A = 2$ and $B = 3$, then $df_B = B - 1 = 2$. Also, $df_{B(a_1)} = df_{B(a_2)} = 2$. To find the MS for a simple effect, divide the simple SS by the same df used for the SS itself.

6.3 EXPECTED MEAN SQUARE COMPONENTS

6.3.1 General Composition of EMS Formulas

Over successive replications, any particular SS varies randomly, because the estimates on which an SS is based are composed in part, at least, of parameters such as the **e**'s whose values are randomly determined over replications. In addition to the **e**'s, the parameters of any key term having a bold-faced random factor symbol are randomly determined over replications. Since an MS is simply an SS divided by a constant, the df, an MS is also a random variable over replications.

An expected mean square (EMS) is the expectation of an MS, that is, the average value over an infinity of replications. There is an EMS for each key term of the model, which is denoted by capital letter subscripts on EMS that correspond to the subscripts on the SS's and MS's. For example, if SS_A is the sum of squares associated with term $\mathbf{a}_a$, then $SS_A/df_A = MS_A$, and $E(SS_A/df_A) = E(MS_A) = EMS_A$. Since df_A is a constant, EMS_A also equals $E(SS_A)/df_A$.

This chapter shows that the EMS for the error term, EMS_E, equals σ_E^2, the variance of **e**. (Of course, we have no SS_E, and thus no EMS_E, unless we have an unconfounded error term in the score model.) The EMS for any other key term consists of σ_E^2 plus other components of the form $k_\tau\sigma_\tau^2$ added together. The letter τ is used as a general symbol referring to some unspecified key term of the design under consideration. When more specificity is desired, τ is replaced by the capital letter or letters corresponding to the bold-faced letters of the key term. For example, instead of a general reference to $k_\tau\sigma_\tau^2$, we may make specific reference to $k_A\sigma_A^2$ or $k_{AB}\sigma_{AB}^2$. The subscripting here corresponds exactly to the subscripting of SS's and df's by capital letters to refer to specific key terms.

A k_τ is a positive constant whose value depends on the upper subscript limits for the design. The σ_τ^2 is a variability associated with the parameters of key term τ. For $A \times B$ we have σ_A^2, σ_B^2, σ_{AB}^2, and σ_E^2. Sometimes σ_τ^2 is technically a variance, as defined in Section 2.2.3, but often it differs from a variance, as will be explained in Section 6.3.3. The same $k_\tau\sigma_\tau^2$ frequently appears as a component of more than one EMS. Before we can say exactly which $k_\tau\sigma_\tau^2$ components appear in an EMS for some design, we have to know which factors are random and which are fixed. If in $A \times B$ factor A is fixed and B is random, then

$$\begin{aligned} \text{EMS}_A &= \sigma_E^2 + E\sigma_{AB}^2 + BE\sigma_A^2 \\ \text{EMS}_B &= \sigma_E^2 + AE\sigma_B^2 \\ \text{EMS}_{AB} &= \sigma_E^2 + E\sigma_{AB}^2 \\ \text{EMS}_E &= \sigma_E^2 \end{aligned}$$

Here $k_{AB} = E$, and $k_{AB}\sigma_{AB}^2$ appears in the formulas for EMS_A and EMS_{AB}. Also, we see that $k_A = B \cdot E$, $k_B = A \cdot E$, and $k_E = 1$. The last equality, for k_E, holds true for every design.

The remainder of Section 6.3 explains the nature of the σ_τ^2's, which depends on whether the factors of τ are random, fixed, or a mixture. Derivation of the EMS formulas will be considered in Sections 6.9 and 6.10.

6.3.2 Random τ Referents

If key term τ contains only random factor referents, σ_τ^2 is a variance, as defined in Section 2.2.3. For example, in $A \times B$, B random, σ_B^2 is a variance of the population of $\mathbf{b}_b$'s corresponding to the population of $\mathbf{b}_b$'s from which we randomly chose B levels for our experiment. Chapter 4 explained that, for the population of $\mathbf{b}_b$'s, $\text{E}(\mathbf{b}_b) = 0$. (It should be remembered that the mean of the B $\mathbf{b}_b$'s of an experiment does not in general equal zero, when B is a random factor.) As pointed out in Section 2.2.3, if the expectation of a random variable is zero, its variance is the expectation of the square of that variable. Therefore, the variance of the $\mathbf{b}_b$'s, σ_B^2, is $\text{E}(\mathbf{b}_b^2)$. If both A and B are random, then σ_{AB}^2 is the variance of the $\mathbf{ab}_{ab}$'s, $\text{E}(\mathbf{ab}_{ab}^2)$.

6.3.3 Fixed τ Referents

When τ has one or more fixed factor referents, σ_τ^2 is not exactly a variance. Since, in general, σ^2 has been used to symbolize a variance (Chapter 2), the use of σ^2 here to denote a near variance is not an ideal arrangement, but this text follows the lead of prominent authorities in using this symbolism. The term Var ( ) is used to emphasize referral to the variance proper for a key term. For example, Var $(\mathbf{a}_a)$ means the variance of the parameters of key term $\mathbf{a}_a$.

When factor A is random, Var $(\mathbf{a}_a) = \sigma_A^2$, the variance referring to the population of $\mathbf{a}_a$'s, as has been pointed out. When factor A is fixed, Var $(\mathbf{a}_a) \neq \sigma_A^2$. Because of the side condition that $\sum \mathbf{a}_a = 0$ when factor A is fixed, the mean of the $\mathbf{a}_a$'s must also equal zero. Therefore, the variance of the $\mathbf{a}_a$'s, in accordance with the definition in Section 2.2.3, is $\sum \mathbf{a}_a^2/A$. However, the symbolism used to describe the EMS formulas is simpler if we let σ_A^2 stand for $\sum_a \mathbf{a}_a^2/(A - 1)$ rather than the variance. We follow a similar procedure for any fixed factor: We sum the squared parameters over the fixed factor subscript and divide by the subscript limit minus one. The division by $A - 1$ is reminiscent of the unbiased estimate of a population variance based on a random sample. The $\mathbf{a}_a$'s are not a random sample, however, when factor A is fixed, and $\sum \mathbf{a}_a^2/(A - 1)$ is neither a variance of the $\mathbf{a}_a$'s nor an unbiased estimate of the variance.

If the term includes several fixed factors, we sum over the square of each and divide by the product of

1. the upper limit minus one for factors represented with non-nest subscripts; or
2. the upper limit undiminished for factors represented with nest subscripts.

For example, if in $A \times B$ both A and B are fixed, then

$$\sigma_{AB}^2 = \frac{\sum \mathbf{ab}_{ab}^2}{(A - 1)(B - 1)}$$

If, in $B(A) \times C$ all factors are fixed, then

$$\sigma_B^2 = \frac{\sum \mathbf{b}_{b(a)}^2}{A(B - 1)}$$

6.3.4 Mixed τ Referents

Comparing the definitions of σ_τ^2 when τ is random or fixed, we see that when τ is random, we take the expectation of the squared parameter, and when τ is fixed, we sum the squared parameters over the factor subscripts and divide by the limits or by the limits each diminished by one, or by the limit itself for a nest subscript. Now, suppose that τ refers to both random and fixed factors. First we take the expectation of the squared model term over the random subscripts; then we sum and divide over the expectation in the usual manner for fixed factor subscripts.

To exemplify, if in $A \times B$ factor A is fixed and B is random, then, by definition,

$$\sigma_{AB}^2 = \frac{\sum_a \mathrm{E}(\mathbf{ab}_{ab}^2)}{A - 1}$$

It is frequently assumed that $\mathrm{E}(\mathbf{ab}_{ab}^2)$ is the same for each level a_a. That is, there is a variance $\mathrm{E}(\mathbf{ab}_{1b}^2)$ of $\mathbf{ab}_{ab}$ at level a_1, a variance $\mathrm{E}(\mathbf{ab}_{2b}^2)$ at level a_2, and so on. The assumption is that these variances at the different a_a levels are the same. Hence the $\sum_a$ in the preceding equation is over a constant relative to level a_a, so we can write

$$\sigma_{AB}^2 = \frac{A}{A - 1} \mathrm{E}(\mathbf{ab}_{ab}^2)$$

In $A \times B \times C$, if factors A and B are fixed but C is random, then, by definition,

$$\sigma^2_{ABC} = \frac{\sum_{ab} \mathrm{E}(\mathbf{abc}^2_{abc})}{(A-1)(B-1)}$$

If we assume that the variance $\mathrm{E}(\mathbf{abc}^2_{abc})$ is the same for each ab_{ab} level, then we have

$$\sigma^2_{ABC} = \frac{AB}{(A-1)(B-1)} \mathrm{E}(\mathbf{abc}^2_{abc})$$

In $B(A) \times C$, if B is random but A and C are fixed, then

$$\sigma^2_B = \frac{\sum_a \mathrm{E}(\mathbf{b}^2_{b(a)})}{A}$$

If we assume that $\mathrm{E}(\mathbf{b}^2_{b(a)})$ is the same at each a_a level, then we have

$$\sigma^2_B = \mathrm{E}(\mathbf{b}^2_{b(a)})$$

6.3.5 Calculating Variances from σ^2_τ's

If all the non-nest subscripts for a term are random, then σ^2_τ for that term is a variance, properly speaking (or possibly a mean of variances if there is a fixed nest subscript or subscripts). If the term has non-nest fixed factor subscripts, then σ^2_τ is not a variance, strictly speaking, but we can compute the variance from σ^2_τ. To do this, we assume that the expectation over the random factor subscripts is the same for each of the levels of the fixed factors. For example, we noted previously that when only factor B of $A \times B$ is random,

$$\sigma^2_{AB} = \frac{A}{A-1} \mathrm{E}(\mathbf{ab}^2_{ab}),$$

assuming that $\mathrm{E}(\mathbf{ab}^2_{ab})$ is the same across levels of fixed factor A. Since the variance of $\mathbf{ab}_{ab}$, $\mathrm{Var}\,(\mathbf{ab}_{ab}) = \mathrm{E}(\mathbf{ab}^2_{ab})$, we have

$$\mathrm{Var}\,(\mathbf{ab}_{ab}) = \frac{A-1}{A} \sigma^2_{AB}$$

Similarly, when only factor C of $A \times B \times C$ is random,

$$\mathrm{Var}\,(\mathbf{abc}_{abc}) = \frac{(A-1)(B-1)}{AB} \sigma^2_{ABC}$$

We never know the σ^2_τ's exactly, but Section 6.9 shows how they can be estimated. We can estimate the variances by using such formulas as are described here, replacing the σ^2_τ's by their estimates.

6.4 SIGNIFICANCE TESTING

6.4.1 The F-Ratio and the Null Hypothesis

The most important significance tests in the analysis of variance are those concerning null hypotheses of the form $\sigma^2_\tau = 0$. Although he may subsequently decide to use other significance tests, an experimenter begins by testing H_N's of the type $\sigma_\tau^2 = 0$, and performs such tests for as many different τ's as the design allows. For some designs it is possible to test all different τ's except σ_E^2; for other designs it is not possible to test H_N that $\sigma_\tau^2 = 0$ for one or more other τ's of the design unless one or more dubious assumptions are made. The basic test for each σ_τ^2 consists of determining the ratio of two different MS's of the design. This is the F-ratio. If the quantity so determined is less than some critical value for F, we *accept the null hypothesis;* that is, we conclude that, indeed, $\sigma_\tau^2 = 0$. If, however, the F-ratio exceeds this critical value, we *reject the null hypothesis;* that is, we conclude that $\sigma_\tau^2 > 0$.

If we accept the null hypothesis that $\sigma_\tau^2 = 0$, it means we conclude that each parameter of key term τ equals zero. In no other way can $\sigma_\tau^2 = 0$. This is true regardless of whether the referent factors of τ are fixed, random, or a mixture. When we say that $\sigma_\tau^2 = 0$ for random factor referents, we mean that the variance is zero for the *population* of parameters, not just for those of the one experiment. But, of course, if the variance is zero for the population, all the parameter values sampled for any experiment must equal zero also. We limit our conclusions regarding fixed factor referents to those levels actually used in the experiment.

To conclude that $\sigma_\tau^2 = 0$ is to conclude that term τ can be eliminated from the score model without affecting the accuracy of the model. For example, suppose that for $A \times B$ $\sigma_A^2 = 0$ but that $\sigma_B^2 \neq 0$ and $\sigma_{AB}^2 \neq 0$. Then instead of model

$$X_{abe} = \mathbf{m} + \mathbf{a}_a + \mathbf{b}_b + \mathbf{ab}_{ab} + \mathbf{e}_{e(ab)},$$

we conclude that the following simpler model suffices:

$$X_{abe} = \mathbf{m} + \mathbf{b}_b + \mathbf{ab}_{ab} + \mathbf{e}_{e(ab)}$$

6.4.2 The *F*-Distribution in Significance Testing

The MS's are random variables over replications because the SS's are random variables. Likewise, the F-ratio, a ratio of two MS's, is a random variable. Assuming that various assumptions hold, including the assumption that $\sigma_\tau^2 = 0$, the F-ratio over replications has a well-known distribution: the F-distribution. Besides the null hypothesis, some other assumptions are required:

1. The error term has a normal distribution over replications. The distribution over replications for each unique set of subscripts on **e** is identical: The expectation is zero for each, and the variance is the same for each; that is, the variance is *homogeneous.*
2. The error parameters with different subscript values are independent; that is, they are uncorrelated over replications.
3. Assumptions comparable to those in (1) or (2) apply to all random factors.

The assumptions of normality and homogeneity of variance are undoubtedly unjustified much of the time. Assumptions and the effects resulting from their failure are discussed at greater length in Section 9.2.

The *F*-distribution under the null hypothesis is more precisely called the *central* F-*distribution*, to distinguish it from the *noncentral* F-*distribution* that occurs when the null hypothesis is false for a fixed effect. Unless otherwise qualified, "*F*-distribution" refers to the central distribution. There are actually many central *F*-distributions. The shape of the distribution depends on the df's of the two MS's forming the *F*-ratio. Therefore, tabled values for the *F*-distribution, such as those in Appendix A.5, differ depending on both df's. Anyone using such a table must know the df's for the numerator MS and the denominator MS. These are symbolized by df_1 and df_2, respectively. To specify a particular *F*-distribution, the notation $F(\mathrm{df}_1, \mathrm{df}_2)$ is used.

Tables of *F* specify what quantity for a particular $F(\mathrm{df}_1, \mathrm{df}_2)$ the *F*-ratio will exceed for different proportions of replications. For example, if we want to know what value the *F*-ratio will exceed for .1 of all replications ("10% of time") for distribution $F(5, 10)$, we look in a table of the *F*-distribution and find that $F_{90}(5, 10)$ is 2.52. In other words, the 90th percentile of $F(5, 10)$ is 2.52. If the null hypothesis is true (that is, if $\sigma_\tau^2 = 0$), and if df_1 and df_2 are 5 and 10, respectively, then we may expect that on .1 (10%) of the replications the *F*-ratio will exceed the value 2.52. (This does not mean, of course, that if we replicated the experiment 10 times, the *F*-ratio under consideration would necessarily exceed 2.52 exactly once.)

The general nature of hypothesis testing for hypotheses of the form $\sigma_\tau^2 = 0$ is the same as that discussed in Section 2.4. There is a *test statistic*, the *F*-ratio (also called the *F*-statistic) that is calculated from the data of the experiment. Under some null hypothesis, over innumerable replications, this statistic would have a *sampling distribution*, in this case the *F*-distribution. The abscissa of the sampling distribution is divided into acceptance and rejection regions; $100\alpha\%$ of the distribution falls in the rejection region, where α is the significance level of the test. In general, there may be a rejection region in each tail of the distribution, or the rejection region may fall in one tail only, but tests of the H_N that $\sigma_\tau^2 = 0$ are always one-tailed, for reasons that will be explained in Section 6.4.4. Rejection of H_N occurs if and only if the *F*-statistic (*F*-ratio) falls along the upper portion of the abscissa along which lies $100\alpha\%$ of the *F*-distribution.

Section 6.4.3 explains which MS's are used in forming the *F*-ratio to test a particular H_N for a particular key term. Given this information, we of course know the corresponding df's, since they were used in calculating the MS's. Normally, a test statistic of any magnitude whatever might occur if the null hypothesis were true (though in the case of the *F*-ratio only non-negative values can occur, because MS's, like SS's, cannot be negative). Therefore, no value of the F-ratio can prove that the null hypothesis is true or false. However, large F-ratios are more likely to occur if the null hypothesis is false, whereas the more moderate F-ratios are more likely when the null hypothesis is true.

Typically, an experimenter screens an F-ratio against critical F's for several α's, and reports the null hypothesis rejection at the smallest α for which rejection occurs. For example, an experimenter might report that "the main effect for factor A was significant at the .05 significance level, and the main effect of factor B was significant at the .01 significance level." In other words, he concludes that $\sigma_A^2 > 0$ and $\sigma_B^2 > 0$; the former conclusion is based on an F-test using $\alpha = .05$, but for $\alpha = .01$ he would have to accept the null hypothesis; the latter conclusion, that $\sigma_B^2 > 0$, can be justified with an F-test based on $\alpha = .01$ (but with an F-test based on $\alpha = .05$ also, since this latter test is less stringent).

6.4.3 Choosing the MS's

We now know how to perform a significance test for the hypothesis that $\sigma_\tau^2 = 0$, given the F-ratio MS_1/MS_2. We know how to find the SS's for a design, and, using the df's, how to calculate the MS's required. We still must determine which MS's to choose as MS_1 and MS_2 in order to test a particular null hypothesis, $\sigma_\tau^2 = 0$.

For MS_1 we choose the MS whose subscript corresponds to the term we wish to test. For example, to test the null hypothesis that $\sigma_A^2 = 0$, $MS_1 = MS_A$. To test the null hypothesis that $\sigma_{AB}^2 = 0$, $MS_1 = MS_{AB}$. For MS_2 the standard procedure is to choose the MS having the same expectation (EMS) as MS_1 would have if the null hypothesis were true. If a simple effect is being tested, we must ask which EMS_2 would be appropriate if the design were really only at that simple level. We then choose the MS from the full design having the appropriate EMS.

Sometimes there is no MS having the required EMS. Then four different actions may be taken, depending on the circumstances:

1. We may simply forget about testing the σ_τ^2, since it may be of little interest, or since we can assume that $\sigma_\tau^2 > 0$ on other grounds.
2. We can assume that some other $\sigma_\tau^2 = 0$, thereby changing some EMS so that the associated MS is now appropriate as MS_2 for the test. Such an approach is typically very dubious.
3. We may start out by performing only certain significance tests for which proper MS's are available. On the basis of such tests it may turn out that we accept H_N for certain σ_τ^2's; that is, we conclude that each of these σ_τ^2's is zero. If we then eliminate such σ_τ^2's from all the EMS formulas, we may be able to locate appropriate MS's for tests that previously

lacked them. Significance tests whose conclusions affect the subsequent tests for the same experiment are known as *preliminary tests.* Preliminary testing is discussed in Section 6.6.

4. We can form a quasi F-ratio, a ratio in which the denominator or numerator or both are quantities obtained by adding and subtracting MS's instead of individual MS's. The numerator and denominator have the same expectation under the H_N that $\sigma_\tau^2 = 0$. Quasi F-ratios are considered in more detail in Section 6.7.

Up to this point we have not considered the possibility that more than one MS may meet our standards for MS_2 for some test. Normally, there would be at most one such MS_2. However, if we take certain σ_τ^2's to be zero, either by assumption or because of preliminary testing, there may be two or more MS's with the same EMS as MS_2 for some test. Under such circumstances, we may wish to combine (or *pool*) these MS's to get a single MS_2 for the F-ratio. In fact, preliminary testing is often performed so that several MS's may be pooled to get a combined MS_2. The combined MS_2 has a df_2 equal to the sum of the component df's. The larger df_2 increases the power of the F-test, if the assumptions behind the pooling are justified. Section 6.6 discusses pooling more extensively.

6.4.4 Why Tests of H_N Are One-Tailed

We have noted that the rejection region for H_N along the F-distribution is in the upper tail only, that is, the test is one-tailed. If H_N is false (if H_A is true), then $\sigma_\tau^2 > 0$ instead of equal to zero. It is impossible for σ_τ^2 to be a negative quantity (though its estimate might be), since it is defined in terms of a sum or the expectation of squared parameters. The square of any number, positive, zero, or negative, can never be negative. Therefore, the sum or expectation of such numbers cannot be negative. For fixed or mixed terms, the sum of the squared parameters is divided by one or more subscript limits minus one. A subscript limit must always be at least 2, so the division must be positive also. Failure of H_N, then, means that MS_1 has an additional component, the $k_\tau\sigma_\tau^2$ for H_N, which would tend to make the F-ratio larger than if H_N were true.

Note the difference between the present situation and the test of H_N that $\mathbf{m} = W$ (Section 2.4.2). Failure of H_N may equally be due to $\mathbf{m}$ being larger or smaller than H_N postulates; that is, the expectation of the t-statistic may either increase or decrease if H_N is false.

6.5 EMS's AND F-RATIOS FOR SPECIFIC DESIGNS

Sections 6.9 and 6.10 show the student how to drive the EMS's for a design. This section illustrates how examination of the EMS's can tell us which MS's to use for the F-ratios. In addition, the computation of F-ratios is exemplified for a variety of designs. The EMS formulas presented here assume that the design is balanced, but the same MS's are used for F-ratios if simple proportionality holds.

6.5.1 Design A

First we consider design A, for which the EMS's are as follows:

Source	EMS
A	$\text{EMS}_A = \sigma_E^2 + E\sigma_A^2$
E	$\text{EMS}_E = \sigma_E^2$

In this simple example, the EMS formulas are the same whether factor A is fixed or random. To test the null hypothesis that $\sigma_A^2 = 0$, we use MS_A for MS_1. For MS_2 we use MS_E, since EMS_E is the same as EMS_A if $\sigma_A^2 = 0$. The hypothesis that $\sigma_E^2 = 0$ cannot be tested, but we can take for granted that $\sigma_E^2 > 0$. However, if σ_E^2 were zero, then MS_E would have to be zero, which it virtually never is. (But MS_E might be zero if σ_E^2 were positive, so we would not conclude that σ_E^2 equals zero even should MS_E be zero.)

6.5.2 Design $A \times B$, A and B Fixed

Now consider design $A \times B$, with both factors fixed, and with more than one score (subject) per condition. The EMS's are:

Source	EMS
A	$\text{EMS}_A = \sigma_E^2 + BE\sigma_A^2$
B	$\text{EMS}_B = \sigma_E^2 + AE\sigma_B^2$
AB	$\text{EMS}_{AB} = \sigma_E^2 + E\sigma_{AB}^2$
E	$\text{EMS}_E = \sigma_E^2$

We can test any of the following three null hypotheses: $\sigma_A^2 = 0$, $\sigma_B^2 = 0$, $\sigma_{AB}^2 = 0$. In each case, MS_2 for the F-ratio is MS_E. A test of the simple effect of A at any b_b or of the simple effect of B at any a_a would also use MS_E as MS_2. It should be noted that, although MS_1 for the test of a simple effect utilizes data from the simple level of interest only, for MS_2 we use MS_E calculated from the design as a whole. We do this because we assume that σ_E^2 is the same for each cell, and we get a better estimate with the larger df_2 using data from all cells rather than from the simple level alone.

Table 6.5-1 Analysis of variance table for $A \times B$, A and B fixed

Source	SS	df	MS	MS_2	F
A	4.5	1	4.5	MS_E	1.20
B	32.0	1	32.0	MS_E	8.53*
AB	.5	1	.5	MS_E	.13
E	15.0	4	3.75	–	–

* $p < .05$.
** $p < .01$.
*** $p < .001$.

A similar situation holds for other completely crossed designs (for example, $A \times B \times C$), if all factors are fixed and $E > 1$, as here. Each EMS except EMS_E consists of only two terms, including σ_E^2. The denominator of the F-ratio is MS_E for each null hypothesis. If there is but one score per condition, however, the error term is confounded with an interaction and we have no separate MS_E. Such a condition occurs infrequently for completely crossed designs, however, with the exception of Class IV designs, which are discussed in Sections 6.5.4 and 6.5.5. Should $E = 1$, however, the way to proceed with hypothesis testing is to assume that $\sigma_{AB}^2 = 0$. With such an assumption, $EMS_{AB} = \sigma_E^2$, so MS_{AB} can be used as MS_2 to test the H_N's that $\sigma_A^2 = 0$ and $\sigma_B^2 = 0$. The assumption that $\sigma_{AB}^2 = 0$ is not likely to be based on any convincing evidence. If the assumption is false, as would be likely, MS_2 would in general be too large, because of σ_{AB}^2, so the F-statistics would be biased negatively, that is, they would tend to be smaller than they should be under the assumptions. It would be less likely, then, for significant F-ratios to occur. In other words, failure of the assumption that $\sigma_{AB}^2 = 0$ would increase the probability of type 2 errors. In order to circumvent the necessity of assuming that $\sigma_{AB}^2 = 0$, we should have at least two subjects in each condition.

Table 6.5-1 gives the basic quantities used in the significance testing for $A \times B$. The SS's were computed in Section 5.6. In the example, $A = 2$ and $B = 2$. We take both factors A and B to be fixed, so MS_2 is MS_E for all three tests.

A table like 6.5-1 frequently appears in experimental reports, except that there are no columns headed SS or MS_2. Since SS = MS·df, the SS entry is redundant for the final report. Likewise, the experimenter does not report MS_2 in his table—presumably, it should be known to discerning students. We can easily check which quantity was used for MS_2, however, since

$MS_2 = MS_1/F$. Therefore, unless by rare coincidence the possible MS's for MS_2 are equal, we can deduce what MS_2 was in each case. Under the column headed "Source" are listed the symbols that would occur as subscripts on SS. These are simply the capital letter equivalents of the bold-faced key term letters. Source AB, then, refers to the interaction term $\mathbf{ab}_{ab}$. Many authors write $A \times B$ instead of AB under "Source."

It is usual, when such a table is used to report results, to indicate conclusions by placing asterisks next to significant F-ratios. Some authors use only one significance level α, some use two, and some use three. When more than one level is used, one asterisk stands for the largest α. The meaning of the different numbers of asterisks is explained in the footnote to Table 6.5-1. In reporting significance levels for F-ratios it is conventional to use the form $p < \alpha$, where p stands for probability. In Table 6.5-1, one asterisk means that the F-ratio fell into the rejection region defined by $\alpha = .05$, but fell short of the rejection region defined by $\alpha = .01$. Two asterisks mean that the F-ratio fell into the rejection region defined by $\alpha = .01$, but fell short of the rejection region defined by $\alpha = .001$. Three asterisks mean that the F-ratio fell into the rejection region defined by $\alpha = .001$. Experimenters often use one, two, or three asterisks to indicate significance levels. The α's might be .05, .01, and .001, or .025, .01, .001, or, as recently appeared in one journal, .10, .01, and .001. (The latter triplet is unusual, since a significance level of .10 is not often used in such tables.) Instead of three α's, there may be two, or one. In practice, the use of multiple asterisks is normally restricted to tables that are intended to report F-ratios at various significance levels. When there is only one significant effect, as in Table 6.5-1, or when all significant F-ratios are reported at the same α level, only the interpretation for one asterisk is normally given at the bottom of the table. In addition to omitting asterisks, an experimenter may label a "not significant" F-ratio with *ns* or *n.s.*

The only significant effect for Table 6.5-1 is B. Since df_1 and df_2 are 1 and 4, respectively, the critical F for $\alpha = .05$ is $F_{95}(1, 4) = 7.71$; for $\alpha = .01$ it is $F_{99}(1, 4) = 21.2$. Factor B is significant at the .05 significance level, but not at the .01 level. Since space in most journals is at a premium, the ANOVA table is often omitted. The experimenter then simply states in the text of his report that "factor B was significant, $F(1, 4) = 8.53$, $p < .05$."

If only factor B is significant, the postanalysis score model would be $X_{abe} = \mathbf{m} + \mathbf{b}_b + \mathbf{e}_{e(ab)}$. There would be no point in estimating the $\mathbf{a}_a$ or $\mathbf{ab}_{ab}$ parameters, since our conclusion is that all these parameters are zero. There would be some point in estimating the $\mathbf{b}_b$'s, but estimates of the model terms are rarely reported as such. Instead, the mean scores for various cells or levels that differ significantly are reported. In the present example, the experimenter would probably report the mean scores at b_1 and b_2 as 9 and 5, respectively. These are estimates of $\mathbf{m} + \mathbf{b}_1$ and $\mathbf{m} + \mathbf{b}_2$.

6.5.3 Design $A \times B$, A Fixed, B Random

Suppose that factor B of $A \times B$ is now random, although A is fixed as before. The EMS's are:

Source	EMS
A	$\sigma_E^2 + E\sigma_{AB}^2 + BE\sigma_A^2$
B	$\sigma_E^2 + AE\sigma_B^2$
AB	$\sigma_E^2 + E\sigma_{AB}^2$
E	σ_E^2

Henceforth the EMS designation will be omitted from the table body, since the proper EMS designation should by now be apparent. The only change from when B was fixed is in EMS_A, which now has $E\sigma_{AB}^2$ in addition to the previous components. Because of this extra component, MS_2 for testing the H_N that $\sigma_A^2 = 0$ is no longer MS_E, but MS_{AB}. Whence this σ_{AB}^2? Recall that Chapter 4 showed how parameter estimates could be interpreted via theoretical components. For design $A \times B$, when B is fixed, $\hat{\mathbf{a}}_a = \mathbf{a}_a + \bar{\mathbf{e}}_{\cdot(a\cdot)} - \bar{\mathbf{e}}_{\cdot(\cdot\cdot)}$, but when B is random, $\hat{\mathbf{a}}_a = \mathbf{a}_a + \overline{\mathbf{ab}}_{a\cdot} + \bar{\mathbf{e}}_{\cdot(a\cdot)} - \bar{\mathbf{e}}_{\cdot(\cdot\cdot)}$. Term $\overline{\mathbf{ab}}_{a\cdot}$ is zero when B is fixed, but not when B is random. Since $\text{SS}_A = BE\sum\hat{\mathbf{a}}_a^2$, the $\overline{\mathbf{ab}}_{a\cdot}$ carries through into the EMS_A. In general, the σ_τ^2 components in the EMS's correspond to the terms in the theoretical formulas for the parameter estimates.

Interpretation of factors as random or fixed does not affect the estimation of parameters, the SS's, the df's for the key terms, or, therefore, the MS's. As we have seen, however, it can affect the EMS components for a term, and hence can affect the MS to be used as MS_2. Therefore, the F-ratio for a term can be affected. Also, because a different MS_2 generally has a different df_2, the critical F for a given significance level will be different. The result is that whether an effect is judged significant can depend on whether design factors are held to be fixed or random.

How would the analysis shown in Table 6.5-1 be affected if B were random instead of fixed? The MS_2 for testing B would not change, so the F-ratio and conclusion for B would be the same as shown. Likewise, the test of AB would not be changed. However, the test of source A now requires MS_{AB} instead of MS_E as MS_2. Now the F-ratio for A would be $\text{MS}_A/\text{MS}_{AB} = 4.5/.5 = 9$, which is much larger than before. But since df_2 now equals 1 instead of 4, the critical F is also much larger, $F_{95}(1, 1) = 161$, so A is still judged not significant.

If A instead of B were random, EMS_B, rather than EMS_A, would have the extra $E\sigma_{AB}^2$ component. The F-ratio for B would then be $\text{MS}_B/\text{MS}_{AB} = 64$. Although this ratio is considerably larger than the ratio of 8.53 shown in Table 6.5-1, B would not be significant since the critical F, $F_{95}(1, 1)$, is now 161.

It is always true that the smaller the df_2, the larger the F-ratio must be to reach significance for any given level of α. Since experimenters normally desire to demonstrate significant effects in their experiments, they prefer to have df_2 larger rather than smaller. If the AB effect is not significant, then $\text{EMS}_{AB} = \sigma_E^2$, just as does EMS_E. Furthermore, if $\sigma_{AB}^2 = 0$, then EMS_B under H_N that $\sigma_B^2 = 0$ equals simply σ_E^2. We might then sum the SS's for AB and E and divide this sum by the total df's for AB and E. This *pooled* MS could then be used as the proper MS_2 for testing $\sigma_A^2 = 0$ and $\sigma_B^2 = 0$, with $\text{df}_2 = \text{df}_E + \text{df}_{AB}$. With this larger df_2, effects will reach significance at smaller F-ratios. Some authorities frown on pooling, however. (Pooling is discussed at greater length in Section 6.6.)

A test of the simple effect $\mathbf{a}_{a(b_b)}$ would use MS_E as MS_2, even though the test of effect $\mathbf{a}_a$ uses MS_{AB} when B is random. As explained in Section 6.5.1, for a one-factor design A we use MS_E as MS_2, and, effectively, for simple effect $\mathbf{a}_{a(b_b)}$, we have a one-factor design at the simple level b_b. Another way to view the situation is that our simple design has no separate σ_A^2 and σ_{AB}^2 components, since $\mathbf{a}_{a(b_b)}$ combines $\mathbf{a}_a$ and $\mathbf{ab}_{ab}$. Hence, as MS_2 we seek an MS whose EMS lacks both σ_A^2 and σ_{AB}^2, but otherwise has the same σ_τ^2 components as EMS_A of the full design.

6.5.4 Design $S \times A$, A Fixed

In the designs considered thus far in this section, we have assumed that there is more than one score per condition, that is, that $E > 1$. Otherwise, there is no MS_E to use as MS_2 in the F-ratio. What happens, then, with Class IV designs such as $S \times A$ and $S \times A \times B$, which are completely crossed like the previous designs, but which have only one score per condition? Assuming that in $S \times A$, S is random and A is fixed, the pattern of the EMS's is the same as for $A \times B$, with B random and A fixed. The only changes are that (1) symbol S replaces B, (2) since $E = 1$ we omit it from our formulas, and (3) since we can calculate no MS_E, we omit the EMS_E. The EMS's for $S \times A$, then, are:

Source	EMS
A	$\sigma_E^2 + \sigma_{AS}^2 + S\sigma_A^2$
S	$\sigma_E^2 + A\sigma_S^2$
AS	$\sigma_E^2 + \sigma_{AS}^2$

Table 6.5-2 Analysis of variance table for $S \times A$, A fixed

Source	SS	df	MS	MS_2	F
A	124.4	3	41.47	MS_{AS}	14.2*
S	115.3	4	28.82	–	–
AS	35.1	12	2.92	–	–

* $p < .001$.

Recall that with $A \times B$, B random, we used MS_{AB} for MS_2 to test H_N that $\sigma_A^2 = 0$. Here, likewise, we use MS_{AS} as MS_2 to test H_N that $\sigma_A^2 = 0$, so the lack of an MS_E is immaterial. However, we did require an MS_E to test H_N that $\sigma_B^2 = 0$ and the H_N that $\sigma_{AB}^2 = 0$. We are faced with a problem: We cannot test H_N that $\sigma_S^2 = 0$ or H_N that $\sigma_{AS}^2 = 0$, since we lack an EMS equaling σ_E^2. If we were willing to assume that $\sigma_{BS}^2 = 0$, the MS_{AS} could be used as MS_2 in testing H_N that $\sigma_S^2 = 0$, but the correctness of such an assumption is dubious. As discussed in Section 4.5, however, design $S \times A$ is used to investigate the effects of A; the experimenter assumes that S has effects, but he normally takes little or no interest in them. Therefore, instead of assuming that $\sigma_{AS}^2 = 0$ to test H_N that $\sigma_S^2 = 0$, the usual procedure is to test only the H_N that $\sigma_A^2 = 0$ and to forget about other σ_τ^2's.

Table 6.5-2 displays the ANOVA table for design $S \times A$, A fixed, using the SS's computed in Section 5.8 for the scores of Section 4.5. Again, the columns headed by SS and MS_2 normally would not appear in a published report. Since $F_{99.9}(3, 12) = 10.8$, factor A is significant at the .001 level. In this table only the one α level is indicated, rather than the three levels shown in Table 6.5-1.

6.5.5 Design $S \times A \times B$, A and B Fixed

The EMS's for design $S \times A \times B$ with A and B fixed are given in Table 6.5-3. Proper MS's are available for testing the H_N's that $\sigma_A^2 = 0$, that

Table 6.5-3 EMS's for $S \times A \times B$, A and B fixed

Source	EMS	MS_2
A	$\sigma_E^2 + B\sigma_{AS}^2 + BS\sigma_A^2$	MS_{AS}
B	$\sigma_E^2 + A\sigma_{BS}^2 + AS\sigma_B^2$	MS_{BS}
S	$\sigma_E^2 + AB\sigma_S^2$	–
AB	$\sigma_E^2 + \sigma_{ABS}^2 + S\sigma_{AB}^2$	MS_{ABS}
AS	$\sigma_E^2 + B\sigma_{AS}^2$	–
BS	$\sigma_E^2 + A\sigma_{BS}^2$	–
ABS	$\sigma_E^2 + \sigma_{ABS}^2$	–

Table 6.5-4 Analysis of variance table for $S \times A \times B$, A and B fixed

Source	SS	df	MS	MS_2	F
A	84	2	42	MS_{AS}	7*
B	288	1	288	MS_{BS}	16
S	156	2	78	–	–
AB	84	2	42	MS_{ABS}	5.25
AS	24	4	6	–	–
BS	36	2	18	–	–
ABS	32	4	8	–	–

* $p < .05$.

$\sigma_B^2 = 0$, and that $\sigma_{AB}^2 = 0$, but not for testing any H_N involving factor S. Again, these latter H_N's are of little interest and may simply be ignored.

We have calculated the SS's for an $S \times A \times B$ design (see Section 5.10). The scores were taken from Section 4.8.5, though there they appeared in an $S(A) \times B$ design. Table 6.5-4 shows the MS's and F-ratios for these data, assuming that A and B are fixed. Again, the columns headed by SS and MS_2 would normally not appear in a published report.

6.5.6 Design $S(A) \times B$, A and B Fixed

The EMS's for $S(A) \times B$, A and B fixed, are shown in Table 6.5-5. It is customary with Class III designs to tabulate between-subject and within-subject sources separately when writing the EMS's or the ANOVA table. Recall from Section 5.12 that key terms including factors with which S is crossed give rise to within-subject SS's, and that other key terms give rise to between-subject SS's. We classify EMS's as within- or between-subject by the same rule. Tabulation according to this classification gives a clearer picture

Table 6.5-5 EMS's for $S(A) \times B$, A and B fixed

Source	EMS	MS_2
	Between subjects	
A	$\sigma_E^2 + B\sigma_S^2 + BS\sigma_A^2$	MS_S
S	$\sigma_E^2 + B\sigma_S^2$	–
	Within subjects	
B	$\sigma_E^2 + \sigma_{BS}^2 + AS\sigma_B^2$	MS_{BS}
AB	$\sigma_E^2 + \sigma_{BS}^2 + S\sigma_{AB}^2$	BS_{BS}
BS	$\sigma_E^2 + \sigma_{BS}^2$	–

of the F-ratios used in significance testing, since the MS's in an F-ratio are both from within- or between-subject sources. In our example, the between-subject source A is tested with $MS_2 = MS_S$. There is no way to test the H_N that $\sigma_S^2 = 0$, but this is no problem, as we can confidently assume that it is false.

The within-subject sources are tested with $MS_2 = MS_{BS}$. There is no way to test the H_N that $\sigma_{BS}^2 = 0$. However, this effect is of little interest, so there is no problem. One advantage of maintaining both $\mathbf{bs}_{bs(a)}$ and $\mathbf{e}_{abs}$ in our model, even though they are confounded, is now apparent. It is clear that tests of within-subject sources are valid even though $\sigma_{BS}^2 > 0$, but a valid test of source S would be possible only if $\sigma_{BS}^2 = 0$. If we had originally assumed $\mathbf{bs}_{bs(a)}$ to be zero, we would not know how this assumption would affect our analysis.

Table 6.5-5 shows why tests of within-subject effects are more powerful than tests of between-subject effects, so that if the designer of the experiment has a choice, factor S should be crossed with the factor or factors whose effects he is most concerned about establishing, not nested within them. Differences in subjects' mean scores are normally rather large because σ_S^2 is large. On the other hand, σ_{BS}^2 represents the $\mathbf{bs}_{bs(a)}$ parameters, which would usually be relatively small, because they are secondary effects. Hence MS_S, which is MS_2 for between-subject effects, is relatively large, making F-ratios based on it small. Since MS_{BS} is relatively small, F-ratios based on it as MS_2 are relatively large. Also, df_2 is usually larger when MS_2 is MS_{BS} rather than MS_S.

The SS's for an $S(A) \times B$ design were calculated in Section 5.9. The data were the same as for design $S \times A \times B$ in the preceding subsection. Recall that when we interpret the same data as in either $S(A) \times B$ or $S \times A \times B$,

Table 6.5-6 Analysis of variance table for $S(A) \times B$, A and B fixed

Source	SS	df	MS	MS_2	F
			Between subjects		
A	84	2	42	MS_S	1.4
S	180	6	30	–	–
			Within subjects		
B	288	1	288	MS_{BS}	25.42*
AB	84	2	42	MS_{BS}	3.71
BS	68	6	11.33	–	–

* $p < .01$.

some of the SS's are the same, and some from $S(A) \times B$ are a sum of SS's from $S \times A \times B$ (Section 5.11). Table 6.5-6 gives the MS's and F-ratios for $S(A) \times B$, assuming that A and B are fixed.

6.5.7 Using the Table of *F*

If three levels of α are used, significance testing requires that three critical values of F be extracted from the table of F for each combination of df_1, df_2. An orderly way to go about collecting these data is first to determine all unique combinations of df_1 and df_2 for the various sources to be tested, then to find the three critical F's and list them in a table. Such a table appropriate for our example is:

df_1, df_2	Source(s)	F_{95}	F_{99}	$F_{99.9}$
1,6	B	5.99	13.7	35.5
2,6	A, AB	5.14	10.9	27.0

By comparing each F-ratio with this table we can quickly see the level of significance for each F-ratio. This systematic approach is simpler than using the F-table with each separate F-ratio, and we are less likely to make an error.

It may be that the critical F's for the exact df's at hand are missing from Table A.5. Usually it suffices to observe the critical F's for the adjacent df's. For example, if we have an F-ratio based on $df_1 = 22$ and $df_2 = 30$, we cannot locate the exact critical F we need because our table has no column for $df_1 = 22$. However, $F_{95}(22, 30)$ must lie between $F_{95}(20, 30) = 1.93$ and $F_{95}(24, 30) = 1.89$, so that if the F-ratio exceeds 1.93, we can safely say that the effect tested is significant at the .05 level, and if the ratio is less than 1.89, it fails to reach significance at the .05 level. Only if the F-ratio is between 1.89 and 1.93 do we have a problem. In such a situation we interpolate using the reciprocals of the df's. In the preceding example we wish to interpolate to find F_c for $df_1 = 22$, knowing the values for $df_1 = 20$ and $df_1 = 24$. Since $1/20 = .05$, $1/22 = .045$, and $1/24 = .042$, we have

$$F_{95}(22, 30) = 1.89 + \frac{.045 - .042}{.05 - .042}(1.93 - 1.89)$$

$$= 1.89 + \frac{.003}{.008}(.04)$$

$$= 1.89 + .015 = 1.905$$

We proceed similarly for other significance levels and for situations where the exact df_2 is missing from the table.

6.6 PRELIMINARY TESTING AND POOLING OF MS's

6.6.1 Introduction

As mentioned in Section 6.4.3, the term *preliminary tests* refers to significance tests whose conclusions affect subsequent tests for the same experiment. In particular, it is normally the determination of MS_2 that is affected by preliminary testing.

These are two important reasons for preliminary testing: (1) to alter the EMS formulas so that a suitable MS becomes available for some subsequent test, when a suitable MS is not available initially; (2) to allow the determination that two or more EMS's are the same, so that they may be pooled to yield an MS_2 with a higher df_2 than would otherwise be used. Sometimes both purposes apply; we may start with no appropriate MS, but after preliminary testing, we may decide to pool several of them.

The larger values of df_2 obtained from pooling are desirable because for a given df_1, the higher the df_2, the larger the power of the F-test. We can see this from examining the F_c's in Table A.5. For any given df_1 and α, the larger the df_2, the smaller the F_c, and hence, the larger is the region of rejection. For example, $F_{95}(6, 4) = 6.16$, $F_{95}(6, 10) = 3.22$, $F_{95}(6, 60) = 2.25$. An observed F-ratio of 2.7 would not be significant if $df_2 = 10$, but it would be significant if $df_2 = 60$. The increase in power with df_2 relates to the assertion in Section 2.5 that power increases with sample size because the greater the sample size, the less variable the test statistic, so the larger is the region of rejection. Pooling, of course, does not increase the sample size, but it makes MS_2 less variable over replications, because with pooling an MS_2 is based on an average of two or more independent SS's rather than merely on one.

The conclusions obtained from preliminary testing can be wrong. If we falsely conclude in preliminary testing that one or more σ_τ^2 components are zero, then subsequent tests based on these conclusions will be biased. In order to decrease the probability of such type 2 errors (concluding that H_N is true when it is false), preliminary tests may use larger significance levels than usual. There is no standard α for preliminary testing, such as the conventional .05 level for final testing, but α's from .20 to .30 are frequently used, and even larger levels have sometimes been recommended. The larger α levels increase the probability of type 1 errors, but decrease the probability of type 2 errors (Section 2.5). In other words, with a larger α, we are more likely to reject an H_N that $\sigma_\tau^2 = 0$, and thus less likely to proceed with subsequent pooling and testing on the erroneous belief that $\sigma_\tau^2 = 0$. Once preliminary testing is accomplished, final testing proceeds using the usual significance level or levels. Even hypotheses already accepted or rejected in preliminary testing will be retested with the conventional significance levels and possibly with different MS_2's; that is, the conclusions about σ_τ^2's that result from preliminary testing can only be used to determine the subsequent steps in testing. The final conclusions are based on conventional α's and pooled MS's, where applicable.

Pooling of MS's occurs only among MS's that have the same EMS's, as determined by the conclusions from preliminary testing, or in some cases, by a priori assumptions. It is not the MS's themselves that are combined. Pooling consists of summing the associated SS's and dividing the total by the sum of the associated df's. For example, suppose that MS_E and MS_{AB} are to be pooled because we have concluded that $\sigma_{AB}^2 = 0$, so that $EMS_E = EMS_{AB}$. Then

$$MS_{pooled} = \frac{SS_E + SS_{AB}}{df_E + df_{AB}}$$

It is conceivable that an experimenter could use preliminary testing without any intention of pooling. The purpose might simply be to alter the EMS formulas so that a suitable MS becomes available for a subsequent test. In actuality, however, an experimenter who performs preliminary testing for such a purpose will be likely to pool MS's also, whenever the preliminary tests indicate that such pooling is justified. For this reason, the examples of preliminary testing presented here also involve pooling.

6.6.2 Numerical Example: $A \times B$

Table 6.5-1 gives the MS's for design $A \times B$ and the F-ratios assuming that A and B are fixed. As previously explained, however (Section 6.5.3), if B is random, the F-ratio for A would be $MS_A/MS_{AB} = 9$, but $F_{95}(1, 1) = 161$, so A is not significant. With $df_2 = 1$, the test has little power. Source AB is not significant, even using a large significance level. If we conclude, then, that $\sigma_{AB}^2 = 0$, then both AB and E have the same EMS, σ_E^2. Furthermore, if $\sigma_{AB}^2 = 0$, the EMS for A is $\sigma_E^2 + BE\sigma_A^2$, so the pooled MS_E can now serve as MS_2. The pooled $MS_E = (SS_E + SS_{AB})/(df_E + df_{AB}) = 15.5/5 = 3.1$. Then $F = 4.5/3.1 = 1.45$. Since $F_{95}(1, 5) = 6.61$, A is still not significant.

Now suppose A is the random factor and B is fixed. We saw that under these circumstances the F-ratio for B is 64, a very large F, but one that is not significant since $F_{95}(1, 1) = 161$. The preliminary test on AB, however, allows us to conclude that $\sigma_{AB}^2 = 0$ and to form a pooled MS_E, exactly as previously. Then the F-ratio for B is $32.0/3.1 = 10.3$. Since $F_{95}(1, 5) = 6.61$, pooling allows us to conclude that B is significant, whereas without pooling, we could not so conclude.

Obviously, whether an experimenter pools can affect whether an effect is judged to be significant. Pooling decisions should never be made, however, simply in order to obtain the desired conclusion. To guard against this possibility, policy regarding pooling should be established before the scores are analyzed. Pooling policy is considered in Section 6.6.4.

6.6.3 Preliminary Testing and Pooling in Stages

Preliminary testing may proceed through several stages, the first stage utilizing as MS_2 the MS whose expectation has the fewest components. For example, an experimenter often begins by testing all possible H_N's using MS_E, whose EMS is σ_E^2. Then he eliminates from the EMS formulas any σ_τ^2's found to equal zero, and pools the MS's now having the same EMS's. In stage (2) he tests all further hypotheses that are possible based on MS_E (the pooled version, if it exists) or any other MS whose expectation has a positive σ_τ^2 component, as confirmed by testing in stage (1). Then he pools the MS's as possible whenever some $\sigma_\tau^2 = 0$. In stage (3) he tests all further hypotheses based on MS_E or any other MS whose expectations have σ_τ^2 components confirmed by previous tests. The experimenter proceeds in this manner through successive stages until the final stage, in which all sources not yet tested become due for testing, or, if any remaining sources still cannot be tested, there is no possibility that further preliminary testing can produce MS_2's for them.

During the preliminary stages no source is retested. Once we conclude for purposes of preliminary testing that some $\sigma_\tau^2 > 0$, we continue to accept that conclusion throughout the stages of preliminary testing, even though, due to pooling of MS's we could change the F-ratio for σ_τ^2 and possibly our conclusion. But in the final stage, all sources that have not been eliminated by preliminary tests are retested. As a result of pooling and a decrease in α to a conventional level, we may finally conclude that some σ_τ^2 is zero after all, even though we previously proceeded on the basis that $\sigma_\tau^2 > 0$. At this point, however, we would not utilize further pooling and retesting.

These stages can be illustrated using design $A \times B \times C$ with B and C random. Table 6.6-1 gives the EMS's and the MS_2's that would be appropriate before any preliminary testing. There is no MS_2 to test $\sigma_A^2 = 0$. Although we have encountered such a situation before with Class IV designs, sources that lacked appropriate MS_2's were of little interest and could be ignored.

Table 6.6-1 EMS's for $A \times B \times C$, B and C random

Source	EMS	MS_2
A	$\sigma_E^2 + E\sigma_{ABC}^2 + BE\sigma_{AC}^2 + CE\sigma_{AB}^2 + BCE\sigma_A^2$	–
B	$\sigma_E^2 + AE\sigma_{BC}^2 + ACE\sigma_B^2$	MS_{BC}
C	$\sigma_E^2 + AE\sigma_{BC}^2 + ABE\sigma_C^2$	MS_{BC}
AB	$\sigma_E^2 + E\sigma_{ABC}^2 + CE\sigma_{AB}^2$	MS_{ABC}
AC	$\sigma_E^2 + E\sigma_{ABC}^2 + BE\sigma_{AC}^2$	MS_{ABC}
BC	$\sigma_E^2 + AE\sigma_{BC}^2$	MS_E
ABC	$\sigma_E^2 + E\sigma_{ABC}^2$	MS_E
E	σ_E^2	–

With the present design, however, the fixed factor A is likely to be of primary importance, so we are not content to ignore it in our testing.

Note that if σ^2_{AB} were zero, a test of $\sigma^2_A = 0$ could proceed with MS_{AC} as MS_2. Likewise, if σ^2_{AC} were zero, a test of $\sigma^2_A = 0$ could proceed with MS_{AB} as MS_2. Although, in general, it would be dubious to assume on a priori grounds that either σ^2_{AB} or σ^2_{AC} were zero, each of these σ^2_τ's can be tested by using MS_{ABC} as MS_2. If, on the basis of a preliminary test, we conclude that either or both of σ^2_{AB} and σ^2_{AC} equal zero, a subsequent test of $\sigma^2_A = 0$ is possible.

If an experimenter's policy is never to pool, he might simply start with preliminary tests on σ^2_{AB} and σ^2_{AC}, and then proceed with the final tests, including a test of σ^2_A if the conclusions from the preliminary test allow it. The preliminary tests mentioned, however, are likely to be lacking in power, for $df_2 = (A - 1)(B - 1)(C - 1)$ may be rather small. If an experimenter's policy is to pool when he can, he begins with preliminary tests using MS_E as MS_2. Only in a second stage of preliminary testing does he consider the H_N's for σ^2_{AB} and σ^2_{AC}. The following table shows how the testing might proceed if

Stage (1)

Sources tested	F-ratio	Conclusion
BC	MS_{BC}/MS_E	$\sigma^2_{BC} > 0$
ABC	MS_{ABC}/MS_E	$\sigma^2_{ABC} = 0$
	Pooling	
E:	$MS_{E(1)} = \dfrac{SS_E + SS_{ABC}}{df_{E(1)}}$	
	$df_{E(1)} = df_E + df_{ABC}$	

preliminary testing and pooling were to be done. The term $MS_{E(1)}$ symbolizes the mean square for error and $df_{E(1)}$ symbolizes the degrees of freedom for error after stage (1) pooling.

The revised EMS's and MS_2's would now be as shown in Table 6.6-2. All components σ^2_{ABC} have been omitted from the EMS's. The term $E(1)$ refers to the error obtained by stage (1) pooling. The entries under the column headed "Stage for testing" show which sources come up for testing in stage (2), and which was previously tested in stage (1). We have already tested source BC; we do not retest it until the final stage, although the result could be different from before because of the new $MS_{E(1)}$. (The result would

Table 6.6-2 EMS's for $A \times B \times C$ after stage (1) testing

Source	Stage for testing	EMS	MS_2 for stage (2)
A	?	$\sigma_E^2 + BE\sigma_{AC}^2 + CE\sigma_{AB}^2 + BCE\sigma_A^2$	–
B	2	$\sigma_E^2 + AE\sigma_{BC}^2 + ACE\sigma_B^2$	MS_{BC}
C	2	$\sigma_E^2 + AE\sigma_{BC}^2 + ABE\sigma_C^2$	MS_{BC}
AB	2	$\sigma_E^2 + CE\sigma_{AB}^2$	$MS_{E(1)}$
AC	2	$\sigma_E^2 + BE\sigma_{AC}^2$	$MS_{E(1)}$
BC	1	$\sigma_E^2 + AE\sigma_{BC}^2$	–
$E(1)$	–	σ_E^2	–

generally be the same, however.) We still have no appropriate MS_2 for testing $\sigma_A^2 = 0$. We proceed to stage (2), in which we perform all tests possible with $MS_{E(1)}$ (except for hypotheses already tested in stage (1)) and with any other MS's whose expectations we have shown to differ from σ_E^2. In our example, the latter specification refers only to MS_{BC}.

Stage (2)

Sources tested	F-ratio	Conclusion
AB	$MS_{AB}/MS_{E(1)}$	$\sigma_{AB}^2 = 0$
AC	$MS_{AC}/MS_{E(1)}$	$\sigma_{AC}^2 > 0$
B	MS_B/MS_{BC}	$\sigma_B^2 > 0$
C	MS_C/MS_{BC}	$\sigma_C^2 = 0$

Pooling

E:

$$MS_{E(2)} = \frac{SS_E + SS_{ABC} + SS_{AB}}{df_{E(2)}}$$

$$df_{E(2)} = df_E + df_{ABC} + df_{AB}$$

BC:

$$MS_{BC(2)} = \frac{SS_{BC} + SS_C}{df_{BC(2)}}$$

$$df_{BC(2)} = df_{BC} + df_C$$

The revised EMS's and MS_2's are now as shown in Table 6.6-3. We have reached the point where all sources not previously tested can now be tested by using MS_2's whose expectation components have all been verified in preliminary testing. Thus we are now ready for final testing. In the final testing, all sources possible must be subjected to F-tests, even those deemed to be significant in the preliminary stages. Because a smaller α is likely to be used for final testing, and because pooling subsequent to the preliminary test

Table 6.6-3 EMS's for $A \times B \times C$ after stage (2) testing

Source	EMS	MS_2 for final testing
A	$\sigma_E^2 + BE\sigma_{AC}^2 + BCE\sigma_A^2$	MS_{AC}
B	$\sigma_E^2 + AE\sigma_{BC}^2 + ACE\sigma_B^2$	MS_{BC}
AC	$\sigma_E^2 + BE\sigma_{AC}^2$	$MS_{E(2)}$
$BC(2)$	$\sigma_E^2 + AE\sigma_{BC}^2$	$MS_{E(2)}$
$E(2)$	σ_E^2	–

may have changed the MS_2 and df_2 used for testing a source, the final conclusion could be changed. Although in preliminary testing we concluded that $\sigma_\tau^2 > 0$, in final testing we may conclude that $\sigma_\tau^2 = 0$. We would not then do more pooling to alter our EMS's, however. Although we concluded in preliminary testing that $\sigma_{AC}^2 > 0$, and developed our subsequent testing procedure on that basis, we now decide that $\sigma_{AC}^2 = 0$ after all. We do not, however, do more pooling.

Stage (3), final

Sources tested	F-ratio	Conclusion
A	MS_A/MS_{AC}	$\sigma_A^2 > 0$
B	MS_B/MS_{BC}	$\sigma_B^2 > 0$
AC	$MS_{AC}/MS_{E(2)}$	$\sigma_{AC}^2 = 0$
BC	$MS_{BC(2)}/MS_{E(2)}$	$\sigma_{BC}^2 > 0$

It is possible for preliminary testing to indicate that $\sigma_{AB}^2 > 0$ and $\sigma_{AC}^2 > 0$, and then we would lack a suitable MS_2 for testing $\sigma_A^2 = 0$, even in the final stage. However, we still use a quasi F-ratio to test $\sigma_A^2 = 0$. (Quasi F-ratios are considered in Section 6.7.)

6.6.4 To Pool or Not to Pool

In regard to pooling, an experimenter has various options:

1. He may pool, basing his decision on a priori assumptions that some σ_τ^2 or σ_τ^2's are zero. No preliminary testing would then be required.
2. He may pool only whenever preliminary testing indicates that two or more sources have the same EMS's.
3. He may follow a policy of never pooling.

Each policy has its adherents and its rationale. As stated in Section 6.6.2, however, basing a decision about whether to pool on the desired conclusion is never justified. To protect himself against the temptation to follow this approach, an experimenter should establish his pooling policy before examining the scores. Preferably, his policy should apply to his research projects in general, and not be reformulated for each experiment. The author of this textbook cannot tell the student what this policy should be. An experimenter tends to follow the customs of his mentors and colleagues in his particular area of research.

Authorities favoring option (1) often stress that an assumption that some $\sigma_\tau^2 = 0$ should be based on evidence from previous experimentation or from expert knowledge of the subject matter. In practice, however, "previous experimentation" and "expert knowledge" normally add up to no more than an educated guess. One reason is that there is seldom a substantial amount of evidence concerning the existence of an interaction for the particular conditions of the present experiment. Furthermore, psychological theory usually does well even to predict whether interactions between two variables should exist, much less to predict them with certitude.

The authorities who favor option (3) critize any pooling of MS's whatsoever because the pooling is not valid unless some $\sigma_\tau^2 = 0$. Even if a preliminary test indicates that this $\sigma_\tau^2 = 0$, the conclusion could be false, leading to a biased F-test. For example, suppose we conclude that $\sigma_{AB}^2 = 0$ in $A \times B$ (with A and B fixed) and pool MS_E and MS_{AB}. If, indeed, $\sigma_{AB}^2 = 0$, then $\text{EMS}_{AB} = \sigma_E^2$, the same as EMS_E; and the EMS of the pooled MS also is σ_E^2. However, if $\sigma_{AB}^2 > 0$, the EMS of the pooled MS is actually greater than σ_E^2, so such pooled MS's will be larger, on the average, than they should be. Then the F-ratios having these pooled MS's as MS_2 will tend to be smaller than they should be by the assumptions, and thus a reduced number of F-ratios will reach significance. Preliminary testing alone, without pooling, is subject to similar criticism, since it may lead to the elimination of some σ_τ^2 from EMS formulas when that σ_τ^2 is not truly zero. The F-ratios formed on the basis of the revised EMS's will then be biased.

Option (2) can be seen as a kind of middle ground between options (1) and (3). A σ_τ^2 may be considered to be zero for the purposes of subsequent testing, but only if evidence from the experiment at hand (that is, from preliminary testing) leads to this conclusion. To guard against biases that would result if σ_τ^2 were not zero, a larger value of α may be used for the preliminary tests. thus reducing the number of type 2 errors that will occur.

Since pooling is motivated primarily by a desire to increase test power by increasing df_2, we may follow a policy of only pooling when df_2 would otherwise be very small, say, less than 10 to 20. If df_2 is already larger than 20, there is relatively little to gain by enlarging df_2, so we may as well avoid the dangers inherent in pooling in such situations.

6.7 QUASI *F*-RATIOS

It may happen, as we have noted, that on first inspection there is no MS appropriate to serve as MS_2 for some possible H_N test. The test sometimes does not interest us, so we may simply forget about it. At other times, however, we may wish to test the effect. What then? As we have just seen, preliminary testing may bring about adjustments in the EMS's so that a suitable MS becomes available. If not, another possibility is to test the effect with a quasi F-ratio, instead of an F-ratio proper. One or both MS's in a quasi F-ratio consist of a sum of MS's instead of single MS's. The summing is on the MS's themselves, not on the SS's, as with pooling.

A prime on F indicates a quasi ratio (F'). The F'-ratio has a distribution approximating F, so F is taken to be the sampling distribution of the F'-ratio. Just as the F-ratio is the ratio of two MS's having the same expectation under H_N, so too is the F'-ratio. With the F'-ratio, however, this condition for equal expectations is met by summing MS's, rather than by taking MS's directly.

Let us see how we could test the H_N that $\sigma_A^2 = 0$ in $A \times B \times C$ (Table 6.6-1) without preliminary testing. Our approach is to find a sum of MS's whose expectation equals EMS_A under H_N. The sum $EMS_{AC} + EMS_{AB} - EMS_{ABC}$ has the form required: It equals $\sigma_E^2 + E\sigma_{ABC}^2 + BE\sigma_{AC}^2 + CE\sigma_{AB}^2$. Therefore, we use $MS_{AC} + MS_{AB} - MS_{ABC}$ as our quasi MS_2; then the quasi F-ratio, $F' = MS_A/MS_2'$, is distributed approximately as F. To form a conclusion with F' we proceed as usual, using the tables of F to find critical F''s. As df_2' we use the nearest integral value to the quantity determined by the formula

$$\frac{(MS_{AC} + MS_{AB} - MS_{ABC})^2}{\dfrac{MS_{AC}^2}{df_{AC}} + \dfrac{MS_{AB}^2}{df_{AB}} + \dfrac{MS_{ABC}^2}{df_{ABC}}}$$

More generally, the formula for df_2' consists of the square of MS_2', however that may be composed, divided by the sum of the individual MS components, each squared and divided by the associated df. Note in the preceding formula that, although MS_{ABC} appears with a minus sign in the numerator, only plus signs appear in the denominator.

Following are some general hints about which MS's to combine to obtain the required MS_2'. The EMS we sought in our example was $\sigma_E^2 + k_{ABC}\sigma_{ABC}^2 + k_{AC}\sigma_{AC}^2 + k_{AB}\sigma_{AB}^2$. Beginning with the simpler σ_τ^2's, σ_{AB}^2 and σ_{AC}^2, we add EMS_{AB} and EMS_{AC} to get a sum including these σ_τ^2's:

$$2\sigma_E^2 + 2k_{ABC}\sigma_{ABC}^2 + k_{AC}\sigma_{AC}^2 + k_{AB}\sigma_{AB}^2$$

In summing EMS's we can have only one more positive than negative entry, since each EMS has a σ_E^2, and we must end up with a formula having σ_E^2, not some multiple of σ_E^2. The sum up to now exceeds the required EMS by

$\sigma_E^2 + k_{ABC}\sigma_{ABC}^2$. We search for an EMS equal to this difference, so we may subtract it. We find the required term, EMS_{ABC}.

The preceding paragraphs illustrated the formation of an F'-ratio by summing MS's to get an MS_2'. We can also combine MS's to get an MS_1' and find df_1' by a formula that is entirely analogous to the one that was given for df_2'. It is possible to form an F'-ratio using both an MS_1' and an MS_2', if they have the same expectations.

6.8 ESTIMATION OF σ_τ^2 AND RELIABILITY

The EMS formulas are important in ANOVA primarily because they tell us from which MS's to form the F-ratios for significance testing. However, they can also be used to estimate the σ_τ^2's. The student might think that it would be enlightening to compute and report the σ_τ^2's for the significant effects found in an experiment; however, this procedure is seldom followed. It is, nevertheless, well for the student and practitioner to understand how such estimates can be obtained.

Consider design A:

$$\text{EMS}_A = \sigma_E^2 + E\sigma_A^2$$
$$\text{EMS}_E = \sigma_E^2$$

Let us solve for σ_A^2 in terms of the EMS's:

$$\text{EMS}_A - \text{EMS}_E = E\sigma_A^2$$

Therefore,

$$\sigma_A^2 = \frac{\text{EMS}_A - \text{EMS}_E}{E}$$

If we knew the EMS's, we could calculate σ_A^2, but we do not. However, we do know the MS's, and each MS is, of course, an unbiased estimate of its corresponding EMS. Hence, we substitute the MS's for the EMS's and calculate an estimate $\hat{\sigma}_A^2$ of σ_A^2:

$$\hat{\sigma}_A^2 = \frac{\text{MS}_A - \text{MS}_E}{E}$$

Since $\sigma_E^2 = \text{EMS}_E$,

$$\hat{\sigma}_E^2 = \text{MS}_E$$

We can use a similar approach to estimate other σ_τ^2's in more complex designs:

1. Solve for σ_τ^2 in terms of various EMS's.
2. For each EMS in the formula so derived, substitute the corresponding MS and compute the value of the formula.
3. The value computed is $\hat{\sigma}_\tau^2$, an unbiased estimate of σ_τ^2.

Recall that for a key term τ, σ_τ^2 is not necessarily a variance, but when it is not, the variance can be calculated from σ_τ^2 (Section 6.3.5). Likewise, we can compute an estimate of the variance from $\hat{\sigma}_\tau^2$.

In general, to solve for some σ_τ^2 in terms of the EMS's, we subtract from EMS_τ the EMS that equals EMS_τ except for $k_\tau\sigma_\tau^2$. In other words, we subtract that EMS whose MS is MS_2 for testing the H_N that $\sigma_\tau^2 = 0$. This gives us

$$\text{EMS}_\tau - \text{EMS}_2 = k_\tau\sigma_\tau^2,$$

so

$$\sigma_\tau^2 = (\text{EMS}_\tau - \text{EMS}_2)/k_\tau$$

Although the MS's must be non-negative, as must σ_τ^2, it is possible for the estimate of σ_τ^2 to be negative. This would happen only when $\text{MS}_\tau < \text{MS}_2$, which means that the F-ratio for term τ would be less than one, so τ would be judged not significant. Under such circumstances, an experimenter would normally conclude that $\sigma_\tau^2 = 0$ and would make no attempt to estimate σ_τ^2 by the methods described here.

Section 1.2 divided the statistical methodology used by investigators into the "experimental" and the "correlational." This book primarily concerns the experimental, but the two methodologies overlap at various points, so occasionally the discussion turns to matters that are of interest to correlationists. For example, the topic under discussion at present, estimation of the σ_τ^2's, relates closely to the topic of test reliability, a central concern of correlationists. Experimentalists normally exhibit no interest in the topic of reliability, though the topic is not without relevance for experimentation.

Reliability is normally discussed with the implicit assumption that all subjects are within one cell. There is no division of subjects into groups receiving different treatments. The correlationist often assumes that each subject is rescored, but under the same condition—not under different conditions, as is true with repeated measurements designs. Nonetheless, the idea of the reliability of a test (scoring instrument) applies to experiments. The experimenter should have a reliable scoring instrument, just as should the correlationist, because an unreliable scoring instrument will result in F-tests of low power.

A basic definition of reliability is the ratio of true score variance to total score variance for some population of subjects, with the understanding that only one cell is involved. By "true score" is meant that part of a subject's score that is the same at each rescoring. Section 4.2.3 contrasted the error term of single measurement designs with those of repeated measurements designs. The "error" of reliability theory is the error of repeated measurements designs. The "error" of single measurement designs also includes, as we may conceive it, a confounded **s** component, a systematic component that would remain the same for that particular subject each time he is scored on the same instrument. We can only estimate reliability for repeated measurements designs, because only in such designs can we estimate true score and error variance separately, as correlationists understand those terms. Furthermore, to estimate reliability for repeated measurements designs, we must be able to estimate σ_E^2, which is only possible if the interaction term confounded with $\mathbf{e}$ is assumed to be zero, or if there is more than one score per condition, which is virtually never the situation.

Let us consider the design $S \times A$. The score model, assuming $\mathbf{as}_{as}$ to be zero, is $X_{as} = \mathbf{m} + \mathbf{a}_a + \mathbf{s}_s + \mathbf{e}_{as}$. The true score is $\mathbf{m} + \mathbf{a}_a + \mathbf{s}_s$. The population variance of the true scores for a particular cell would be simply the variance of the population of $\mathbf{s}_s$'s, since $\mathbf{m}$ and $\mathbf{a}_a$ would be constant. The variance of the $\mathbf{s}_s$'s is σ_S^2. The population variance of a cell is $\sigma_S^2 + \sigma_E^2$. Reliability, then, is $\sigma_S^2/(\sigma_S^2 + \sigma_E^2)$. We assume, as we have all along, that these variances are the same for each cell, so we can make a common estimate of them based on the scores from all cells in an experiment. The EMS's for $S \times A$ were given in Section 6.5.4, where it was said that $\text{EMS}_S = \sigma_E^2 + A\sigma_S^2$, and $\text{EMS}_{AS} = \sigma_E^2 + \sigma_{AS}^2$. Assuming that $\sigma_{AS}^2 = 0$, then, we have

$$\sigma_E^2 = \text{EMS}_{AS}$$

$$\sigma_S^2 = (\text{EMS}_S - \text{EMS}_{AS})/A$$

We do not know the EMS's, but we can estimate them with MS's, which we can compute from experimental scores. By substituting the MS's for the corresponding EMS's in the preceding formulas we can estimate σ_E^2 and σ_S^2:

$$\hat{\sigma}_E^2 = \text{MS}_{AS}$$

$$\hat{\sigma}_S^2 = (\text{MS}_S - \text{MS}_{AS})/A$$

We estimate the reliability r_{XX} as

$$\hat{r}_{XX} = \frac{\hat{\sigma}_S^2}{\hat{\sigma}_E^2 + \hat{\sigma}_S^2}$$

The MS's for the $S \times A$ scores of Section 4.5 are shown in Table 6.5-2. For these data,

$$\hat{\sigma}_E^2 = \text{MS}_{AS} = 2.92$$

$$\begin{aligned}\hat{\sigma}_S^2 &= (\text{MS}_S - \text{MS}_{AS})/A \\ &= (28.82 - 2.92)/4 \\ &= 25.9/4 = 6.48\end{aligned}$$

Hence,

$$\hat{r}_{XX} = \frac{6.48}{2.92 + 6.48} = \frac{6.48}{9.40} = .69$$

For design $S(A) \times B$, estimation of the reliability requires that we assume σ_{BS}^2 to be zero (if there is one score per condition). The EMS's are given in Table 6.5-5, and from them we determine that:

$$\hat{\sigma}_E^2 = \text{MS}_{BS}$$

$$\hat{\sigma}_S^2 = \frac{\text{MS}_S - \text{MS}_{BS}}{B}$$

As before, the reliability is estimated as $\hat{\sigma}_S^2/(\hat{\sigma}_S^2 + \hat{\sigma}_E^2)$. If the reliability is high, σ_E^2 is small compared with σ_S^2, so within-subject F-tests are particularly powerful relative to between-subject tests. If the reliability is small, within-subject F-tests have less advantage.

6.9 GENERAL RULES FOR FINDING EMS FORMULAS

Preceding sections have shown how to determine which F-ratios to use for significance testing, given the EMS formulas, and also how to estimate the σ_τ^2 components of these formulas. If the EMS formulas are not readily available for the design of interest, the experimenter can quickly derive them using the simple rules that follow. Use of these rules is exemplified using design $S(A) \times B$, A and B fixed.

Step 1. Make a table in which the model terms (except **m** and **e**) read down the rows and the subscripts of the factors head the columns. If error is not confounded with an interaction, include a column for subscript e. Normally, such a column is required only for single measurement designs having more than one score per cell. Indicate whether the factor heading each column is fixed or random.

Step 2. Fill in the table column by column. For fixed factor columns, three runs down each column are needed; for random factor columns, two runs are needed.

Fixed factor columns:

2a. Enter 0 for each key term having the column subscript, if it is not within a nest.

2b. Enter 1 for each key term having the column subscript within a nest.

2c. Enter the upper limit of the column subscript in the remaining cells of the column.

Random factor columns:

2a. Enter 1 for each key term in which the column subscript appears, regardless of whether it is within a nest.

2b. Enter the upper limit of the subscript in the remaining cells of the column.

Table 6.9-1 shows the table so constructed for $S(A) \times B$.

Step 3. Write the EMS for each key term. Beginning with the first key term, $\mathbf{a}_a$, listed along the left edge of Table 6.9-1, write its EMS, and then proceed downward, writing the EMS for each key term. Each EMS except EMS_E is written according to the substeps that follow. After all other EMS's

Table 6.9-1 Derivation of EMS formulas: steps (1) and (2) completed, for $S(A) \times B$

	Fixed a	Fixed b	Random s
$\mathbf{a}_a$	0	B	S
$\mathbf{b}_b$	A	0	S
$\mathbf{s}_{s(a)}$	1	B	1
$\mathbf{ab}_{ab}$	0	0	S
$\mathbf{bs}_{bs(a)}$	1	0	1

have been derived, simply write $\text{EMS}_E = \sigma_E^2$, if an unconfounded error term appears in the score model.

3a. First, note which subscripts appear in the model term whose EMS is being derived; cover the columns corresponding to such subscripts with narrow strips of paper. Cover subscripts appearing in a nest as well as non-nested ones.

3b. The first component of the EMS formula is always σ_E^2.

3c. Examine each key term, beginning at the bottom ($\mathbf{bs}_{bs(a)}$); if it has all the subscripts of the key term whose EMS is sought, then it may contribute a σ_τ^2 component to the formula. For example, since $\mathbf{bs}_{bs(a)}$ has all the subscripts of $\mathbf{a}_a$, σ_{BS}^2 is considered a possible component of EMS_A.

3d. The coefficient of the σ_τ^2 being considered in (3c) is obtained by multiplying those row entries for the possible component σ_τ^2 that were not covered in (3a). For this design, the row entries for $\mathbf{bs}_{bs(a)}$ are 0 and 1. The product is 0, so k_{BS} is 0, and therefore σ_{BS}^2 is not a component of EMS_A after all. (Had B been random, it would have been a component.)

3e. Continue working up the list of model terms, until the term is reached whose EMS is being sought. This term will certainly contribute a $k_\tau \sigma_\tau^2$ to the EMS, but no term higher on the list need be considered.

If all columns are covered in (3a), we are seeking the EMS for the interaction term that has all design subscripts. Its EMS is simply $\sigma_E^2 + \sigma_\tau^2$, where τ stands for the interaction term.

To find EMS_A, we cover column a with a strip of paper. Since subscript a appears in $\mathbf{bs}_{bs(a)}$, $\mathbf{ab}_{ab}$, and $\mathbf{s}_{s(a)}$, we can consider σ_{BS}^2, σ_{AB}^2, and σ_S^2 as possible components to EMS_A. We find that $k_{BS} = 0$, $k_{AB} = 0$, $k_S = B$, and $k_A = BS$, so

$$\text{EMS}_A = \sigma_E^2 + B\sigma_S^2 + BS\sigma_A^2$$

Table 6.9-2 Derivation of EMS formulas: steps (1) and (2) completed, for $A \times B \times C$, B and C random

	Fixed a	Random b	Random c	Random e
$\mathbf{a}_a$	0	B	C	E
$\mathbf{b}_b$	A	1	C	E
$\mathbf{c}_c$	A	B	1	E
$\mathbf{ab}_{ab}$	0	1	C	E
$\mathbf{ac}_{ac}$	0	B	1	E
$\mathbf{bc}_{bc}$	A	1	1	E
$\mathbf{abc}_{abc}$	0	1	1	E

We proceed to find EMS_B. Possible σ_τ^2 components are σ_{BS}^2, σ_{AB}^2, and σ_B^2. Now column b is covered, so we find that $k_{BS} = 1$, $k_{AB} = 0$, and $k_B = AS$, and thus

$$\text{EMS}_B = \sigma_E^2 + \sigma_{BS}^2 + AS\sigma_B^2$$

Next we find EMS_S by covering columns a and s. We consider component σ_{BS}^2, but we find that $k_{BS} = 0$, so we omit this component. We do not consider σ_{AB}^2, since $\mathbf{ab}_{ab}$ lacks subscript s. We find that $k_S = B$, so we have

$$\text{EMS}_S = \sigma_E^2 + B\sigma_S^2$$

To find EMS_{AB}, we cover columns a and b. We find that $k_{BS} = 1$ and $k_{AB} = S$, so

$$\text{EMS}_{AB} = \sigma_E^2 + \sigma_{BS}^2 + S\sigma_{AB}^2$$

For EMS_{BS} we can consider only σ_{BS}^2. All columns are covered, but we know that in such circumstances the coefficient is 1, so

$$\text{EMS}_{BS} = \sigma_E^2 + \sigma_{BS}^2$$

We have no unconfounded error term in $S(A) \times B$, so we do not consider EMS_E. (Recall that we have no separate MS_E in such a situation.) All the EMS's we have derived agree with those given in Table 6.5-5.

Table 6.9-2 shows steps (1) and (2) completed for design $A \times B \times C$, B and C random. The formula for EMS_A contains σ_E^2 plus possible σ_τ^2 components for τ of ABC, AC, and AB, plus a σ_A^2 component. To get the k_τ's, we cover column a and multiply in the respective rows, obtaining

$$\text{EMS}_A = \sigma_E^2 + E\sigma_{ABC}^2 + BE\sigma_{AC}^2 + BCE\sigma_A^2$$

Derivation of the other EMS's is left to the student; the results can be checked using Table 6.6-1, which shows all the EMS's.

6.10 DERIVATION OF EMS FORMULAS FOR SPECIFIC DESIGNS

6.10.1 General Approach

By using the simple rules given in Section 6.9, it is easy to obtain the $k_\tau \sigma_\tau^2$ components for the EMS's. When these components are not given in a readily available textbook, an experimenter normally uses these rules. Although the $k_\tau \sigma_\tau^2$ components can be easily obtained by using these rules, they provide no understanding of how such components are derived mathematically.

In the belief that the mathematical derivations provide intellectual satisfaction as well as a better comprehension of ANOVA in general, this text presents such derivations for a few simple designs. Students who are not inspired by such vague motivations may be assured that the techniques demonstrated here need not be mastered, either as a prerequisite to further study in this book, or as necessary tools for the analysis of data in actual experimentation.

The general procedure followed in this section is to begin with the calculational formulas for the SS elements, obtained according to procedures given in Section 5.4. Recall that each such formula involves X's summed over one or more subscripts and then squared (except that for one element the scores are squared without preliminary summing). For each such sum over X, we substitute the theoretical values obtained by summing over the score model. Then we evaluate the expectation (expected value) of the SS element. By adding and subtracting these expectations of the SS elements for a key term instead of the elements themselves, we obtain a formula for the expected SS (ESS), expressed via model parameters. Whereas an MS is an SS divided by the appropriate df, an EMS may be obtained by dividing an ESS by the appropriate df. Since the ESS is expressed via model parameters, so also is the EMS. We therefore have our desired end result. The procedure is now demonstrated for designs A and $S(A) \times B$.

6.10.2 Design A, A Fixed

The score model is:

$$X_{ae} = \mathbf{m} + \mathbf{a}_a + \mathbf{e}_{e(a)}$$

There is an SS element for each term of the model: SS_1, SS_a, and SS_{ae}. We require the corresponding ESS elements: ESS_1, ESS_a, and ESS_{ae}. We derive them in reverse order.

ESS_{ae}:

$$ESS_{ae} = E\left[\sum_{ae} X_{ae}^2\right]$$

$$= \mathrm{E}\left[\sum_{ae} (\mathbf{m} + \mathbf{a}_a + \mathbf{e}_{e(a)})^2\right]$$

$$= \mathrm{E}\left[\sum_{ae} (\mathbf{m}^2 + \mathbf{a}_a^2 + \mathbf{e}_{e(a)}^2 + \underset{\text{(I)}}{2\mathbf{m}\mathbf{a}_a} + \underset{\text{(II)}}{2\mathbf{m}\mathbf{e}_{e(a)}} + \underset{\text{(III)}}{2\mathbf{a}_a\mathbf{e}_{e(a)}})\right]$$

We arrived at the last stage by squaring the score model, as indicated. The result consists of model terms squared plus cross products of model terms. The cross products are labeled I, II, and III. After we sum with $\sum_{ae}$ and take the expectation, we find that each cross product equals zero. Let us see why, term by term.

(I) $$\mathrm{E}\left(\sum_{ae} 2\mathbf{m}\mathbf{a}_a\right) = 2\mathbf{m}\mathrm{E}\left(\sum_e \sum_a \mathbf{a}_a\right)$$

But $\sum_a \mathbf{a}_a = 0$, by the side condition, so the whole term equals zero.

(II) $$\mathrm{E}\left(\sum_{ae} 2\mathbf{m}\mathbf{e}_{e(a)}\right) = 2\mathbf{m}\mathrm{E}\left(\sum_{ae} \mathbf{e}_{e(a)}\right) = 2\mathbf{m}\sum_{ae} \mathrm{E}(\mathbf{e}_{e(a)})$$

It should be noted that we have reversed the order for applying the $\sum$ and E operations; such a reversal is always allowed. The expectation of $\mathbf{e}$ is always zero, by the usual assumptions, so term II equals zero.

(III) $$\mathrm{E}\left(\sum_{ae} 2\mathbf{a}_a\mathbf{e}_{e(a)}\right) = 2\sum_{ae} \mathbf{a}_a\mathrm{E}(\mathbf{e}_{e(a)})$$

We not only reversed the order for applying $\sum$ and E, but we moved $\mathbf{a}_a$ to the left of E, a legitimate operation when factor A is fixed so that $\mathbf{a}_a$ is a constant with respect to replications. Since $\mathrm{E}(\mathbf{e}_{e(a)}) = 0$, the whole term equals zero.

Since the cross products equal zero, we are left with:

$$\mathrm{ESS}_{ae} = \mathrm{E}\sum_{ae} (\mathbf{m}^2 + \mathbf{a}_a^2 + \mathbf{e}_{e(a)}^2)$$

$$= \underset{\text{(I)}}{\mathrm{E}\left(\sum_{ae} \mathbf{m}^2\right)} + \underset{\text{(II)}}{\mathrm{E}\left(\sum_{ae} \mathbf{a}_a^2\right)} + \underset{\text{(III)}}{\mathrm{E}\left(\sum_{ae} \mathbf{e}_{e(a)}^2\right)}$$

Since $\mathbf{m}^2$ is a constant, both with respect to subscripts a and e and with respect to replications, term (I) equals $AE\mathbf{m}^2$. Since $\sum_{ae}\mathbf{a}_a^2 = \sum_e\sum_a\mathbf{a}_a^2$, since $\sigma_A^2 = \sum_a\mathbf{a}_a^2/(A-1)$ when factor A is fixed, and since the expectation of a constant equals that constant, term (II) equals $(A-1)E\sigma_A^2$. Since $\mathrm{E}(\mathbf{e}_{e(a)}) = 0$, $\mathrm{E}(\mathbf{e}_{e(a)}^2) = \sigma_E^2$, the variance of $\mathbf{e}$, which we assume is the same (is constant) regardless of the value of the subscripts on $\mathbf{e}$. It follows, then, that $\mathrm{E}(\sum_{ae}\mathbf{e}_{e(a)}^2) = \sum_{ae}\mathrm{E}(\mathbf{e}_{e(a)}^2) = \sum_{ae}\sigma_E^2 = AE\sigma_E^2$.

The net result is that

$$\mathrm{ESS}_{ae} = AE\mathbf{m}^2 + (A-1)E\sigma_A^2 + AE\sigma_E^2$$

ESS_a:

$$\mathrm{ESS}_a = \mathrm{E}\left[\frac{\sum_a X_{a.}^2}{E}\right]$$

$$X_{a.} = E\mathbf{m} + E\mathbf{a}_a + \mathbf{e}_{.(a)}$$

$$\text{ESS}_a = \text{E}\left[\frac{\sum_a (E\mathbf{m} + E\mathbf{a}_a + \mathbf{e}_{.(a)})^2}{E}\right]$$

$$= \text{E}\left[\frac{\sum_a (E\mathbf{m})^2}{E}\right] + \text{E}\left[\frac{\sum_a (E\mathbf{a}_a)^2}{E}\right] + \text{E}\left[\frac{\sum_a \mathbf{e}^2_{.(a)}}{E}\right]$$
$$\text{(I)} \qquad\qquad \text{(II)} \qquad\qquad \text{(III)}$$

We did not include the cross products, since they drop out as before.

(I) $$\text{E}\left[\frac{\sum_a (E\mathbf{m})^2}{E}\right] = \text{E}\left[\frac{\sum_a E^2\mathbf{m}^2}{E}\right] = \text{E}[AE\mathbf{m}^2] = AE\mathbf{m}^2$$

(II) $$\text{E}\left[\frac{\sum_a (E\mathbf{a}_a)^2}{E}\right] = \text{E}\left[E\sum_a \mathbf{a}_a^2\right] = \text{E}[(A-1)E\sigma_A^2] = (A-1)E\sigma_A^2$$

The derivation here depends on the definition that $\sigma_A^2 = \sum_a \mathbf{a}_a^2/(A-1)$ for fixed A and the fact that the expectation of a constant over replications equals that constant.

(III) $$\text{E}\left[\frac{\sum_a \mathbf{e}^2_{.(a)}}{E}\right] = \frac{\sum_a E(\mathbf{e}^2_{.(a)})}{E}$$

Since $\text{E}(\mathbf{e}^2_{e(a)}) = \sigma_E^2$, and since $\mathbf{e}_{.(a)}$ is the sum of E independent and identically distributed random variables, $\text{E}(\mathbf{e}^2_{.(a)}) = E\sigma_E^2$. Therefore, term III equals $\sum_a E\sigma_e^2/E = \sum_a \sigma_E^2 = A\sigma_E^2$.

The net result is that

$$\text{ESS}_a = AE\mathbf{m}^2 + (A-1)E\sigma_A^2 + A\sigma_E^2$$

ESS_1:

$$\text{ESS}_1 = \text{E}\left[\frac{X^2_{..}}{AE}\right]$$

$$X_{..} = AE\mathbf{m} + 0 + \mathbf{e}_{.(.)}$$

The zero stands for $\sum_{ae} a_a = \sum_e \sum_a \mathbf{a}_a$, which equals zero because of the side condition for fixed factor A.

$$\text{ESS}_1 = \text{E}\left[\frac{(AE\mathbf{m} + \mathbf{e}^2_{.(.)})}{AE}\right]$$

$$= \text{E}\left[\frac{(AE\mathbf{m})^2}{AE}\right] + \text{E}\left[\frac{\mathbf{e}^2_{.(.)}}{AE}\right]$$

$$= AE\mathbf{m}^2 + \sigma_E^2$$

Again, the cross product is zero, so it is omitted. Since $\text{E}(\mathbf{e}^2_{e(a)}) = \sigma_E^2$, and since $\mathbf{e}_{.(.)}$ is the sum of $A \cdot E$ independent observations from identical $\mathbf{e}$ distributions,

$$\text{E}[\mathbf{e}^2_{.(.)}/AE] = \text{E}[\mathbf{e}^2_{.(.)}]/AE = AE\sigma_E^2/AE = \sigma_E^2$$

To summarize, we have

$$\text{ESS}_{ae} = AE\mathbf{m}^2 + (A-1)E\sigma_A^2 + AE\sigma_E^2$$
$$\text{ESS}_a = AE\mathbf{m}^2 + (A-1)E\sigma_A^2 + A\sigma_E^2$$
$$\text{ESS}_1 = AE\mathbf{m}^2 + \sigma_E^2$$

Just as there is an SS for each key term (in this case SS_A and SS_E), there is an ESS for each key term, ESS_A and ESS_E. To get the ESS from the ESS elements, we add and subtract the ESS elements in the same way that we add and subtract the SS elements to get the SS's. For design A,

$$\text{SS}_A = \text{SS}_a - \text{SS}_1$$
$$\text{SS}_E = \text{SS}_{ae} - \text{SS}_a$$

Therefore,

$$\text{ESS}_A = \text{ESS}_a - \text{ESS}_1$$
$$\text{ESS}_E = \text{ESS}_{ae} - \text{ESS}_a$$

Using the previously given formulas for the ESS elements, we find that

$$\text{ESS}_A = (A-1)\sigma_E^2 + (A-1)E\sigma_A^2$$
$$\text{ESS}_E = A(E-1)\sigma_E^2$$

Since $\text{EMS}_A = \text{ESS}_A/\text{df}_A$, and $\text{EMS}_E = \text{ESS}_E/\text{df}_E$, and since $\text{df}_A = A - 1$ and $\text{df}_E = A(E-1)$, we have the following EMS formulas we have sought:

$$\text{EMS}_A = \sigma_E^2 + E\sigma_A^2$$
$$\text{EMS}_E = \sigma_E^2$$

It is customary, in writing the EMS formulas, to reverse the order we use in writing the model terms; that is, σ_E^2 comes first in the EMS formula, though **e** comes last in writing the model.

Although in the present example, combining the ESS elements to find the ESS's is simple, for more complicated designs it is convenient to proceed more systematically by constructing a table of ESS element coefficients. Such a table (see Table 6.10-1) has as column headings the σ_τ^2's, in reverse order, beginning with σ_E^2. The ESS elements are listed vertically as row headings, also in reverse order. The body of the table gives the coefficients of the σ_τ^2's in the formulas for the ESS elements.

Table 6.10-1 Coefficients of σ_τ^2's in the ESS element formulas for design A

	σ_E^2	σ_A^2	$\mathbf{m}^2$
ESS_{ae}	AE	$(A-1)E$	AE
ESS_a	A	$(A-1)E$	AE
ESS_1	1	0	AE

We use the table to find k_τ coefficients of the σ_τ^2 in the EMS formulas as follows:

EMS_A:

$$\text{EMS}_A = \text{ESS}_A/\text{df}_A = (\text{ESS}_a - \text{ESS}_1)/(A-1)$$

To obtain the k_τ's, we combine the ESS elements for each column separately as required, and divide each result by df_A, which equals $A - 1$:

$$k_E = (A-1)/\text{df}_A = 1$$
$$k_A = (A-1)E/\text{df}_A = E$$
$$k_M = (AE - AE)/\text{df}_A = 0$$

EMS_E:

$$\text{EMS}_E = \text{ESS}_E/\text{df}_E = (\text{ESS}_{ae} - \text{ESS}_a)/A(E-1)$$
$$k_E = (AE - A)/\text{df}_E = A(E-1)/A(E-1) = 1$$
$$k_A = ((A-1)E - (A-1)E)/\text{df}_E = 0$$
$$k_M = (AE - AE)/\text{df}_E = 0$$

The k_τ's agree with those in the preceding EMS formulas. The coefficient for $\mathbf{m}^2$ is always zero, so $\mathbf{m}^2$ is never found in an EMS formula. Therefore, the $\mathbf{m}^2$ column will henceforth be omitted from the coefficients table.

6.10.3 Design $S(A) \times B$, A and B Fixed

We now derive the EMS's for design $S(A) \times B$. Factors A and B are assumed to be fixed, but factor S is, as usual, random. The derivation is considerably more complicated than the derivation for design A, though the procedures are the same. However, fewer intermediary steps are shown than were shown with design A, since the student should now require less guidance.

The score model is:

$$X_{abs} = \mathbf{m} + \mathbf{a}_a + \mathbf{b}_b + \mathbf{s}_{s(a)} + \mathbf{ab}_{ab} + \mathbf{bs}_{bs(a)} + \mathbf{e}_{abs}$$

We assume that, as usual with this design, there is only one score per condition. Since $\mathbf{e}_{abs}$ is confounded with $\mathbf{bs}_{bs(a)}$, it has no separate estimate, SS, or EMS. There is an SS element for each term of the model except $\mathbf{e}_{abs}$. We require the corresponding ESS elements: ESS_{abs}, ESS_{ab}, ESS_{as}, ESS_b, ESS_a, and ESS_1. They are derived as follows.

ESS_{abs}:

$$\begin{aligned}\text{ESS}_{abs} &= \text{E}\left[\sum_{abs} X_{abs}^2\right] \\ &= \text{E}\left[\sum_{abs} (\mathbf{m} + \mathbf{a}_a + \mathbf{b}_b + \mathbf{s}_{s(a)} + \mathbf{ab}_{ab} + \mathbf{bs}_{bs(a)} + \mathbf{e}_{abs})^2\right] \\ &= ABS\mathbf{m}^2 + (A-1)BS\sigma_A^2 + A(B-1)S\sigma_B^2 + ABS\sigma_S^2 \\ &\qquad + (A-1)(B-1)S\sigma_{AB}^2 + A(B-1)S\sigma_{BS}^2 + ABS\sigma_E^2\end{aligned}$$

The preceding derivation is based on our knowledge that $\sigma_S^2 = \mathrm{E}(\mathbf{s}_{s(a)}^2)$, and since σ_S^2 is assumed to be a constant for different values of a, $\sum_a \mathrm{E}(\mathbf{s}_{s(a)}^2) = A\sigma_S^2$. In addition, for B fixed and S random, $\sigma_{BS}^2 = \sum_b \mathrm{E}(\mathbf{bs}_{bs(a)})/(B-1)$.

ESS_{ab}:

$$\mathrm{ESS}_{ab} = \mathrm{E}\left[\frac{\sum_{ab} X_{ab}^2}{S}\right]$$

$$= \mathrm{E}\left[\frac{\sum_{ab}(S\mathbf{m} + S\mathbf{a}_a + S\mathbf{b}_b + \mathbf{s}_{\cdot(a)} + S\mathbf{ab}_{ab} + \mathbf{bs}_{b\cdot(a)} + \mathbf{e}_{ab\cdot}^2)^2}{S}\right]$$

$$= ABS\mathbf{m}^2 + (A-1)BS\sigma_A^2 + A(B-1)S\sigma_B^2 + AB\sigma_S^2$$
$$+ (A-1)(B-1)S\sigma_{AB}^2 + A(B-1)\sigma_{BS}^2 + AB\sigma_E^2$$

The basis for the preceding derivation is that $\mathrm{E}(\mathbf{s}_{\cdot(a)}^2) = S E(\mathbf{s}_{s(a)}^2) = S\sigma_S^2$, and that $\mathrm{E}(\mathbf{bs}_{b\cdot(a)}^2) = SE(\mathbf{bs}_{bs(a)}^2) = S\sigma_{BS}^2$.

ESS_{as}:

$$\mathrm{ESS}_{as} = \mathrm{E}\left[\frac{\sum_{as} X_{a\cdot s}^2}{B}\right]$$

$$= \mathrm{E}\left[\frac{\sum_{as}(B\mathbf{m} + B\mathbf{a}_a + B\mathbf{s}_{s(a)} + \mathbf{e}_{a\cdot s})^2}{B}\right]$$

$$= ABS\mathbf{m}^2 + (A-1)BS\sigma_A^2 + ABS\sigma_S^2 + AS\sigma_E^2$$

ESS_b:

$$\mathrm{ESS}_b = \mathrm{E}\left[\frac{\sum_b X_{\cdot b\cdot}^2}{AS}\right]$$

$$= \mathrm{E}\left[\frac{\sum_b(AS\mathbf{m} + AS\mathbf{b}_b + \mathbf{s}_{\cdot(\cdot)} + \mathbf{bs}_{b\cdot(\cdot)} + \mathbf{e}_{\cdot b\cdot}^2)^2}{AS}\right]$$

$$= ABS\mathbf{m}^2 + A(B-1)S\sigma_B^2 + B\sigma_S^2 + (B-1)\sigma_{BS}^2 + B\sigma_E^2$$

ESS_a:

$$\mathrm{ESS}_a = \mathrm{E}\left[\frac{\sum_a X_{a\cdot\cdot}^2}{BS}\right]$$

$$= \mathrm{E}\left[\frac{\sum_a(BS\mathbf{m} + BS\mathbf{a}_a + B\mathbf{s}_{\cdot(a)} + \mathbf{e}_{a\cdot\cdot})^2}{BS}\right]$$

$$= ABS\mathbf{m}^2 + (A-1)BS\sigma_A^2 + AB\sigma_S^2 + A\sigma_E^2$$

ESS_1:

$$\mathrm{ESS}_1 = \mathrm{E}\left[\frac{X_{\cdots}^2}{ABS}\right]$$

$$= \mathrm{E}\left[\frac{(ABS\mathbf{m} + B\mathbf{s}_{\cdot(\cdot)} + \mathbf{e}_{\cdots})^2}{ABS}\right]$$

$$= ABS\mathbf{m}^2 + B\sigma_S^2 + \sigma_E^2$$

Table 6.10-2 Coefficients of σ_τ^2 in the ESS element formulas for design $S(A) \times B$

	σ_E^2	σ_{BS}^2	σ_{AB}^2	σ_S^2	σ_B^2	σ_A^2
ESS_{abs}	ABS	$A(B-1)S$	$(A-1)(B-1)S$	ABS	$A(B-1)S$	$(A-1)BS$
ESS_{ab}	AB	$A(B-1)$	$(A-1)(B-1)S$	AB	$A(B-1)S$	$(A-1)BS$
ESS_{as}	AS	0	0	ABS	0	$(A-1)BS$
ESS_b	B	$B-1$	0	B	$A(B-1)S$	0
ESS_a	A	0	0	AB	0	$(A-1)BS$
ESS_1	1	0	0	B	0	0

The column for $\mathbf{m}^2$ was omitted from Table 6.10-2, which shows the ESS coefficients, since, as we previously noted, the $\mathbf{m}^2$ coefficient drops out for every ESS.

We proceed now to find the EMS's.

EMS_A:

$$\begin{aligned}\text{EMS}_A &= \text{ESS}_A/\text{df}_A \\ &= (\text{ESS}_a - \text{ESS}_1)/(A-1)\end{aligned}$$

We use Table 6.10-2 in deriving the EMS k_τ coefficients, beginning with k_E:

$$\begin{aligned}k_E &= (A-1)/\text{df}_A = 1 \\ k_{BS} &= k_{AB} = k_B = 0 \\ k_S &= (AB - B)/\text{df}_A = B(A-1)/(A-1) = B \\ k_A &= (A-1)BS/\text{df}_A = BS\end{aligned}$$

Therefore,

$$\text{EMS}_A = \sigma_E^2 + B\sigma_S^2 + BS\sigma_A^2$$

EMS_B:

$$\begin{aligned}\text{EMS}_B &= \text{ESS}_B/\text{df}_B \\ &= (\text{ESS}_b - \text{ESS}_1)/(B-1)\end{aligned}$$

Again using the table of ESS component coefficients, we find:

$$\begin{aligned}k_E &= (B-1)/\text{df}_B = 1 \\ k_{BS} &= (B-1)/\text{df}_B = 1 \\ k_{AB} &= k_S = 0 \\ k_B &= A(B-1)/\text{df}_B = AS \\ k_A &= 0\end{aligned}$$

Thus,

$$\text{EMS}_B = \sigma_E^2 + \sigma_{BS}^2 + AS\sigma_B^2$$

EMS_S:

$$\begin{aligned}
\text{EMS}_S &= \text{ESS}_S/\text{df}_S \\
&= (\text{ESS}_{as} - \text{ESS}_a)/A(S-1) \\
k_E &= (AS - A)/\text{df}_S = A(S-1)/A(S-1) = 1 \\
k_{BS} &= k_{AB} = 0 \\
k_S &= (ABS - AB)/\text{df}_S = AB(S-1)/A(S-1) = B \\
k_B &= k_A = 0
\end{aligned}$$

Consequently,

$$\text{EMS}_S = \sigma_E^2 + B\sigma_S^2$$

EMS_{AB}:

$$\begin{aligned}
\text{EMS}_{AB} &= \text{ESS}_{AB}/\text{df}_{AB} \\
&= (\text{ESS}_{ab} - \text{ESS}_b - \text{ESS}_a + \text{ESS}_1)/(A-1)(B-1) \\
k_E &= (AB - B - A + 1)/\text{df}_{AB} \\
&= (A-1)(B-1)/(A-1)(B-1) = 1 \\
k_{BS} &= [A(B-1) - (B-1)]/\text{df}_{AB} \\
&= (A-1)(B-1)/(A-1)(B-1) = 1 \\
k_{AB} &= (A-1)(B-1)S/\text{df}_{AB} = S \\
k_S &= k_B = k_A = 0
\end{aligned}$$

Therefore,

$$\text{EMS}_{AB} = \sigma_E^2 + \sigma_{BS}^2 + S\sigma_{AB}^2$$

EMS_{BS}:

$$\begin{aligned}
\text{EMS}_{BS} &= \text{ESS}_{BS}/\text{df}_{BS} \\
&= (\text{ESS}_{abs} - \text{ESS}_{ab} - \text{ESS}_{as} + \text{ESS}_a)/A(B-1)(S-1) \\
k_E &= (ABS - AB - AS + A)/\text{df}_{BS} \\
&= A(B-1)(S-1)/A(B-1)(S-1) = 1 \\
k_{BS} &= [A(B-1)S - A(B-1)]/\text{df}_{BS} \\
&= A(B-1)(S-1)/A(B-1)(S-1) = 1 \\
k_{AB} &= k_S = k_B = k_A = 0
\end{aligned}$$

Thus,

$$\text{EMS}_{BS} = \sigma_E^2 + \sigma_{BS}^2$$

Table 6.5-5 summarizes the EMS's for $S(A) \times B$.

CHAPTER SEVEN

Latin Square Designs

7.1 INTRODUCTION

Although the class of designs that can be symbolized by the crossing and nesting relations is used most often in research, an arrangement of factor levels is used fairly often that cannot be so symbolized. This is the Latin square design, the topic of this chapter. Single measurement Latin square designs are discussed in Sections 7.2 to 7.4. Repeated measurements Latin square designs are discussed in Sections 7.5 to 7.7.

7.2 THE BASIC SINGLE MEASUREMENT LATIN SQUARE DESIGN

7.2.1 The Layout

The basic Latin square design has three fixed factors, each with an equal number of levels. These three factors are symbolized by A, B, and C, and the common number of levels is denoted by L. This section considers only the single measurement version, which is a completely randomized design. Unlike the repeated measurements version, the single measurement version is seldom used in psychological research because the assumptions required for a valid analysis are normally very dubious. However, an examination of this version in some detail is worthwhile as an introduction to variations of it that have proved to be more useful.

As with design $A \times B \times C$, we can say of the Latin square design that $A \times B$, $B \times C$, and $A \times C$. But these three relations do not assure that

$A \times B \times C$, and we cannot assert of the Latin square design that $A \times B \times C$. Although each level a_a occurs with each level b_b and with each level c_c, each level of a_a does not occur with each combination bc_{bc}, which is a requirement if we assert that $A \times B \times C$. A similar statement holds for level b_b relative to levels a_a and c_c, and for level c_c relative to levels a_a and b_b. Instead, each a_a occurs with only L of the L^2 possible bc_{bc} combinations, and similarly for each level of b_b and c_c. In all, whereas design $A \times B \times C$ with L levels per factor would have L^3 conditions, the Latin square design with A, B, and C has only L^2 conditions, or $1/L$ as many. Hence, assuming that E subjects are to be assigned to each cell for either design, the Latin square design requires only $1/L$ as many subjects. This is its most appealing feature.

The layout for the basic single measurement Latin square design is usually diagrammed as in Table 7.2-1 (top). The diagram shows a Latin square based on factors A, B, and C, each of which has $L = 4$ levels. A Latin square has L^2 conditions—in this example, 16. A particular condition is indicated by the level shown within the table (for factor A) and along its row and column. The 16 conditions are explicitly enumerated in the bottom part of Table 7.2-1. A particular one of these 16 conditions is the same as some condition of $A \times B \times C$, assuming that the same factors and levels are used. In other words, abc_{243} would be the same for a subject exposed to it, and presumably his score would be the same, regardless of whether the complete design were the Latin square or $A \times B \times C$. The only difference is that the Latin square uses only a fraction, $1/L$, of the conditions present in $A \times B \times C$.

In the Latin square, a particular level of each factor occurs in conjunction exactly once with each level of both of the other two factors. For example, in

Table 7.2-1 The basic Latin square design

	b_1	b_2	b_3	b_4
c_1	a_1	a_2	a_3	a_4
c_2	a_2	a_3	a_4	a_1
c_3	a_3	a_4	a_1	a_2
c_4	a_4	a_1	a_2	a_3

	b_1	b_2	b_3	b_4
c_1	abc_{111}	abc_{221}	abc_{331}	abc_{441}
c_2	abc_{212}	abc_{322}	abc_{432}	abc_{142}
c_3	abc_{313}	abc_{423}	abc_{133}	abc_{243}
c_4	abc_{414}	abc_{124}	abc_{234}	abc_{344}

Table 7.2-1, a_3 appears exactly once with each b_b and with each c_c. Likewise, b_2 appears exactly once with each a_a and c_c, and c_1 appears exactly once with each a_a and b_b. Given this restriction, there are many ways that the levels of A, B, and C can be combined. For the 4×4 Latin square, there are 576 arrangements. One of these should be chosen at random for use in an experiment.

7.2.2 Choosing a Random Square

A Latin square that is diagrammed as in the top part of Table 7.2-1 is called a *standard square* if the first row and first column are arranged in the regular, ascending order. The square of Table 7.2-1 is a standard square because both the first row and the first column are in the order a_1, a_2, a_3, a_4. If L equals 2 or 3, there is only one standard square. For $L = 4$, there are 4 standard squares, for $L = 5$ there are 56, and for $L = 6$ there are 9408.

In principle, to choose a random Latin square we proceed as follows:

1. Choose one of the standard squares at random.
2. Randomly permute the columns.
3. Randomly permute the rows.
4. Randomly permute the symbols in the body of the diagram.

The procedure can be illustrated with an example. Table 7.2-2 shows the 4 standard squares for $L = 4$. The row and column labels have been omitted; they would be the same as in Table 7.2-1. We choose one of these four squares at random (for example, the first square shown in Table 7.2-2, which is the same as the top square in Table 7.2-1). We must permute first the columns, then the rows, and then the symbols themselves. For each permutation we require a list of 4 random numbers, 1 to 4. Such lists can be obtained from a table of random digits such as Table A.1. In practice, it is adequate to enter the table at some haphazard location.

Table 7.2-2 The four standard Latin squares for $L = 4$

a_1	a_2	a_3	a_4	a_1	a_2	a_3	a_4
a_2	a_3	a_4	a_1	a_2	a_1	a_4	a_3
a_3	a_4	a_1	a_2	a_3	a_4	a_2	a_1
a_4	a_1	a_2	a_3	a_4	a_3	a_1	a_2
a_1	a_2	a_3	a_4	a_1	a_2	a_3	a_4
a_2	a_4	a_1	a_3	a_2	a_1	a_4	a_3
a_3	a_1	a_4	a_2	a_3	a_4	a_1	a_2
a_4	a_3	a_2	a_1	a_4	a_3	a_2	a_1

Suppose we have the following blocks of random numbers:

83035	92350
97662	24822
88824	71013
12544	41035
22716	19792

Random numbers are printed in blocks (in this example, blocks of size 5 × 5) rather than as continuous digits, for ease of use. In proceeding through the numbers, it is easier to proceed block by block instead of along a straight line. Starting at the beginning of the first block shown, we list the numbers 1 to 4 in the order in which they appear, not repeating a number already listed. The numbers we derive are (3, 2, 4, 1), shown in italics in the display. We then proceed to find another set (2, 4, 1, 3), and the third set is also (2, 4, 1, 3). Such random sets could also be found by drawing poker chips marked 1 to 4 from a bag.

Using the set (3, 2, 4, 1), we rearrange the columns of the a_a's in the top part of Table 7.2-1 by putting column 3 first, column 2 second, column 4 third, and column 1 last, which gives us

$$\begin{array}{cccc} a_3 & a_2 & a_4 & a_1 \\ a_4 & a_3 & a_1 & a_2 \\ a_1 & a_4 & a_2 & a_3 \\ a_2 & a_1 & a_3 & a_4 \end{array}$$

Next, we rearrange the rows of this second square according to the random numbers (2, 4, 1, 3), putting row 2 on top, row 4 next, and so on, which gives us

$$\begin{array}{cccc} a_4 & a_3 & a_1 & a_2 \\ a_2 & a_1 & a_3 & a_4 \\ a_3 & a_2 & a_4 & a_1 \\ a_1 & a_4 & a_2 & a_3 \end{array}$$

Finally, we resubscript the symbols according to the random numbers (2, 4, 1, 3); that is, we resubscript a_1 as a_2, a_2 as a_4, a_3 as a_1, and a_4 as a_3, yielding

$$\begin{array}{cccc} a_3 & a_1 & a_2 & a_4 \\ a_4 & a_2 & a_1 & a_3 \\ a_1 & a_4 & a_3 & a_2 \\ a_2 & a_3 & a_4 & a_1 \end{array}$$

By implication, the columns of the preceding squares are headed b_1, b_2, b_3, and b_4, and the rows are labeled c_1, c_2, c_3, c_4, as in Table 7.2-1. There is no need to rearrange the listing of the b_b's and c_c's, since the preceding randomization of the a_a's of the square suffices to give us a random selection of a Latin square from all the possible squares of size L^2.

The theory of hypothesis testing for effects in the Latin square assumes that, for each replication of the experiment, the particular square to be used is randomly chosen anew. The major difficulty with this procedure is choosing one standard square at random when $L > 4$. Fisher and Yates[1] give all standard squares for L of 4, 5, and 6, and sample squares for L of up to 12. Thus a standard square can be chosen at random from this source. The occasional user of a Latin square design, however, can hardly be faulted if, when $L > 4$, he begins with any standard square and then randomizes the columns, rows, and symbols. It is easy to construct a standard square for any size L. Simply write the first column and row in order, as the definition of a standard square requires. Then fill in each row beginning at the initial symbol, and continuing in order. When a_L is reached, the next symbol is a_1, and the symbols continue in order. When $L = 6$, for example, it is easy to construct the following standard square:

$$\begin{matrix} a_1 & a_2 & a_3 & a_4 & a_5 & a_6 \\ a_2 & a_3 & a_4 & a_5 & a_6 & a_1 \\ a_3 & a_4 & a_5 & a_6 & a_1 & a_2 \\ a_4 & a_5 & a_6 & a_1 & a_2 & a_3 \\ a_5 & a_6 & a_1 & a_2 & a_3 & a_4 \\ a_6 & a_1 & a_2 & a_3 & a_4 & a_5 \end{matrix}$$

It is not as easy to write out a standard square haphazardly, but it can be done. Start out with the standard first row and column and fill in each cell with a symbol that is, at present, missing from the row and column. Unfortunately, a point may be reached where there is no such symbol, so a revision of past entries may be required. The following square for $L = 6$ was constructed in this way.

$$\begin{matrix} a_1 & a_2 & a_3 & a_4 & a_5 & a_6 \\ a_2 & a_6 & a_4 & a_1 & a_3 & a_5 \\ a_3 & a_1 & a_6 & a_5 & a_2 & a_4 \\ a_4 & a_5 & a_2 & a_3 & a_6 & a_1 \\ a_5 & a_3 & a_1 & a_6 & a_4 & a_2 \\ a_6 & a_4 & a_5 & a_2 & a_1 & a_3 \end{matrix}$$

We would permute the columns, rows, and symbols of this standard square before using it.

[1] R. A. Fisher and F. Yates, *Statistical Tables for Biological, Agricultural and Medical Research* (Edinburgh: Oliver & Boyd, 1955).

7.2.3 The Score Model

The score model for the basic single measurement Latin square design is patterned after that for $A \times B \times C$. The only difference is that no separate interaction terms are included. In fact, valid significance testing requires that all interaction terms be zero. The score model may be written

$$X_{abce} = \mathbf{m} + \mathbf{a}_a + \mathbf{b}_b + \mathbf{c}_c + \mathbf{res}_{abc} + \mathbf{e}_{e(abc)}$$

Term $\mathbf{res}_{abc}$ represents the sum of all interaction terms from $A \times B \times C$; it is a specific number for a cell, but it varies among the cells. Actually, $\mathbf{res}_{abc}$ must equal zero in order to have valid significance tests for main effects. However, if $L > 2$, we can estimate $\mathbf{res}_{abc}$, calculate an SS_{res}, and test an H_N that $\mathbf{res}_{abc} = 0$, so there is some basis for carrying $\mathbf{res}_{abc}$ in the model.

7.2.4 Estimation of Model Terms

As with the completely crossed designs, a main effect is estimated as the difference between the mean of all scores having a main effect parameter as a component and the mean of all scores in the experiment:

$$\hat{\mathbf{a}}_a = \bar{X}_{a...} - \bar{X}_{....}$$

$$\hat{\mathbf{b}}_b = \bar{X}_{.b..} - \bar{X}_{....}$$

$$\hat{\mathbf{c}}_c = \bar{X}_{..c.} - \bar{X}_{....}$$

With the Latin square design, both dot and $\sum$-notation require special interpretation. Consider $\bar{X}_{....}$ Previously, this notation would have meant $(1/EL^3)\sum_{abce} X_{abce}$. However, with respect to a Latin square, there are only $E \cdot L^2$ scores over which we can sum, and hence the proper divisor for computing $\bar{X}_{....}$ is $E \cdot L^2$. Similarly, $\bar{X}_{a...}$ means the sum of all scores in the design at level a_a divided by $E \cdot L$, not by $E \cdot L^2$.

To determine the theoretical components of $\hat{\mathbf{a}}_1$, say, we need to find $\bar{X}_{1...}$ and $\bar{X}_{....}$ in terms of the score model. More specifically, assume that we have data from the following 4×4 Latin square:

	b_1	b_2	b_3	b_4
c_1	a_3	a_1	a_2	a_4
c_2	a_1	a_4	a_3	a_2
c_3	a_2	a_3	a_4	a_1
c_4	a_4	a_2	a_1	a_3

To find $\bar{X}_{1...}$ we find the mean of each term in the score model for X_{abce} for scores at level a_1. Hence,

$$\bar{X}_{1...} = \mathbf{m} + \mathbf{a}_1 + \bar{\mathbf{b}}_. + \bar{\mathbf{c}}_. + \overline{\mathbf{res}}_{1..} + \bar{\mathbf{e}}_{.(1..)}$$

Because B and C are fixed, $\bar{\mathbf{b}}. = \bar{\mathbf{c}}. = 0$, so

$$\bar{X}_{1...} = \mathbf{m} + \mathbf{a}_1 + \overline{\mathbf{res}}_{1..} + \bar{\mathbf{e}}_{.(1..)}$$

Let us assume for now that the triple interaction term of $A \times B \times C$ is zero, so that we can analyze the two-factor composition of $\overline{\mathbf{res}}_{1...}$ The four cells with a_1, those with the scores over which we are averaging, are abc_{112}, abc_{121}, abc_{134}, and abc_{143}. The two-factor interaction parameters in the score model for scores in each of these cells are:

	$\mathbf{ab}_{ab}$	$\mathbf{ac}_{ac}$	$\mathbf{bc}_{bc}$
abc_{112}	$\mathbf{ab}_{11}$	$\mathbf{ac}_{12}$	$\mathbf{bc}_{12}$
abc_{121}	$\mathbf{ab}_{12}$	$\mathbf{ac}_{11}$	$\mathbf{bc}_{21}$
abc_{134}	$\mathbf{ab}_{13}$	$\mathbf{ac}_{14}$	$\mathbf{bc}_{34}$
abc_{143}	$\mathbf{ab}_{14}$	$\mathbf{ac}_{13}$	$\mathbf{bc}_{43}$

In summing across scores at a_1, each interaction parameter tabled would occur E times. The sum of the $\mathbf{ab}_{ab}$'s would be $E\mathbf{ab}_{1.}$, which is zero because of the side condition for two-factor interaction when B is fixed. The sum of the $\mathbf{ac}_{ac}$'s would be $E\mathbf{ac}_{1.}$, which is likewise zero. However, no side condition requires that the sum of the $\mathbf{bc}_{bc}$'s tabled must equal zero, since it is not a sum over one subscript with the other subscript constant. If, then, $\mathbf{abc}_{abc}$ were zero, $\overline{\mathbf{res}}_{1..}$ would consist only of $(1/L)$ $(\mathbf{bc}_{12} + \mathbf{bc}_{21} + \mathbf{bc}_{34} + \mathbf{bc}_{43})$; the other two-factor interaction terms would be absent.

Now consider $\bar{X}_{....}$, the mean over all $E \cdot L^2$ scores:

$$\bar{X}_{....} = \mathbf{m} + \overline{\mathbf{res}}_{...} + \bar{\mathbf{e}}_{.(...)}$$

Here $\overline{\mathbf{res}}_{...}$ consists only of three-factor interaction parameters, and if they equal zero, $\overline{\mathbf{res}}_{...} = 0$. Although the $\mathbf{bc}_{bc}$'s do not sum to zero, in general, within a given level a_a, the 16 parameters across all levels of a_a do sum to zero because of the side conditions.

Since $\hat{\mathbf{a}}_1 = \bar{X}_{1...} - \bar{X}_{....}$, we have

$$\hat{\mathbf{a}}_1 = \mathbf{a}_1 + \overline{\mathbf{res}}_{1..} - \overline{\mathbf{res}}_{...} + \bar{\mathbf{e}}_{.(1..)} - \bar{\mathbf{e}}_{.(...)}$$

Recall that for designs A, $A \times B$, or $A \times B \times C$ with fixed factors, an estimate of $\mathbf{a}_a$ deviates from its true value only as a result of error parameters. Here there is a discrepancy as a result of $\mathbf{res}_{abc}$ also. If the three-factor interaction is zero, then the discrepancy due to $\mathbf{res}_{abc}$ is simply $(1/L)$ $(\mathbf{bc}_{12} + \mathbf{bc}_{21} + \mathbf{bc}_{34} + \mathbf{bc}_{43})$. The particular $\mathbf{bc}_{bc}$ parameters averaged would depend on the randomized square used, and hence would vary from replication to replication. For no randomization, however, would the mean equal zero because of a side condition. For each $\hat{\mathbf{a}}_a$, a different set of $\mathbf{bc}_{bc}$'s would be averaged.

If $\mathbf{bc}_{bc}$ is not zero, then SS_A and EMS_A will be inflated and invalidate the test of H_N that $\sigma_A^2 = 0$. However, if $\mathbf{ab}_{ab}$ and $\mathbf{ac}_{ac}$ exist, they will in no way invalidate the test that $\sigma_A^2 = 0$. If $\mathbf{ab}_{ab}$ is not zero, its parameters will enter into the estimate $\hat{\mathbf{c}}_c$ and invalidate the test of $\sigma_C^2 = 0$. If $\mathbf{ac}_{ac}$ is not zero, its parameters will enter into the estimate $\hat{\mathbf{b}}_b$ and invalidate the test of $\sigma_B^2 = 0$. In short, if any two-factor interaction is not zero, it will only invalidate the significance test for the main effect of the third factor. If three-factor interaction occurs, however, it will give a positive bias to the F-ratios for all three main terms.

The residual term can be estimated as the discrepancy between the mean score for a cell and the value for that cell estimated by the main effects model; that is,

$$\widehat{\mathbf{res}}_{abc} = \bar{X}_{abc.} - (\hat{\mathbf{m}} + \hat{\mathbf{a}}_a + \hat{\mathbf{b}}_b + \hat{\mathbf{c}}_c),$$

or, in terms of $\bar{X}$ quantities,

$$\widehat{\mathbf{res}}_{abc} = \bar{X}_{abc.} - \bar{X}_{a...} - \bar{X}_{.b..} - \bar{X}_{..c.} + 2\bar{X}_{....}$$

An error term is estimated as the difference between an individual score and the cell mean, as is usual where there is more than one score per condition:

$$\hat{\mathbf{e}}_{e(abc)} = X_{abce} - \bar{X}_{abc.}$$

The residual term is the same for all scores within the same condition, so differences among such scores are due entirely to the error terms. Regardless of the presence or absence of a residual term, we can get an unbiased estimate of error variance.

7.2.5 Sums of Squares

The rules given in Chapter 5 for developing SS formulas do not apply to the Latin square design, although there are similarities. We can describe the breakdown of SS's in the Latin square design as follows:

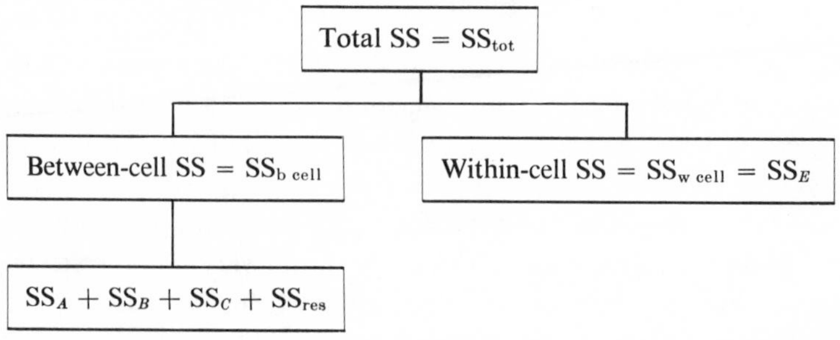

In equation form,

$$SS_{tot} = SS_{b\ cell} + SS_{w\ cell} = SS_A + SS_B + SS_C + SS_{res} + SS_E$$

We have, as previously, an SS for each term of the model except **m**.

Chapter 5 showed how to calculate an SS element for each model term, and how to combine these elements to form the SS's. Likewise, for the Latin square there is an SS element for each model term, including **m** and $\mathbf{res}_{abc}$, which are summed to provide the SS's. The elements are defined in much the same way as previously, except that summations are only over such subscript combinations as occur, and the divisor has L in place of L^2, and L^2 in place of L^3. Table 7.2-3 shows the SS elements formulas and the table for combining the elements. The quantity found by the SS elements formula is entered in the appropriate row wherever there is a plus or minus sign. To find the SS's proper we sum the columns. The double plus sign for SS_1 under SS_{res} means that SS_1 must be added twice in finding SS_{res}. The rules given in Chapter 5 for combining SS elements into SS's based on the ESP's do not work for the Latin square design. Table 7.2-3 should be used as given. As before, we can check the addition within the table by adding the SS's for the model terms to

Table 7.2-3 SS elements formulas and table

Model term	SS element
$\mathbf{e}_{e(abc)}$	$SS_{abce} = \sum X^2_{abcd}$
$\mathbf{res}_{abc}$	$SS_{abc} = (\sum X^2_{abc.})/E$
$\mathbf{c}_c$	$SS_c = (\sum X^2_{..c.})/EL$
$\mathbf{b}_b$	$SS_b = (\sum X^2_{.b..})/EL$
$\mathbf{a}_a$	$SS_a = (\sum X^2_{a...})/EL$
$\mathbf{m}$	$SS_1 = (X^2_{....})/EL^2$

	SS_A	SS_B	SS_C	SS_{res}	SS_E	SS_{tot}
SS_{abce}					+	+
SS_{abc}				+	−	
SS_c			+	−		
SS_b		+		−		
SS_a	+			−		
SS_1	−	−	−	++		−
Sum						

see that they sum to SS_{tot}. As before, however, this check does not protect against making errors in computing the SS elements except insofar as such errors lead to negative SS's. Since each SS must be positive, any negative outcome indicates that an error was made.

7.2.6 Expected Mean Squares and Hypothesis Testing

To calculate the mean squares, we divide the SS's by the appropriate df's. The df's for Latin square designs cannot be obtained by following the rules given in Section 6.2.1. The df's and the EMS's are as given in Table 7.2-4. The term σ^2_{res} is the component of EMS_{res} due to the two-factor and three-factor interactions. Each two-factor interaction contributes to σ^2_{res} unless it equals zero. By $\sigma^2_{res\ BC}$ we mean the component of EMS_A due to the BC and ABC interactions. It is free of contribution from the AB and AC interactions, even if they are not zero. Analogous statements apply to $\sigma^2_{res\ AC}$ and $\sigma^2_{res\ AB}$. If $\sigma^2_{res} = 0$, then $\sigma^2_{res\ BC}$, $\sigma^2_{res\ AC}$, and $\sigma^2_{res\ AB}$ also equal zero. Then, and only then, is there a valid test for the H_N's that $\sigma^2_A = 0$, $\sigma^2_B = 0$, and $\sigma^2_C = 0$. If $\sigma^2_{res} = 0$, we can pool SS_{res} and SS_E to get a pooled MS_E for testing the three main effects.

Sometimes textbook writers take σ^2_{res} to be zero from the start, because of previous evidence, or by a priori reasoning, or simply because of necessity. Then the design model, SS's, and EMS's may lack any reference to a residual term, and the SS_E and df_E described in their books are already pooled. Alternatively, an experimenter can follow the method given here and do preliminary testing of H_N that $\sigma^2_{res} = 0$. If σ^2_{res} tests greater than zero, however, the other tests are not valid. If σ^2_{res} tests greater than zero when the original scores are used, it might test zero when transformed scores are used (Section 9.3), so an analysis could proceed with the transformed scores. However, there are great risks both in assuming σ^2_{res} to be zero and in proceeding in the hope that it will test out zero. Therefore, the basic single measurement Latin square design is seldom used in psychological research.

Note from Table 7.2-4 that if $L = 2$, the df for res is zero. In such a situation, no test for σ^2_{res} is possible. Only for squares of size 3×3 or larger

Table 7.2-4 EMS's and df's for the Latin square design

Source	df	EMS
A	$L - 1$	$\sigma^2_E + \sigma^2_{res\ BC} + EL\sigma^2_A$
B	$L - 1$	$\sigma^2_E + \sigma^2_{res\ AC} + EL\sigma^2_B$
C	$L - 1$	$\sigma^2_E + \sigma^2_{res\ AB} + EL\sigma^2_C$
res	$(L - 1)(L - 2)$	$\sigma^2_E + \sigma^2_{res}$
E	$(E - 1)L^2$	σ^2_E

is a test of H_N that $\sigma^2_{\text{res}} = 0$ possible. In a 2×2 square, there are four cell means, so a total of four df's can be used to describe these means. The overall mean uses one, and each of the main effects uses one, so there are no df left. In other words, $\hat{\mathbf{m}} + \hat{\mathbf{a}}_a + \hat{\mathbf{b}}_b + \hat{\mathbf{c}}_c$ must exactly equal the cell mean for a 2×2 Latin square design. There can be no discrepancy between this estimate and the cell mean with which to estimate $\mathbf{res}_{abc}$. For 3×3 squares there are nine cell means but only $1 + 3(L - 1) = 7$ df are used in estimating the overall mean and the main effects, leaving 2 df for the residual. In general, the df for the residual term is $L^2 - (1 + 3(L - 1)) = L^2 - 1 - 3L + 3 = L^2 - 3L + 2 = (L - 1)(L - 2)$, which is the formula given in Table 7.2-4.

7.3 GRECO-LATIN SQUARE DESIGN

This section and the remaining sections of this chapter show how the Latin square can be used in the construction of more complicated designs, beginning with the basic *Greco-Latin square design.* This design and the one described in Section 7.4 are single measurement designs. Repeated measurements versions of the Latin square design are considered in Sections 7.5 to 7.7.

Two Latin squares of the same size are *orthogonal* if, when superimposed, no combination of levels appears more than once. The two squares would be based on the levels of different factors. For example, the following two squares, which are based on the levels of factors A and D, are orthogonal:

$$
\begin{matrix} a_1 & a_2 & a_3 & a_4 \\ a_2 & a_1 & a_4 & a_3 \\ a_3 & a_4 & a_1 & a_2 \\ a_4 & a_3 & a_2 & a_1 \end{matrix}
\qquad\qquad
\begin{matrix} d_1 & d_2 & d_3 & d_4 \\ d_3 & d_4 & d_1 & d_2 \\ d_4 & d_3 & d_2 & d_1 \\ d_2 & d_1 & d_4 & d_3 \end{matrix}
$$

Superimposed, two orthogonal squares can be combined with factors B and C, to form the four-factor Greco-Latin square design. The preceding squares give us the following layout for the design:

	b_1	b_2	b_3	b_4
c_1	ad_{11}	ad_{22}	ad_{33}	ad_{44}
c_2	ad_{23}	ad_{14}	ad_{41}	ad_{32}
c_3	ad_{34}	ad_{43}	ad_{12}	ad_{21}
c_4	ad_{42}	ad_{31}	ad_{24}	ad_{13}

No combination ad_{ad} repeats itself in the design. It is not possible to construct Greco-Latin squares of all dimensions. For example, there are no such squares of size 6×6 or 10×10.

If all interactions among the four factors of a Greco-Latin square are zero, we can evaluate the main effects very economically, because only L^2

conditions are required, rather than the L^4 required by the completely crossed design. For example, if $L = 4$, we need only 16 conditions instead of 256 conditions—a tremendous saving in subject-hours. The score model and analysis of variance for a Greco-Latin square are similar to those for the Latin square, except that four fixed factors are used instead of three. The Greco-Latin square design suffers from the requirements that factors have the same number of levels and that all interactions must equal zero, so it is seldom used. Various designs can be constructed using the basic Greco-Latin square as a component, but they are not considered in this book.

7.4 CROSSING THE LATIN SQUARE DESIGN

Another way of extending the Latin square design is to cross the Latin square with fixed factor D. In other words, the Latin square is repeated D times; all conditions for a particular square include that square's characteristic level d_d. This single measurement design can be diagrammed as follows:

		b_1	b_2	b_3
d_1	c_1	a_3	a_1	a_2
	c_2	a_1	a_2	a_3
	c_3	a_2	a_3	a_1

		b_1	b_2	b_3
d_2	c_1	a_3	a_1	a_2
	c_2	a_1	a_2	a_3
	c_3	a_2	a_3	a_1

Like the Greco-Latin square, this design extends the Latin square from three factors to four. Unlike the Greco-Latin square, it increases the number of conditions D times, and hence increases the number of subjects required D times (for a constant number of subjects per condition). In compensation, however, it allows assessment not only of the main effects of factor D but of the two-factor interactions of D with each of A, B, and C. These interactions, then, need not equal zero in order that tests of the main effects be valid. All other interactions must be zero, however, if the tests of the main effects and two-factor interactions with **d** are to be valid.

The score model is

$$X_{abcde} = \mathbf{m} + \mathbf{a}_a + \mathbf{b}_b + \mathbf{c}_c + \mathbf{d}_d + \mathbf{ad}_{ad} + \mathbf{bd}_{bd} + \mathbf{cd}_{cd} + \mathbf{res}_{abcd} + \mathbf{e}_{e(abcd)}$$

The EMS's and df's are given in Table 7.4-1. The $\mathbf{res}_{abcd}$ components of sources A to CD have been omitted from the EMS formulas, though such components will be present if $\mathbf{res}_{abcd}$ does not equal zero. If $\mathbf{res}_{abcd}$ does not equal zero, these sources have no valid test, since an appropriate MS_2 is not available. A preliminary test is possible for the hypothesis that $\mathbf{res}_{abcd}$ is zero, using MS_{res} as MS_1 and MS_E as MS_2. If the hypothesis is confirmed, SS_{res} may be pooled with MS_E for the final testing. If the hypothesis is rejected, a

Table 7.4-1 EMS's and df's for the crossed Latin square design

Source	df	EMS
A	$L-1$	$\sigma_E^2 + DEL\sigma_A^2$
B	$L-1$	$\sigma_E^2 + DEL\sigma_B^2$
C	$L-1$	$\sigma_E^2 + DEL\sigma_C^2$
D	$D-1$	$\sigma_E^2 + EL^2\sigma_D^2$
AD	$(D-1)(L-1)$	$\sigma_E^2 + EL\sigma_{AD}^2$
BD	$(D-1)(L-1)$	$\sigma_E^2 + EL\sigma_{BD}^2$
CD	$(D-1)(L-1)$	$\sigma_E^2 + EL\sigma_{CD}^2$
res	$D(L-1)(L-2)$	$\sigma_E^2 + E\sigma_{\text{res}}^2$
E	$DL^2(E-1)$	σ_E^2

transformation of the scores (Section 9.3) might yield new scores for which the residual term would be zero.

Table 7.4-2 gives the formulas for the SS elements and shows how these elements are combined into SS's.

7.5 THE BASIC REPEATED MEASUREMENTS LATIN SQUARE DESIGN

7.5.1 The Layout

As we have noted, the completely randomized Latin square design of Section 7.2 receives little use. However, the repeated measurements version of the basic Latin square design that will now be discussed requires less stringent assumptions and therefore has more use. The greater practicability of the repeated measurements design comes at a price: One of the three Latin square factors cannot be used as such.

The 3×3 version of the design is diagrammed in Table 7.5-1. The arrangement of factors A, B, and C forms a Latin square, and the particular arrangement of the a_a's within the body of the table is determined randomly for each replication, as before. Some number S of subjects is randomly chosen from a single population of subjects and assigned to each level of C. The diagram shows three different subjects at each level of C, and literal labeling is used to emphasize that different subjects appear at each level of C. Although the diagram shows $S = 3$ subjects per group, any number could be used—it need not equal L.

All subjects in a group are exposed to the same L cells of the design. For example, subjects s_4, s_5, and s_6 all receive ab_{13}, ab_{22}, and ab_{31}. The diagram does not imply that the subjects in a group receive their treatments in the same

Table 7.4-2 SS elements formulas and table for the crossed Latin square design

Model term	SS element
$\mathbf{e}_{e(abcd)}$	$SS_{abcde} = \sum X^2_{abcde}$
$\mathbf{res}_{abcd}$	$SS_{res} = (\sum X^2_{abcd.})/E$
$\mathbf{cd}_{cd}$	$SS_{cd} = (\sum X^2_{..cd.})/EL$
$\mathbf{bd}_{bd}$	$SS_{bd} = (\sum X^2_{.b.d.})/EL$
$\mathbf{ad}_{ad}$	$SS_{ad} = (\sum X^2_{a..d.})/EL$
$\mathbf{d}_{d}$	$SS_{d} = (\sum X^2_{...d.})/EL^2$
$\mathbf{c}_{c}$	$SS_{c} = (\sum X^2_{..c..})/DEL$
$\mathbf{b}_{b}$	$SS_{b} = (\sum X^2_{.b...})/DEL$
$\mathbf{a}_{a}$	$SS_{a} = (\sum X^2_{a....})/DEL$
$\mathbf{m}$	$SS_{1} = X^2_{.....}/DEL^2$

	SS_A	SS_B	SS_C	SS_D	SS_{AD}	SS_{BD}	SS_{CD}	SS_{res}	SS_E	SS_{tot}
SS_{abcde}									+	+
SS_{res}								+	−	
SS_{cd}							+	−		
SS_{bd}						+		−		
SS_{ad}					+			−		
SS_{d}				+	−	−	−	++		
SS_{c}			+				−			
SS_{b}		+				−				
SS_{a}	+				−					
SS_{1}	−	−	−	−	+	+	+			−
Sum										

Table 7.5-1 Layout for the basic repeated measurements Latin square design

		b_1	b_2	b_3
c_1	s_1 s_2 s_3	a_2	a_1	a_3
c_2	s_4 s_5 s_6	a_3	a_2	a_1
c_3	s_7 s_8 s_9	a_1	a_3	a_2

order. The order would normally be individually randomized for each subject in the experiment, unless B were a trial factor, in which case all subjects in a group would have the same sequence of treatments.

Factor C in this design is a *dummy factor;* that is, the c_c's do not refer to different situations, stimulation, or categories of subjects; they are simply labels for the groups of subjects. Factor C, then, is not available as an experimental factor; only A and B, which we assume to be fixed, are available. In this respect, then, the design is an alternative to other designs we have considered with a basic purpose of investigating factors A and B. Such designs include the completely randomized design $A \times B$ and the repeated measurements designs $S(A) \times B$ (or $S(B) \times A$) and $S \times A \times B$. All these designs include among their conditions all possible ab_{ab}'s an equal number of times. Likewise, our Latin square design includes all possible ab_{ab}'s S times. The Latin square design is like $S(A) \times B$ in that each subject is exposed to more than one but less than all of the ab_{ab} combinations.

Why use one design rather than the other? The nested design is relatively insensitive for detecting differences among the levels of the nest factor, since the denominator EMS includes a component that is due to systematic individual differences, σ_S^2. The design is more sensitive for detecting differences in the crossed factor levels, since the denominator MS is free of σ_S^2. The Latin square design, on the other hand, has relatively sensitive tests for both A and B. If an experimenter has approximately equal interest in detecting the effects of A and B, but cannot expose or feels it unwise to expose each subject to all combinations of ab_{ab}, then the basic repeated measurements Latin square design is worth considering. The Latin square design, however, does require somewhat stronger assumptions than the nested design, and in this sense is less desirable. Where C is a dummy factor, however, the assumptions required are considerably milder than those required by the completely randomized Latin square design.

When factor B is a trial factor, the repeated measurements Latin square design under consideration is basically an alternative to $S \times A$. The main purpose of factor B is to balance out the effect of order of presentation, but factor B can also be tested to see if scores change systematically with the order of presentation. Use of the Latin square for such purposes is considered further in Section 8.1.

7.5.2 The Score Model

The score model is

$$X_{abcs} = \mathbf{m} + \mathbf{a}_a + \mathbf{b}_b + \mathbf{c}_c + \mathbf{s}_{s(c)} + \mathbf{ab}'_{ab} + \mathbf{e}_{abcs}$$

We have no separate subscript e, since there is only one score per condition, and the other subscripts suffice to specify a particular score and its error parameter. In this model, however, error is not confounded with any other model term—it has a separate estimate and MS, thanks to certain assumptions we make thus eliminating some other conceivable terms.

In contrast to the model for the completely randomized design, this model has a two-factor interaction term $\mathbf{ab}'_{ab}$ and, of course, a term $\mathbf{s}_{s(c)}$ to account for systematic differences among subjects, but it has no residual term. The inclusion of an $\mathbf{ab}'_{ab}$ term here implies that we can test for the presence of AB interaction, which we could not do with the completely randomized design. The prime in the term $\mathbf{ab}'_{ab}$ is included because the term is defined somewhat differently from $\mathbf{ab}_{ab}$. It is a within-subject component of $\mathbf{ab}_{ab}$ and is discussed more fully in the next subsection.

A proper analysis of the main effects of the single measurement Latin square design requires that we assume or demonstrate that all interactions are zero. The differences for the present design depend largely on the difference in factor C. If all levels of factor C are the same, then any conceivable model terms having $\mathbf{c}$ must equal zero. We cannot attribute differences in scores to differences in c_c levels if all those levels are the same. Hence, we can assume that $\mathbf{c}_c$, $\mathbf{ac}_{ac}$, $\mathbf{bc}_{bc}$, and $\mathbf{abc}_{abc}$ of the present design are all zero. The $\mathbf{res}_{abc}$ term of the completely randomized model consisted of a confounding of all the interaction terms between A, B, and C. Of those terms only $\mathbf{ab}_{ab}$ remains in the repeated measurements design, so whereas before we had a residual estimate and a test of $\sigma^2_{\text{res}} = 0$, here we have an $\mathbf{ab}_{ab}$ estimate and a test of $\sigma^2_{AB} = 0$.

Our model includes a main term $\mathbf{c}_c$. Since we have determined that it is zero, why include it? Because it serves as a handle for certain calculations we wish to make. Specifically, we can still estimate $\mathbf{c}_c$ and calculate an SS_C, just as we can estimate $\mathbf{a}_a$ and calculate an SS_A, even though we may find that $\sigma_A^2 = 0$. Although SS_C is part of the breakdown of SS_{tot} for the design, it normally plays no role in hypothesis testing.

When we compare the present model with the score model for $S \times A \times B$, we see that interaction terms involving S with A and B are missing from the former. But is it not reasonable to suppose here, as we did for $S \times A \times B$, that, besides having different mean scores, the various subjects differ in their patterns of scores, depending on a_a, b_b, and ab_{ab}? In other words, can we not imagine that AS, BS, and ABS interactions can occur? The answer is yes, but the present design does not allow for separate consideration of such effects. We assume that they equal zero.

7.5.3 Estimation of Model Terms

This section presents the formulas for estimating the parameters. First the formula for calculating a parameter by the method of successive extraction is presented, followed by the formula based on $\bar{X}$'s and the formula giving the parametric components of an estimate.

$\hat{\mathbf{m}}$:

$$\begin{aligned}\hat{\mathbf{m}} &= \bar{X}_{....}\\ &= \frac{1}{L^2S}\sum_{abcs}(\mathbf{m} + \mathbf{a}_a + \mathbf{b}_b + \mathbf{c}_c + \mathbf{s}_{s(c)} + \mathbf{ab}_{ab} + \mathbf{e}_{abcs})\\ &= \mathbf{m} + \bar{\mathbf{s}}_{.(.)} + \bar{\mathbf{e}}_{....}\end{aligned}$$

$\hat{\mathbf{a}}_a$:

$$\begin{aligned}\hat{\mathbf{a}}_a &= \bar{X}_{a...} - \hat{\mathbf{m}}\\ &= \bar{X}_{a...} - \bar{X}_{....}\end{aligned}$$

$$\begin{aligned}\bar{X}_{a...} &= \frac{1}{LS}\sum_{bcs}(\mathbf{m} + \mathbf{a}_a + \mathbf{b}_b + \mathbf{c}_c + \mathbf{s}_{s(c)} + \mathbf{ab}_{ab} + \mathbf{e}_{abcs})\\ &= \mathbf{m} + \mathbf{a}_a + \bar{\mathbf{s}}_{.(.)} + \bar{\mathbf{e}}_{a...}\\ \hat{\mathbf{a}}_a &= \mathbf{a}_a + \bar{\mathbf{e}}_{a...} - \bar{\mathbf{e}}_{....}\end{aligned}$$

$\hat{\mathbf{b}}_b$:

The formulas for $\hat{\mathbf{b}}_b$ are similar to those for $\hat{\mathbf{a}}_a$; for example,

$$\hat{\mathbf{b}}_b = \mathbf{b}_b + \bar{\mathbf{e}}_{.b..} - \bar{\mathbf{e}}_{....}$$

$\hat{\mathbf{c}}_c$:

$$\begin{aligned}\hat{\mathbf{c}}_c &= \bar{X}_{..c.} - \hat{\mathbf{m}}\\ &= \bar{X}_{..c.} - \bar{X}_{....}\end{aligned}$$

$$\begin{aligned}\bar{X}_{..c.} &= \frac{1}{LS}\sum_{abs}(\mathbf{m} + \mathbf{a}_a + \mathbf{b}_b + \mathbf{c}_c + \mathbf{s}_{s(c)} + \mathbf{ab}_{ab} + \mathbf{e}_{absc})\\ &= \mathbf{m} + \mathbf{c}_c + \bar{\mathbf{s}}_{.(c)} + \overline{\mathbf{ab}}_{..(c_c)} + \bar{\mathbf{e}}_{..c.}\\ \hat{\mathbf{c}}_c &= \mathbf{c}_c + \bar{\mathbf{s}}_{.(c)} - \bar{\mathbf{s}}_{.(.)} + \overline{\mathbf{ab}}_{..(c_c)} + \bar{\mathbf{e}}_{..c.} - \bar{\mathbf{e}}_{....}\end{aligned}$$

$\hat{\mathbf{s}}_{s(c)}$:

$$\hat{\mathbf{s}}_{s(c)} = \bar{X}_{..cs} - (\hat{\mathbf{m}} + \hat{\mathbf{c}}_c)$$
$$= \bar{X}_{..cs} - \bar{X}_{..c.}$$
$$\bar{X}_{..cs} = \mathbf{m} + \mathbf{s}_{s(c)} + \overline{\mathbf{ab}}_{..(c_c)} + \bar{\mathbf{e}}_{..cs}$$

Using the previously found $\bar{X}_{..c.}$, we conclude that

$$\hat{\mathbf{s}}_{s(c)} = \mathbf{s}_{s(c)} - \bar{\mathbf{s}}_{.(c)} + \bar{\mathbf{e}}_{..cs} - \bar{\mathbf{e}}_{..c.}$$

$\widehat{\mathbf{ab}}_{ab}$:

$$\widehat{\mathbf{ab}}_{ab} = \bar{X}_{ab..} - (\hat{\mathbf{m}} + \hat{\mathbf{a}}_a + \hat{\mathbf{b}}_b)$$
$$= \bar{X}_{ab..} - \bar{X}_{.b..} - \bar{X}_{a...} + \bar{X}_{....}$$

For a specific pair of subscripts a and b, it is not possible to average over different values of subscript c, since a specific pair a and b occur only in conjunction with one value of c. Hence we can replace $\bar{X}_{ab..}$ with $\bar{X}_{abc..}$ These two forms mean the same thing, but the latter can be grasped more readily. By averaging over the model terms separately as we did previously, we find that

$$\bar{X}_{abc.} = \mathbf{m} + \mathbf{a}_a + \mathbf{b}_b + \mathbf{c}_c + \bar{\mathbf{s}}_{.(c)} + \mathbf{ab}_{ab} + \bar{\mathbf{e}}_{abc.}$$
$$\widehat{\mathbf{ab}}_{ab} = \mathbf{c}_c + \bar{\mathbf{s}}_{.(c)} - \bar{\mathbf{s}}_{.(.)} + \mathbf{ab}_{ab} + \bar{\mathbf{e}}_{abc.} - \bar{\mathbf{e}}_{a...} - \bar{\mathbf{e}}_{.b..} + \bar{\mathbf{e}}_{....}$$

$\widehat{\mathbf{ab}}'_{ab}$:

Term $\mathbf{ab}_{ab}$ proper does not appear in the score model, nor can the MS based on $\mathbf{ab}_{ab}$ be used in an F-ratio for testing the H_N that $\sigma^2_{AB} = 0$, for there is no suitable MS_2. The test of significance for $\mathbf{ab}_{ab}$ is, instead, based on an adjusted $\mathbf{ab}_{ab}$ estimate $\widehat{\mathbf{ab}}'_{ab}$. We estimate $\mathbf{ab}'_{ab}$ by the following formula:

$$\widehat{\mathbf{ab}}'_{ab} = \widehat{\mathbf{ab}}_{ab} - \hat{\mathbf{c}}_c$$
$$= \mathbf{ab}_{ab} - \overline{\mathbf{ab}}_{..(c_c)} + \bar{\mathbf{e}}_{abc.} - \bar{\mathbf{e}}_{a...} - \bar{\mathbf{e}}_{.b..} - \bar{\mathbf{e}}_{..c.} + 2\bar{\mathbf{e}}_{....}$$

By definition, $\mathbf{ab}'_{ab} = \mathbf{ab}_{ab} - \overline{\mathbf{ab}}_{..(c_c)}$, so

$$\widehat{\mathbf{ab}}'_{ab} = \mathbf{ab}'_{ab} + \bar{\mathbf{e}}_{abc.} - \bar{\mathbf{e}}_{a...} - \bar{\mathbf{e}}_{.b..} - \bar{\mathbf{e}}_{..c.} + 2\bar{\mathbf{e}}_{....}$$

In other words, $\mathbf{ab}'_{ab}$ is the difference between an $\mathbf{ab}_{ab}$ parameter and the mean of the $\mathbf{ab}_{ab}$ parameters for level c_c. The mean $\overline{\mathbf{ab}}_{..(c)}$, however, is the same for all subjects in a group. Hence, we can think of $\mathbf{ab}'_{ab}$ as the difference between a subject's $\mathbf{ab}_{ab}$ parameter for a particular score and the mean of that subject's $\mathbf{ab}_{ab}$ parameters across all his scores. For this reason, the adjusted interaction term $\mathbf{ab}'_{ab}$ is also called the within-subject component of the $\mathbf{ab}_{ab}$ interaction term. We use the MS for this within-subject component, $\text{MS}_{AB'}$, to test for the presence of an AB interaction.

7.5.4 Sums of Squares

In previous designs we had an SS for each key term of the score model. Here the situation is similar, but instead of an SS_{AB} we have an $SS_{AB'}$. As before, we have an SS element for each model term. These elements are SS_1, SS_a, SS_b, SS_c, SS_{cs}, SS_{ab}, and SS_{abcs}, corresponding to $\mathbf{m}$, $\mathbf{a}_a$, $\mathbf{b}_b$, $\mathbf{c}_c$, $\mathbf{ab}_{ab}$, $\mathbf{s}_{s(c)}$, and $\mathbf{e}_{abcs}$, respectively. The SS's we require are obtained by adding and subtracting various of these elements. Although the score model has $\mathbf{ab}'_{ab}$, not $\mathbf{ab}_{ab}$, element SS_{ab} is defined in relation to $\mathbf{ab}_{ab}$. Table 7.5-2 shows the formulas for the SS elements and how to combine them to obtain the SS's. In calculating the SS elements, remember that not all combinations of subscripts exist; the summations by dots or $\sum$'s refer only to such combinations as occur in the design. The double plus sign for SS_1 under $SS_{AB'}$ means that SS_1 must be added twice. Symbol S stands for the number of subjects within a group (that is, at one level c_c), not the total number of subjects in the experiment.

Table 7.5-2 SS elements formulas and table

Model term	SS element
$\mathbf{e}_{abcs}$	$SS_{abcs} = \sum X^2_{abcs}$
$\mathbf{ab}_{ab}$	$SS_{ab} = (\sum X^2_{ab..})/S$
$\mathbf{s}_{s(c)}$	$SS_{cs} = (\sum X^2_{..cs})/L$
$\mathbf{c}_c$	$SS_c = (\sum X^2_{..c.})/LS$
$\mathbf{b}_b$	$SS_b = (\sum X^2_{.b..})/LS$
$\mathbf{a}_a$	$SS_a = (\sum X^2_{a...})/LS$
$\mathbf{m}$	$SS_1 = X^2_{....}/LS^2$

	SS_A	SS_B	SS_C	SS_S	$SS_{AB'}$	SS_E	SS_{tot}
SS_{abcs}						+	+
SS_{ab}					+	−	
SS_{cs}				+	+	−	
SS_c			+	−	−	+	
SS_b		+			−		
SS_a	+				−		
SS_1	−	−	−		++		−
Sum							

Table 7.5-3 EMS's and df's

Source	df	EMS
C	$L - 1$	$\sigma_E^2 + L\sigma_S^2 + LS\sigma_{\overline{AB}}^2 + LS\sigma_C^2$
S	$L(S - 1)$	$\sigma_E^2 + L\sigma_S^2$
A	$(L - 1)$	$\sigma_E^2 + LS\sigma_A^2$
B	$(L - 1)$	$\sigma_E^2 + LS\sigma_B^2$
AB'	$(L - 1)(L - 2)$	$\sigma_E^2 + S\sigma_{AB'}^2$
E	$L(L - 1)(S - 1)$	σ_E^2

7.5.5 Expected Mean Squares and Hypothesis Testing

The EMS's and df's are shown in Table 7.5-3. To test any of the H_N's that $\sigma_A^2 = 0$, $\sigma_B^2 = 0$, or $\sigma_{AB'}^2 = 0$, $\text{MS}_2 = \text{MS}_E$. Although there is an $LS\sigma_C^2$ in the formula for EMS_C, according to our assumption that C is a dummy factor, $\sigma_C^2 = 0$. Hence we could test the H_N that $\sigma_{\overline{AB}}^2 = 0$ by using MS_C as MS_1 and MS_S as MS_2.

Recall that, by definition, $\mathbf{ab}'_{ab} = \mathbf{ab}_{ab} - \overline{\mathbf{ab}}_{..(c_c)}$, where $\overline{\mathbf{ab}}_{..(c_c)}$ is the mean of the $\mathbf{ab}_{ab}$ parameters for level c_c (or for a subject at level c_c). Again by definition,

$$\sigma_{AB'}^2 = \frac{\text{E}(\sum \mathbf{ab}'^2_{ab})}{(L - 1)(L - 2)}$$

$$\sigma_{AB}^2 = \frac{\text{E}(\sum \mathbf{ab}^2_{ab})}{(L - 1)^2}$$

$$\sigma_{\overline{AB}}^2 = \frac{\text{E}(\sum_c \overline{\mathbf{ab}}^2_{..(cc)})}{L - 1}$$

The three quantities $\sigma_{AB'}^2$, σ_{AB}^2, and $\sigma_{\overline{AB}}^2$ are related as follows:

$$(L - 2)\sigma_{AB'}^2 = (L - 1)\sigma_{AB}^2 - L\sigma_{\overline{AB}}^2$$

The quantity $\sigma_{AB'}^2$ represents variation in the $\mathbf{ab}'_{ab}$ parameters for individual subjects. If and only if the different $\mathbf{ab}_{ab}$ parameters for each subject in each replication are equal can all the $\mathbf{ab}'_{ab}$ parameters, and hence $\sigma_{AB'}^2$, be zero. But such a state of affairs can exist only if all $\mathbf{ab}_{ab}$ parameters equal zero. Hence, to conclude that $\sigma_{AB'}^2 = 0$ is to conclude that $\sigma_{AB}^2 = 0$. Likewise, we cannot have $\sigma_{AB'}^2 > 0$ unless the $\mathbf{ab}_{ab}$ parameters for a subject differ, that is, unless $\sigma_{AB}^2 > 0$ also. Hence to conclude that $\sigma_{AB'}^2 > 0$ is to conclude that $\sigma_{AB}^2 > 0$. The conclusion from the test of $\sigma_{AB'}^2$, then, applies not just to $\mathbf{ab}'_{ab}$ parameters, but to the $\mathbf{ab}_{ab}$ parameters of our score model as well. We could likewise argue that the conclusions from the test of $\sigma_{\overline{AB}}^2$ apply to σ_{AB}^2. The tests on $\sigma_{AB'}^2$ and $\sigma_{\overline{AB}}^2$ can give conflicting results. The normal procedure would be to test only $\sigma^2_{AB'}$, since the test on $\sigma^2_{\overline{AB}}$ would be less powerful (to see why, compare the EMS_2's).

7.5.6 When C Is an Experimental Factor

We could let C be an experimental factor; but then we could no longer assume that $\sigma^2_C = 0$, or that interactions with C are zero. This design would have essentially the same dangers and require the same type of dubious assumptions as the completely randomized design. In order to have valid tests for the three main effects, we would have to assume that all the two- and three-factor interactions were zero.

The SS's and MS's would be calculated exactly as when C is a dummy factor. In this case, however, what was called AB' is interpreted as a residual term comparable to that of the completely randomized design. We can test H_N's that $\sigma^2_A = 0$, $\sigma^2_B = 0$, and $\sigma^2_{\text{res}} = 0$ using MS_E as the denominator of the F-ratio. The test of H_N that $\sigma^2_C = 0$ uses MS_S as the denominator. If we reject H_N that $\sigma^2_{\text{res}} = 0$, significant main effects found for A, B, or C could be due to confounded interaction effects.

7.6 CROSSING THE REPEATED MEASUREMENTS LATIN SQUARE, USING DIFFERENT SUBJECTS

7.6.1 The Layout

Chapter 7 considers two extensions of the basic repeated measurements Latin square design. Both require crossing the design with an additional fixed factor D. This section considers the situation where different subjects appear in squares at different d_d levels. The subsequent section considers the situation where the same subjects appear in all squares, that is, S is crossed with D.

Consider the following layout. Literal labeling is used for the subjects, though in subsequent symbols labeling is initialized for each square.

			b_1	b_2	b_3
	c_1	s_1, s_2	a_2	a_1	a_3
d_1	c_2	s_3, s_4	a_3	a_2	a_1
	c_3	s_5, s_6	a_1	a_3	a_2

			b_1	b_2	b_3
	c_1	s_7, s_8	a_2	a_1	a_3
d_2	c_2	s_9, s_{10}	a_3	a_2	a_1
	c_3	s_{11}, s_{12}	a_1	a_3	a_2

The arrangement of the a_a's forming the Latin square is determined randomly, as explained in Section 7.2.3, but the same square is used under each level of D. The diagram shows two subjects per c_c level, though any number is possible. Likewise, though $D = 2$ in the diagram, it may equal any value. All subjects in the experiment may be drawn randomly from the same population and randomly assigned to positions in the design. However, D may be a blocking factor; then subjects assigned to d_1 would come from one population and subjects assigned to d_2 would come from another. The diagram does not imply that subjects receive the treatments in the order shown. Normally, the order of presentation for the treatments would be randomized separately for each subject, but if B is a trial factor, the order of presentation would accord with the a_a's of the Latin square, so subjects at the same c_c level would have the same order. We let S be the number of subjects in a group (at the same cd_{cd} combination), so the total number of subjects in an experiment would be $D \cdot L \cdot S$. Each subject has L scores, so the total number of scores would be $D \cdot L^2 \cdot S$.

Factor C is a dummy variable, as in Section 7.5, so it has no main effect or interactions with other factors. Hence there are only three experimental factors in this design, A, B, and D. It is possible to test for the presence of all three main effects, as well as the presence of all interactions involving them.

The design is a competitor to other designs with three experimental factors allowing assessment of the interactions, including $A \times B \times D$, $S \times A \times B \times D$, and $S(D) \times A \times B$. Its particular restriction is that A and B must have the same number of levels. Its advantage over $A \times B \times D$ is that most of the effects tested are within-subject effects, and hence the tests are more powerful. If it were undesirable to require that each subject have all treatments abd_{abd}, then the design would have an advantage over $S \times A \times B \times D$, though its test for the main effect of D would be less powerful. The design's tests would be comparable in power to those for $S(D) \times A \times B$ but would require fewer treatments per subject. Our analysis of the present design requires the assumption that all design factors except S are fixed.

7.6.2 The Score Model and Estimation of Model Terms

The score model is

$$X_{abcds} = \mathbf{m} + \mathbf{a}_a + \mathbf{b}_b + \mathbf{c}_c + \mathbf{d}_d + \mathbf{s}_{s(cd)} + \mathbf{ab}'_{ab} + \mathbf{ad}_{ad} + \mathbf{bd}_{bd} + \mathbf{cd}_{cd} + \mathbf{abd}'_{abd} + \mathbf{e}_{abcds}$$

We include $\mathbf{c}_c$ and $\mathbf{cd}_{cd}$, though both terms equal zero, since the estimates of these terms prove useful for certain derivations. We keep terms $\mathbf{c}_c$ and $\mathbf{cd}_{cd}$ throughout the derivation of the EMS's, but we recognize that σ_C^2 and σ_{CD}^2 equal zero.

The estimates for the model terms are as follows. A calculational formula based on the method of successive extraction and the parametric components of the estimates are given for each term. The intermediary steps in deriving the latter from the former are not shown. The student should be able to fill in the gaps by this time, as the steps are very similar to those shown in Chapter 4.

$$\hat{\mathbf{m}} = \bar{X}_{....}$$
$$= \mathbf{m} + \bar{\mathbf{s}}_{.(..)} + \bar{\mathbf{e}}_{.....}$$
$$\hat{\mathbf{a}}_a = \bar{X}_{a....} - \hat{\mathbf{m}}$$
$$= \mathbf{a}_a + \bar{\mathbf{e}}_{a....} - \mathbf{e}_{.....}$$
$$\hat{\mathbf{b}}_b = \bar{X}_{.b...} - \hat{\mathbf{m}}$$
$$= \mathbf{b}_b + \bar{\mathbf{e}}_{.b...} - \bar{\mathbf{e}}_{.....}$$
$$\hat{\mathbf{c}}_c = \bar{X}_{..c..} - \hat{\mathbf{m}}$$
$$= \mathbf{c}_c + \bar{\mathbf{s}}_{.(c.)} - \bar{\mathbf{s}}_{.(..)} + \overline{\mathbf{ab}}_{..(c_c)} + \bar{\mathbf{e}}_{..c..} - \bar{\mathbf{e}}_{.....}$$
$$\hat{\mathbf{d}}_d = \bar{X}_{...d.} - \hat{\mathbf{m}}$$
$$= \mathbf{d}_d + \bar{\mathbf{s}}_{.(.d)} - \bar{\mathbf{s}}_{.(..)} + \bar{\mathbf{e}}_{...d.} - \bar{\mathbf{e}}_{.....}$$
$$\widehat{\mathbf{ab}}_{ab} = \bar{X}_{ab...} - (\hat{\mathbf{m}} + \hat{\mathbf{a}}_a + \hat{\mathbf{b}}_b)$$
$$= \mathbf{c}_c + \mathbf{ab}_{ab} + \bar{\mathbf{s}}_{.(c.)} - \bar{\mathbf{s}}_{.(..)} + \bar{\mathbf{e}}_{ab...} - \bar{\mathbf{e}}_{.b...} - \bar{\mathbf{e}}_{a.....} + \bar{\mathbf{e}}_{.....}$$
$$\widehat{\mathbf{ad}}_{ad} = \bar{X}_{a..d.} - (\hat{\mathbf{m}} + \hat{\mathbf{a}}_a + \hat{\mathbf{d}}_d)$$
$$= \mathbf{ad}_{ad} + \bar{\mathbf{e}}_{a..d.} - \bar{\mathbf{e}}_{...d.} - \bar{\mathbf{e}}_{a....} + \bar{\mathbf{e}}_{.....}$$
$$\widehat{\mathbf{bd}}_{bd} = \bar{X}_{.b.d.} - (\hat{\mathbf{m}} + \hat{\mathbf{b}}_b + \hat{\mathbf{d}}_d)$$
$$= \mathbf{bd}_{bd} + \bar{\mathbf{e}}_{.b.d.} - \bar{\mathbf{e}}_{...d.} - \bar{\mathbf{e}}_{.b...} + \bar{\mathbf{e}}_{.....}$$
$$\widehat{\mathbf{cd}}_{cd} = \bar{X}_{..cd.} - (\hat{\mathbf{m}} + \hat{\mathbf{c}}_c + \hat{\mathbf{d}}_d)$$
$$= \mathbf{cd}_{cd} + \bar{\mathbf{s}}_{.(cd)} - \bar{\mathbf{s}}_{.(.d)} - \bar{\mathbf{s}}_{.(c.)} + \bar{\mathbf{s}}_{.(..)} + \overline{\mathbf{abd}}_{..d(c_c)}$$
$$+ \bar{\mathbf{e}}_{..cd.} - \bar{\mathbf{e}}_{...d.} - \bar{\mathbf{e}}_{..c..} + \bar{\mathbf{e}}_{.....}$$
$$\hat{\mathbf{s}}_{s(cd)} = \bar{X}_{..cds} - (\hat{\mathbf{m}} + \hat{\mathbf{c}}_c + \hat{\mathbf{d}}_d + \widehat{\mathbf{cd}}_{cd})$$
$$= \bar{X}_{..cds} - \bar{X}_{..cd.}$$
$$= \mathbf{s}_{s(cd)} - \bar{\mathbf{s}}_{.(cd)} + \bar{\mathbf{e}}_{..cds} - \bar{\mathbf{e}}_{..cd.}$$
$$\widehat{\mathbf{abd}}_{abd} = \bar{X}_{ab.d.} - (\hat{\mathbf{m}} + \hat{\mathbf{a}}_a + \hat{\mathbf{b}}_b + \hat{\mathbf{d}}_d + \widehat{\mathbf{ab}}_{ab} + \widehat{\mathbf{ad}}_{ad} + \widehat{\mathbf{bd}}_{bd})$$
$$= \mathbf{abd}_{abd} + \mathbf{cd}_{cd} + \bar{\mathbf{s}}_{.(cd)} - \bar{\mathbf{s}}_{.(.d)} - \bar{\mathbf{s}}_{.(c.)} + \bar{\mathbf{s}}_{.(..)} + \bar{\mathbf{e}}_{ab.d.} - \bar{\mathbf{e}}_{a..d.}$$
$$- \bar{\mathbf{e}}_{.b.d.} - \bar{\mathbf{e}}_{ab...} + \bar{\mathbf{e}}_{...d.} + \bar{\mathbf{e}}_{a....} + \bar{\mathbf{e}}_{.b...} - \bar{\mathbf{e}}_{.....}$$
$$\hat{\mathbf{e}}_{abcds} = X_{abcds} - (\hat{\mathbf{m}} + \hat{\mathbf{a}}_a + \hat{\mathbf{b}}_b + \hat{\mathbf{c}}_c + \hat{\mathbf{d}}_d + \widehat{\mathbf{ab}}_{ab} + \widehat{\mathbf{ad}}_{ad} + \widehat{\mathbf{bd}}_{bd}$$
$$+ \widehat{\mathbf{cd}}_{cd} + \hat{\mathbf{s}}_{s(cd)} + \widehat{\mathbf{abd}}_{abd})$$
$$= X_{abcds} - \bar{X}_{ab.d.}$$
$$= \mathbf{e}_{abcds} - \bar{\mathbf{e}}_{ab.d.}$$

As with the basic design of Section 7.5, there is no way to test source AB per se. Instead, we define $\mathbf{ab}'_{ab} = \mathbf{ab}_{ab} - \overline{\mathbf{ab}}_{..(c_c)}$, which is a within-subject effect since it is the deviation of $\mathbf{ab}_{ab}$ from the mean of $\mathbf{ab}_{ab}$ for a subject, $\overline{\mathbf{ab}}_{..(c_c)}$. This mean is the same for each subject at level c_c; hence we can index it by c_c. We estimate $\mathbf{ab}'_{ab}$ by

$$\begin{aligned}\widehat{\mathbf{ab}}'_{ab} &= \widehat{\mathbf{ab}}_{ab} - \hat{\mathbf{c}}_c \\ &= \mathbf{ab}'_{ab} + \bar{\mathbf{e}}_{ab...} - \bar{\mathbf{e}}_{..c..} - \bar{\mathbf{e}}_{.b...} - \bar{\mathbf{e}}_{a....} + 2\bar{\mathbf{e}}_{.....}\end{aligned}$$

The student can verify the latter formula from the equations for $\widehat{\mathbf{ab}}_{ab}$ and $\hat{\mathbf{c}}_c$.

In addition, there is no way to test source ABD per se. Instead, we define $\mathbf{abd}'_{abd} = \mathbf{abd}_{abd} - \overline{\mathbf{abd}}_{..d(c_c)}$, which is a within-subject effect also, since it is the deviation of $\mathbf{abd}_{abd}$ from the mean of $\mathbf{abd}_{abd}$ for a subject, $\overline{\mathbf{abd}}_{..d(c_c)}$. This mean is the same for each subject at a given combination cd_{cd}, that is, within the same level c_c for a given square (level d_d). We estimate $\mathbf{abd}'_{abd}$ by the following formula:

$$\begin{aligned}\widehat{\mathbf{abd}}'_{abd} &= \widehat{\mathbf{abd}}_{abd} - \widehat{\mathbf{cd}}_{cd} \\ &= \mathbf{abd}'_{abd} + \bar{\mathbf{e}}_{ab.d.} - \bar{\mathbf{e}}_{..cd.} - \bar{\mathbf{e}}_{.b.d.} - \bar{\mathbf{e}}_{a..d.} - \bar{\mathbf{e}}_{ab...} \\ &\qquad + 2\bar{\mathbf{e}}_{...d.} + \bar{\mathbf{e}}_{..c..} + \bar{\mathbf{e}}_{.b...} + \bar{\mathbf{e}}_{a....} - 2\bar{\mathbf{e}}_{.....}\end{aligned}$$

Again, the student can verify this latter formula from the equations for $\widehat{\mathbf{abd}}_{abd}$ and $\widehat{\mathbf{cd}}_{cd}$.

Estimates $\mathbf{ab}'_{ab}$ and $\mathbf{abd}'_{abd}$ differ from their true values only because of error parameters, so tests of H_N's that $\mathbf{ab}'_{ab}$ and $\mathbf{abd}'_{abd}$ are zero use MS_E as MS_2.

Table 7.6-1 SS elements formulas and table

Model term	SS element
$\mathbf{e}_{abcds}$	$\mathrm{SS}_{abcds} = \sum X^2_{abcds}$
$\mathbf{abd}_{abd}$	$\mathrm{SS}_{abd} = (\sum X^2_{ab.d.})/S$
$\mathbf{s}_{s(cd)}$	$\mathrm{SS}_{cds} = (\sum X^2_{..cds})/L$
$\mathbf{cd}_{cd}$	$\mathrm{SS}_{cd} = (\sum X^2_{..cd.})/LS$
$\mathbf{bd}_{bd}$	$\mathrm{SS}_{bd} = (\sum X^2_{.b.d.})/LS$
$\mathbf{ad}_{ad}$	$\mathrm{SS}_{ad} = (\sum X^2_{a..d.})/LS$
$\mathbf{ab}_{ab}$	$\mathrm{SS}_{ab} = (\sum X^2_{ab...})/DS$
$\mathbf{d}_d$	$\mathrm{SS}_d = (\sum X^2_{...d.})/L^2S$
$\mathbf{c}_c$	$\mathrm{SS}_c = (\sum X^2_{..c..})/DLS$
$\mathbf{b}_b$	$\mathrm{SS}_b = (\sum X^2_{.b...})/DLS$
$\mathbf{a}_a$	$\mathrm{SS}_a = (\sum X^2_{a....})/DLS$
$\mathbf{m}$	$\mathrm{SS}_1 = X^2_{.....}/DL^2S$

7.6.3 Sums of Squares

We have an SS element for each term of the model, except that we have elements for $\mathbf{ab}_{ab}$ and $\mathbf{abd}_{abd}$ rather than for $\mathbf{ab}'_{ab}$ and $\mathbf{abd}'_{abd}$. We have an SS for each key term. Table 7.6-1 gives the formulas for the SS elements and shows how to combine them into SS's.

7.6.4 Expected Mean Squares and Hypothesis Testing

The EMS's and df's are shown in Table 7.6-2. All within-subject sources are tested using MS_E as MS_2. The between-subject sources can be tested using MS_S as MS_2. Normally, however, only source D is tested using MS_S. We know that σ_C^2 and σ_{CD}^2 are zero because C is a dummy factor. But by examining the EMS for source C, we see that MS_C could be used as MS_1 to test the H_N that $\sigma_{\overline{AB}}^2 = 0$. It is preferable, however, simply to test $\sigma_{AB'}^2$ using $MS_{AB'}$, since this test has more power. The conclusion vis-à-vis source AB' applies to the AB interaction also; that is, if we conclude that $\sigma_{AB'}^2 > 0$, then we conclude that there is an AB interaction; if we conclude that $\sigma_{AB'}^2 = 0$, then we conclude that the AB interaction is zero. Likewise, instead of using MS_{CD} to test $\sigma_{\overline{ABD}}^2$, we would normally test only $\sigma_{ABD'}^2$ and apply the conclusion to ABD.

7.6.5 Numerical Example

Table 7.6-3 shows hypothetical scores from an experiment. This subsection illustrates how the analysis of variance would proceed with such data. Table 7.6-4 gives the various summations required for computing the SS elements. These elements, the summations required for them, and their formulas were given in Table 7.6-1. Once the $X_{ab.d.}$ and $X_{..cds}$ quantities have been computed from the original data, subsequent sums can be computed more easily from previously computed sums than from the original data directly. For example, the $X_{..cd.}$'s can be found by adding adjacent entries in the $X_{..cds}$ table. The "sum" entries in the $X_{..cd.}$ table show the $X_{..c..}$'s and $X_{...d.}$'s. Likewise, the "sum" entries in the $X_{ab...}$ table show the $X_{a....}$'s and $X_{.b...}$'s. Since errors in addition will be carried into subsequent tables, each table should be computed twice to assure accuracy. Then the total of the table entries should be computed, and it should equal the total of all scores in the original table. For example, the sum of all nine $X_{ab...}$quantities equals 152, the sum of all scores in the experiment. (The $X_{a....}$'s and $X_{.b...}$'s should be ignored in doing this summation.)

Table 7.6-5 gives the SS elements. For each, the numerator is simply the sum of the squares of all entries in the corresponding subtable. For example,

$$\begin{aligned} SS_{bd} &= (X_{.b.d.}^2)/LS \\ &= (26^2 + 28^2 + 22^2 + 32^2 + 13^2 + 31^2)/6 \end{aligned}$$

Table 7.6-2 EMS's and df's

Source	df	EMS
	Between subjects	
D	$D - 1$	$\sigma_E^2 + L\sigma_S^2 + L^2S\sigma_D^2$
C	$L - 1$	$\sigma_E^2 + L\sigma_S^2 + DLS\sigma_{\overline{AB}}^2 + DLS\sigma_C^2$
CD	$(L - 1)(D - 1)$	$\sigma_E^2 + L\sigma_S^2 + LS\sigma_{\overline{ABD}}^2 + LS\sigma_{CD}^2$
S	$DL(S - 1)$	$\sigma_E^2 + L\sigma_S^2$
	Within subjects	
A	$L - 1$	$\sigma_E^2 + DLS\sigma_A^2$
B	$L - 1$	$\sigma_E^2 + DLS\sigma_B^2$
AB'	$(L - 1)(L - 2)$	$\sigma_E^2 + DS\sigma_{AB'}^2$
AD	$(D - 1)(L - 1)$	$\sigma_E^2 + LS\sigma_{AD}^2$
BD	$(D - 1)(L - 1)$	$\sigma_E^2 + LS\sigma_{BD}^2$
ABD'	$(D - 1)(L - 1)(L - 2)$	$\sigma_E^2 + S\sigma_{ABD'}^2$
E	$DL(L - 1)(S - 1)$	σ_E^2

	SS_A	SS_B	SS_C	SS_D	$SS_{AB'}$	SS_{AD}	SS_{BD}	SS_{CD}	SS_S	$SS_{ABD'}$	SS_E	SS_{tot}
SS_{abcds}											+	+
SS_{abd}										+	−	
SS_{cds}									+		−	
SS_{cd}								+	−	−	+	
SS_{bd}							+			−		
SS_{ad}						+				−		
SS_{ab}					+					−		
SS_d				+		−	−	−		++		
SS_c			+		−			−		+		
SS_b		+			−		−			+		
SS_a	+				−	−				+		
SS_1	−	−	−	−	++	+	+	+		−−		−
Sum												

Table 7.6-3 Hypothetical scores

		d_1			d_2		
		b_1	b_2	b_3	b_1	b_2	b_3
		a_2	a_1	a_3	a_2	a_1	a_3
c_1	s_1	4	1	3	2	3	6
	s_2	5	2	2	6	5	10
		a_3	a_2	a_1	a_3	a_2	a_1
c_2	s_1	7	6	1	5	4	4
	s_2	5	2	3	5	7	3
		a_1	a_3	a_2	a_1	a_3	a_2
c_3	s_1	3	6	1	4	8	3
	s_2	2	5	3	6	5	5

The squaring and summing should also be done twice to guard against error; then the division should be done twice.

Next we enter the SS elements into the SS elements table (Table 7.6-6), and sum down the columns to get the SS's. As a check on our addition, we

Table 7.6-4 Summations required for computing the SS elements

	$X_{ab.d.}$		
d_1	$X_{21.1.} = 9$	$X_{12.1.} = 3$	$X_{33.1.} = 5$
	$X_{31.1.} = 12$	$X_{22.1.} = 8$	$X_{13.1.} = 4$
	$X_{11.1.} = 5$	$X_{32.1.} = 11$	$X_{23.1.} = 4$
d_2	$X_{21.2.} = 8$	$X_{12.2.} = 8$	$X_{33.2.} = 16$
	$X_{31.2.} = 10$	$X_{22.2.} = 11$	$X_{13.2.} = 7$
	$X_{11.2.} = 10$	$X_{32.2.} = 13$	$X_{23.2.} = 8$

$X_{..cds}$	
$X_{..111} = 8$	$X_{..121} = 11$
$X_{..112} = 9$	$X_{..122} = 21$
$X_{..211} = 14$	$X_{..221} = 13$
$X_{..212} = 10$	$X_{..222} = 15$
$X_{..311} = 10$	$X_{..321} = 15$
$X_{..312} = 10$	$X_{..322} = 16$

	$X_{..cd.}$		
			Sum
	$X_{..11.} = 17$	$X_{..12.} = 32$	$X_{..1..} = 49$
	$X_{..21.} = 24$	$X_{..22.} = 28$	$X_{..2..} = 52$
	$X_{..31.} = 20$	$X_{..32.} = 31$	$X_{..3..} = 51$
Sum	$X_{...1.} = 61$	$X_{...2.} = 91$	

Table 7.6-4 (continued)

$X_{.b.d.}$			
d_1	$X_{.1.1.} = 26$	$X_{.2.1.} = 22$	$X_{.3.1.} = 13$
d_2	$X_{.1.2.} = 28$	$X_{.2.2.} = 32$	$X_{.3.2.} = 31$

$X_{a..d.}$			
d_1	$X_{1..1.} = 12$	$X_{2..1.} = 21$	$X_{3..1.} = 28$
d_2	$X_{1..2.} = 25$	$X_{2..2.} = 27$	$X_{3..2.} = 39$

$X_{ab...}$

	b_1	b_2	b_3	Sum
a_1	$X_{...11} = 15$	$X_{12...} = 11$	$X_{13...} = 11$	$X_{1....} = 37$
a_2	$X_{...21} = 17$	$X_{22...} = 19$	$X_{23...} = 12$	$X_{2....} = 48$
a_3	$X_{...31} = 22$	$X_{32...} = 24$	$X_{33...} = 21$	$X_{3....} = 67$
Sum	$X_{.1...} = 54$	$X_{.2...} = 54$	$X_{.3...} = 44$	

$X_{...d.}$	$X_{..c..}$	$X_{.b...}$	$X_{a....}$
$X_{...1.} = 61$	$X_{..1..} = 49$	$X_{.1...} = 54$	$X_{1....} = 37$
$X_{...2.} = 91$	$X_{..2..} = 52$	$X_{.2...} = 54$	$X_{2....} = 48$
	$X_{..3..} = 51$	$X_{.3...} = 44$	$X_{3....} = 67$
$X_{.....} = 152$			

make sure that the SS's total to SS_{tot}, 150.22, as determined by adding the entries in the SS_{tot} column.

Table 7.6-7 is the ANOVA table for our example. Normally, there would be no MS_2 column, but it is included here for pedagogical reasons. First we

Table 7.6-5 SS elements

$SS_{abcds} = 792$	$SS_{ab} = 2762/4 = 690.5$
$SS_{abd} = 1448/2 = 744$	$SS_d = 12{,}002/18 = 666.78$
$SS_{cds} = 2078/3 = 692.67$	$SS_c = 7706/12 = 642.17$
$SS_{cd} = 4034/6 = 672.33$	$SS_b = 7768/12 = 647.33$
$SS_{bd} = 4098/6 = 683$	$SS_a = 8162/12 = 680.17$
$SS_{ad} = 4244/6 = 707.33$	$SS_1 = 23{,}104/36 = 641.78$

Table 7.6-6 SS elements table

	SS_A	SS_B	SS_C	SS_D	$SS_{AB'}$	SS_{AD}	SS_{BD}	SS_{CD}	SS_S	$SS_{ABD'}$	SS_E	SS_{tot}
SS_{abcds}											+792	+792
SS_{abd}										+744	−744	
SS_{cds}									+692.67		−692.67	
SS_{cd}								+672.33	−672.33	−672.33	+672.33	
SS_{bd}							+683			−683		
SS_{ad}						+707.33				−707.33		
SS_{ab}					+690.5					−690.5		
SS_d				+666.78		−666.78	−666.78	−666.78		++666.78		
SS_c			+642.17		−642.17			−642.17		+642.17		
SS_b		+647.33			−647.33		−647.33			+647.33		
SS_a	+680.17				−680.17	−680.17				+680.17		
SS_1	−641.78	−641.78	−641.78	−641.78	++641.78	+641.78	+641.78	+641.78		−−641.78		−641.78
Sum	38.39	5.55	.39	25.00	4.39	2.16	10.67	5.16	20.34	10.51	27.66	150.22

Table 7.6-7 Analysis of variance table

Source	SS	df	MS	MS_2	F
		Between subjects			
D	25	1	25.00	MS_S	7.37*
C	.39	2	.20	MS_S	.06
CD	5.16	2	2.58	MS_S	.76
S	20.34	6	3.39	–	–
		Within subjects			
A	38.39	2	19.20	MS_E	8.35**
B	5.55	2	2.78	MS_E	1.21
AB'	4.39	2	2.20	MS_E	.96
AD	2.16	2	1.08	MS_E	.47
BD	10.67	2	5.34	MS_E	2.32
ABD'	10.51	2	5.26	MS_E	2.29
E	27.66	12	2.30	–	–

* $p < .05$.
** $p < .01$.

enter the SS's and df's in the table. Then we divide each SS by its df to find the MS's. Finally, for each MS having an MS_2, we compute $F = MS/MS_2$.

Rather than comparing each separate F-ratio with those in the F-table (Table A.5), we prepare Table 7.6-8 showing the critical F's for all combinations of df_1 and df_2 appearing in the ANOVA table. Comparison of the obtained F-ratios with these critical F's shows that only the main effects A and D reached significance.

It should be noted that we carried two decimal places in computing our SS elements. Although the final digits in an SS element may seem of little importance, they can have critical effects on the final conclusions, so we must be very conservative about rounding off SS elements or SS's.

Table 7.6-8 Critical F's

df_1,df_2	Source(s)	F_{95}	F_{99}	$F_{99.9}$
1,6	D	5.99	13.74	35.51
2,6	C, CD	5.14	10.92	27.00
2,12	A, B, AB' AD, BD, ABD'	3.88	6.93	12.97

7.7 CROSSING THE REPEATED MEASUREMENTS LATIN SQUARE, USING THE SAME SUBJECTS

This design is the same as the preceding one except that the same subjects are used across the levels of D, that is, in the different squares. The layout, which uses literal labeling for the subjects, is as follows:

			b_1	b_2	b_3
d_1	c_1	s_1, s_2	a_2	a_1	a_3
	c_2	s_3, s_4	a_3	a_2	a_1
	c_3	s_5, s_6	a_1	a_3	a_2

			b_1	b_2	b_3
d_2	c_1	s_1, s_2	a_2	a_1	a_3
	c_2	s_3, s_4	a_3	a_2	a_1
	c_3	s_5, s_6	a_1	a_3	a_2

Factors A, B, and D are assumed to be fixed, as before. Subject s_1 under d_1 is exactly the same individual as s_1 under d_2. Although the diagram shows two subjects per c_c level, any number is possible. The arrangement of the a_a's forming the Latin square is determined randomly, as explained in Section 7.2.3, but the same square is used under each level of D. As previously, the diagram does not imply that subjects receive the treatments in the order shown in the layout. Normally, the order of the treatments would be randomized separately for each subject.

Factor C is a dummy factor, as before. Hence this design, like the preceding one, has only three experimental factors, all assumed to be fixed. For planning purposes, the main differences between this design and the preceding one is that this design requires a subject to receive D times as many treatments, and here factor D has a within- instead of a between-subject effect. Therefore, in choosing between the designs, the experimenter should consider the amount of time each treatment takes, the total amount of time a subject will be available, possible effects of fatigue, and the need for a sensitive test for effect D.

The following model is suitable for the present design:

$$X_{abcds} = \mathbf{m} + \mathbf{a}_a + \mathbf{b}_b + \mathbf{c}_c + \mathbf{d}_d + \mathbf{s}_{s(c)} + \mathbf{ab}'_{ab} + \mathbf{ad}_{ad} + \mathbf{bd}_{bd} + \mathbf{cd}_{cd} + \mathbf{bs}_{bs(c)} + \mathbf{ds}_{ds(c)} + \mathbf{abd}'_{abd} + \mathbf{bds}_{bds(c)} + \mathbf{e}_{abcds}$$

No separate estimate for $\mathbf{e}_{abcds}$ is possible. As before, $\mathbf{ab}'_{ab}$ is the within-subject component of $\mathbf{ab}_{ab}$, that is, $\mathbf{ab}'_{ab} = \mathbf{ab}_{ab} - \overline{\mathbf{ab}}_{..(c_c)}$. Also, $\mathbf{abd}'_{abd}$ is the within-

Table 7.7-1 SS elements formulas

Model term	SS element formula
$\mathbf{bds}_{bds(c)}$	$SS_{bcds} = \sum X^2_{abcds}$
$\mathbf{abd}_{abd}$	$SS_{abd} = (\sum X^2_{abcd.})/S$
$\mathbf{ds}_{ds(c)}$	$SS_{cds} = (\sum X^2_{..cds})/L$
$\mathbf{bs}_{bs(c)}$	$SS_{bcs} = (\sum X^2_{abc.s})/D$
$\mathbf{cd}_{cd}$	$SS_{cd} = (\sum X^2_{..cd.})/LS$
$\mathbf{bd}_{bd}$	$SS_{bd} = (\sum X^2_{.b.d.})/LS$
$\mathbf{ad}_{ad}$	$SS_{ad} = (\sum X^2_{a..d.})/LS$
$\mathbf{ab}_{ab}$	$SS_{ab} = (\sum X^2_{abc..})/DS$
$\mathbf{s}_{s(c)}$	$SS_{cs} = (\sum X^2_{..c.s})/DL$
$\mathbf{d}_{d}$	$SS_{d} = (\sum X^2_{...d.})/L^2S$
$\mathbf{c}_{c}$	$SS_{c} = (\sum X^2_{..c..})/DLS$
$\mathbf{b}_{b}$	$SS_{b} = (\sum X^2_{.b...})/DLS$
$\mathbf{a}_{a}$	$SS_{a} = (\sum X^2_{a....})/DLS$
$\mathbf{m}$	$SS_{1} = X^2_{.....}/DL^2S$

Table 7.7-2 SS elements table

	SS_A	SS_B	SS_C	SS_D	SS_S	$SS_{AB'}$	SS_{AD}	SS_{BD}	SS_{CD}	SS_{BS}	SS_{DS}	$SS_{ABD'}$	SS_{BDS}	SS_{tot}
SS_{bcds}													+	+
SS_{abd}												+	−	
SS_{cds}											+		−	
SS_{bcs}										+			−	
SS_{cd}									+		−	−	+	
SS_{bd}								+				−		
SS_{ad}							+					−		
SS_{ab}						+				−		−	+	
SS_{cs}					+					−	−		+	
SS_d				+			−	−	−			++		
SS_c			+		−	−			−	+	+	+	−	
SS_b		+				−		−				+		
SS_a	+					−	−					+		
SS_1	−	−	−	−		++	+	+	+			−−		−
Sum														

Table 7.7-3 df's and MS_2's for F-ratios

Source	df	MS_2
	Between subjects	
C	$L - 1$	MS_S
S	$L(S - 1)$	—
	Within subjects	
A	$L - 1$	MS_{BS}
B	$L - 1$	MS_{BS}
AB'	$(L - 1)(L - 2)$	MS_{BS}
BS	$L(L - 1)(S - 1)$	—
D	$D - 1$	MS_{DS}
DC	$(D - 1)(L - 1)$	MS_{DS}
DS	$(D - 1)L(S - 1)$	—
AD	$(D - 1)(L - 1)$	MS_{BDS}
BD	$(D - 1)(L - 1)$	MS_{BDS}
ABD'	$(D - 1)(L - 1)(L - 2)$	MS_{BDS}
BDS	$(D - 1)L(L - 1)(S - 1)$	—

subject component of $\mathbf{abd}_{abd}$, that is, $\mathbf{abd}'_{abd} = \mathbf{abd}_{abd} - \overline{\mathbf{abd}}_{..d(c_c)}$. For purposes of estimation,

$$\widehat{\mathbf{ab}}'_{ab} = \widehat{\mathbf{ab}}_{ab} - \hat{\mathbf{c}}_c$$

$$\widehat{\mathbf{abd}}'_{abd} = \widehat{\mathbf{abd}}_{abd} - \widehat{\mathbf{cd}}_{cd}$$

The SS element formulas are shown in Table 7.7-1; Table 7.7-2 is the SS elements table. The df's and F-ratios are shown in Table 7.7-3. For the analysis shown, three different MS's are used for MS_2. Should the experimenter conclude that all three associated EMS's do not differ, the three MS's may be pooled to form a single MS for testing all within-subject effects.

CHAPTER EIGHT

Miscellaneous Design Considerations

8.1 SEQUENCE AND ORDER EFFECTS

8.1.1 Introduction

In a repeated measurements experiment, each subject experiences more than one treatment. We say that a particular one of those treatments has an *order* of n for a subject if it is the nth treatment a subject receives. A listing, showing the order for each treatment a subject receives, is called a *sequence*. If, for example, a subject receives a_3, then a_1, then a_2, his treatment sequence is a_3, a_1, a_2.

We can refer to the successive treatments in a repeated measurements experiment as trials, except that there is no implication that the design therefore includes a trial factor. Even when there is a trial factor, the number of levels T may be less than the total number of treatments (orders) for a subject, because what is called the trial factor is often crossed with another treatment factor (such as sessions), so that a subject receives t_1 to t_T for each level of this other factor.

The experimenter must pay careful attention to the order in which treatments are administered to subjects. It is entirely inadequate to ignore this matter and to use whatever order seems convenient. For example, in design $S(A) \times B$, where B is a treatment factor, each subject has B treatments. Simply to let all subjects receive ab_{a1} first, ab_{a2} second, and so on, is insufficient, unless B is a trial factor. The danger is that exposure to previous treatments, per se, may affect the scores. Suppose, for example, that the subjects must solve a series of differing but related problems, the score for each problem being the time taken to solve it. Over successive problems, it is likely that the time taken to solve a problem will decrease, even if the problems are inherently of equal difficulty, because with practice, a subject becomes more skillful. If the problems were administered to all subjects in the same order, an analysis of variance of the scores would seem to show that the problems administered last were easier, even if all problems were of equal difficulty.

In order to avoid such distortions in the conclusions, several options are available to the experimenter. The technique most commonly employed—one that has been mentioned here many times already—is to randomize the order of presentation separately for each subject, insofar as this is possible. Alternatively, various forms of *counterbalancing* are possible, all of which aim at reducing or eliminating possible distortions due to the order of presenting treatments by using treatment sequences aimed at balancing out such distortions.

The influence that preceding treatments can have on a score for the current treatment is often referred to as a *carry-over* effect. Carry-over effects depend not only on how many preceding treatments a subject received, but on the specific treatments received. In general, the immediately preceding treatment has the greatest carry-over effect, and the more remote treatments are less critical. Besides paying attention to the order for administering treatments, the experimenter can protect himself from distortions due to carry-over by spacing the successive treatments adequately over time. In general, the longer the time between treatments, the smaller will be the carry-over effects.

8.1.2 Randomized Orders

Although randomizing the order of presentation has been mentioned previously in this text, a detailed example is given in this section in order to ensure that the student thoroughly comprehends the procedure. Randomizing the order of presentation for a subject is equivalent to choosing at random a sequence from all possible treatment sequences of a given length.

Suppose the design is $S(A) \times B$, with $A = 3$, $B = 6$, and $S = 4$. Then we have $S \cdot A = 12$ subjects, each receiving 6 treatments (but not necessarily the same 6, because a_a differs among the subjects, and A may be a treatment factor). Using Table A.1 of random digits, we select 12 randomizations of the digits 1 to 6, in the manner illustrated in Section 7.2.3. One such randomization is assigned to each subject, as follows:

s_1	3 6 4 5 2 1
s_2	3 2 6 5 1 4
s_3	6 3 2 4 5 1
s_4	5 2 6 1 3 4
s_5	3 4 1 5 6 2
s_6	1 6 5 2 3 4
s_7	1 5 3 2 4 6
s_8	2 6 3 5 1 4

s_9	1 2 3 4 5 6
s_{10}	5 1 4 2 3 6
s_{11}	2 3 5 1 6 4
s_{12}	2 5 4 6 1 3

The layout of the design is normally diagrammed as follows; literal labeling is used to emphasize the 12 different subjects used.

		b_1	b_2	b_3	b_4	b_5	b_6
a_1	s_1						
	s_2						
	s_3						
	s_4						
a_2	s_5						
	s_6						
	s_7						
	s_8						
a_3	s_9						
	s_{10}						
	s_{11}						
	s_{12}						

We associate the numbers in the random sequence with the subscripts of b_b. Hence, subject s_1, whose random sequence is 364521, receives his treatments in the order ab_{13}, ab_{16}, ab_{14}, ab_{15}, ab_{12}, ab_{11}. The treatments for the 12 subjects are listed in order as follows:

	Order					
	1	2	3	4	5	6
s_1	ab_{13}	ab_{16}	ab_{14}	ab_{15}	ab_{12}	ab_{11}
s_2	ab_{13}	ab_{12}	ab_{16}	ab_{15}	ab_{11}	ab_{14}
s_3	ab_{16}	ab_{13}	ab_{12}	ab_{14}	ab_{15}	ab_{11}
s_4	ab_{15}	ab_{12}	ab_{16}	ab_{11}	ab_{13}	ab_{14}
s_5	ab_{23}	ab_{24}	ab_{21}	ab_{25}	ab_{26}	ab_{22}
s_6	ab_{21}	ab_{26}	ab_{25}	ab_{22}	ab_{23}	ab_{24}
s_7	ab_{21}	ab_{25}	ab_{23}	ab_{22}	ab_{24}	ab_{26}
s_8	ab_{21}	ab_{26}	ab_{23}	ab_{25}	ab_{21}	ab_{24}
s_9	ab_{31}	ab_{32}	ab_{33}	ab_{34}	ab_{35}	ab_{36}
s_{10}	ab_{35}	ab_{31}	ab_{34}	ab_{32}	ab_{33}	ab_{36}
s_{11}	ab_{32}	ab_{33}	ab_{35}	ab_{31}	ab_{36}	ab_{34}
s_{12}	ab_{32}	ab_{35}	ab_{34}	ab_{36}	ab_{31}	ab_{33}

If the number of orders is 10, we can proceed to find random sequences in the same way, letting the digit 0 stand for 10. If, however, the number of sequences is, say, 12, we must examine successive pairs of digits in the table of random digits, rather than individual digits. The method is not very efficient, since most pairs of digits are irrelevant, that is, they are not 01 to 12. A simpler method would be to write the numbers 1 to 12 on 12 white poker chips, put the poker chips in a bag or box and shake well, then draw out the chips one at a time without looking, and copy the numbers in order as they are drawn. (Once a chip is drawn, it is not replaced in the bag.) The process is repeated for each subject, and the chips are thoroughly mixed before each sequence is observed. Although the numbers could be written on slips of paper (or on anything else), care must be taken because slips of paper may stick together and thus be difficult to mix properly. Of course, randomized sequences can also be generated on a computer, and in some instances, for some experimenters, this might be the most convenient method.

If S is crossed with more than one factor, as with $S \times A \times B$ or $S(A) \times B \times C$, we cannot simply identify the sequence numbers with the subscripts of a single factor. Instead, we must assign to each combination of treatments a subject might receive a number corresponding to a sequence number. For example, if, in $S \times A \times B$, $A = 2$ and $B = 3$, we could assign the numbers 1 to 6 to the ab_{ab} combinations as follows:

Sequence number:	1	2	3	4	5	6
Combination:	ab_{11}	ab_{12}	ab_{13}	ab_{21}	ab_{22}	ab_{23}

Then the combinations would be ordered for a subject according to the random sequence of the digits 1 to 6 assigned to him.

8.1.3 Counterbalancing

Rather than using randomly selected sequences, the experimenter may use a specially arranged combination of sequences aimed at balancing out carry-over effects so that they will not distort the conclusions. (The arrangements will not prevent the *occurrence* of carry-over effects, however.) Three ways of counterbalancing are discussed here: (1) opposite sequences, (2) all sequences, and (3) Latin square sequences.

Opposite sequences. According to this approach, half the subjects use one sequence, and half use the reverse sequence. For example, if, in $S \times A$, there are six a_a treatments, half the subjects might receive, in order, a_1, a_2, a_3, a_4, a_5, and a_6, whereas the other half would receive a_6, a_5, a_4, a_3, a_2, a_1. Although this method of counterbalancing was used in former times, it cannot be recommended. It is mentioned here only to illustrate how carry-over effects can distort conclusions.

Suppose that some sort of warm-up effect were operating that caused a gradual increase in scores over successive treatments, but that the treatments per se had no effect on the scores. Then, if $A = 6$, the expected scores might be:

	Order					
	1	2	3	4	5	6
Sequence	a_1	a_2	a_3	a_4	a_5	a_6
Expected score	3.0	3.1	3.2	3.3	3.4	3.5
Sequence	a_6	a_5	a_4	a_3	a_2	a_1
Expected score	3.0	3.1	3.2	3.3	3.4	3.5

The mean of the expected scores for each a_a would be the same, 3.25, which correctly reflects the absence of a true treatment effect. However, in the preceding example, the increase in expected scores is linear: The scores increase by .1 for each successive order. A more realistic example shows an increase that is largest for the first few trials and then diminishes—that is, the increase is nonlinear. We might have, for example, the following situation:

	Order					
	1	2	3	4	5	6
Sequence	a_1	a_2	a_3	a_4	a_5	a_6
Expected score	3.0	3.3	3.5	3.6	3.6	3.6
Sequence	a_6	a_5	a_4	a_3	a_2	a_1
Expected score	3.0	3.3	3.5	3.6	3.6	3.6

Then the mean expected scores for the a_a's would be:

a_1	a_2	a_3	a_4	a_5	a_6
3.3	3.45	3.55	3.55	3.45	3.3

The mean expected scores are not equal in this example. It would appear that the scores depend on the particular level a_a, but actually the differences result from a nonlinear order effect. Use of opposite sequences, then, cannot be recommended to balance out carry-over effects. The student should note, however, that the use of opposite orders does produce some leveling of the scores in comparison with the use of one order only. With opposite orders, the expected scores for the different treatments range only from 3.3 to 3.55, whereas with one order only, the range is from 3.0 to 3.6.

All sequences. With opposite orders, only two of the possible treatment sequences are employed. With A treatments, there are $A!$ different possible sequences. The term $A!$ (A factorial) means $A(A-1)(A-2)\ldots(2)$. For example, if $A = 6$, then $A! = 6 \cdot 5 \cdot 4 \cdot 3 \cdot 2 = 720$. For $A = 3$, $A! = 3 \cdot 2 = 6$. The six possible sequences are:

Sequence	Order 1	Order 2	Order 3
1	a_1	a_2	a_3
2	a_1	a_3	a_2
3	a_2	a_1	a_3
4	a_2	a_3	a_1
5	a_3	a_1	a_2
6	a_3	a_2	a_1

If the experiment consists only of administering three a_a's to each subject, then we would use some multiple of six subjects and randomly assign an equal number of subjects to each sequence. If there was but one subject per sequence, the design would be analyzed as an $S \times A$ design. If there were two or more subjects per sequence, the design would include a *sequence factor* B, and the design structure would be $S(B) \times A$. If C was a category or treatment factor and the design was $S(C) \times A$ to begin with, we would ideally have a set of all possible sequences for each group of subjects (each c_c level). If this was not possible, sequences could be randomly assigned to the various c_c levels. If there were more than one subject per sequence, the design would be analyzed as $S(B(C)) \times A$, and B would be a sequence factor as before.

By using all sequences, the experimenter can avoid the kind of distortion that has been previously exemplified as occurring with opposite orders, as well as other distortions due to carry-over effects. However, in many repeated measurements experiments it is impossible to use all sequences, because too many subjects are required. We noted previously, for example, that $6! = 720$. In other words, if an experiment requires each subject to receive 6 treatments, the experimenter needs 720 subjects to have one subject per sequence. We also note that $4! = 24$, and $5! = 120$. It is clear that the use of all sequences will generally be impossible for designs requiring more than 3 or 4 treatments per subject.

Latin square sequences. Repeated measurements Latin square designs can be used to balance out pure order effects such as those discussed in connection with opposite sequences. For example, the basic repeated measurements Latin square design of Section 7.5 can be used to determine the sequences of the a_a's. The layout for this design is as follows:

		b_1	b_2	b_3
c_1	S_1 S_2 S_3	a_2	a_1	a_3
c_2	S_4 S_5 S_6	a_3	a_2	a_1
c_3	S_7 S_8 S_9	a_1	a_3	a_2

When B is a treatment factor, the different subjects in a group use separately randomized treatment sequences. However, when there is only one treatment factor A, we can use the same design, by letting B be a trial (order) factor. Then all subjects in a group (c_c level) receive the a_a treatments in the same order. Of course, the particular sequences of a_a's illustrated here may not be used, since the arrangement of the a_a's is determined by a random process, as explained in Section 7.2.3.

The use of the preceding Latin square with B as a trial factor is basically an alternative to $S \times A$. It is analyzed as explained in Section 7.5. An advantage of the Latin square arrangement is that it allows assessment of score changes over trials; that is, the effect of B, as well as the AB interaction, is analyzed. No such information can be obtained from the $S \times A$ analysis. On the other hand, the Latin square arrangement is not a perfect protection against carry-over distortions. Since each treatment occurs at each order, the design does offer a protection against the pure order effect that was illustrated in the discussion of opposite sequences. However, there are other carry-over effects against which it does not offer protection. Specifically, if a score is differentially affected by the particular treatments preceding it, apart from the number of such treatments, then such carry-over effects can distort the conclusions. The use of $S \times A$ and randomized sequences offers a better protection in such circumstances.

The Latin square design of Section 7.6 can also be used with B as a trial factor. It would be an alternative to using $S(D) \times A$ with individually randomized treatment sequences.

8.2 THE ANALYSIS OF COVARIANCE

8.2.1 Introduction

In the analysis of covariance each subject has another measure associated with himself besides the score (or scores). This other measure is called a *covariate* (or, sometimes, a concomitant variable), and is symbolized by Y. Like the score itself, the covariate is a measure along a continuum, or an approximation thereof. (In this book the term covariate is used for both the variable and the particular value of the variable observed on a subject.) The covariate, if it is to serve its intended purpose in an analysis of covariance, must correlate with the criterion. Innumerable covariates might be used in any experiment but unless one is chosen that correlates substantially with the criterion, there is little to be gained from the extra work that covariance analysis entails. In general, there are two purposes for collecting covariates and doing an analysis of covariance: (1) to increase the power of significance tests, and (2) to obtain more accurate estimates of treatment means. In a specific experiment, one purpose or the other normally predominates.

An important consideration in the analysis of covariance is whether the covariates are affected by the experimental treatments. Although the interpretation of the results from an analysis of covariance is frequently problematical, it is particularly precarious if the covariates are affected by the experimental treatments. If the covariates are collected before the experiment begins, and subjects are then randomly assigned to treatments, the covariates are certainly unaffected by the treatments. Such a covariate might be the subject's IQ, available from college records; or it might be a measure from a pretest given to the subject just previous to experimentation. If, however, the covariate, like the score, is taken after a subject is exposed to his treatment, then these covariates, as well as the scores, may be affected. Simplistic application of the analysis of covariance is then fraught with danger. Of course, the covariates will not necessarily be affected by the treatments just because the treatments precede observation of the covariate.

Analysis of covariance is frequently used when intact groups must be used in place of groups that should actually be formed by random assignment of subjects. At times, however, random assignment can be so inconvenient as to preclude experimentation altogether. For example, educational research often requires that subjects exposed to one treatment or another compose an intact class. It is administratively impossible to regroup all the experimental subjects randomly into new classes. Under such circumstances, an experimenter may have to compromise with the statistical ideal in order to proceed with experimentation at all. The major problem with intact groups is the likely possibility that they will have different scoring tendencies on the criterion apart from any treatments received—tendencies that are not in accord with the statistical theory behind the analysis. These natural tendencies may be mistaken for treatment effects. The analysis of covariance is used in an attempt to adjust mean group scores to remove the bias. Although analysis of covariance has value in this context, it is not really a solution. The technique provides a kind of equating of the groups along the covariate, but the adjustment is not perfect.

It is well to clear up at this point a common misconception about random assignment of subjects. Certainly treatment groups formed from random assignment differ in natural scoring tendency just as do intact groups. Such differences do not invalidate the inferential process of the ANOVA, which takes account of them. The differences among intact groups, however, are not incorporated into ANOVA theory. When an experimenter randomly assigns subjects to groups and then finds that these groups are not perfectly matched according to IQ, age, or other variables, he should not then reshuffle the assignments to force a match, as such reassignments invalidate the analysis.

The real problem with the completely randomized designs and with between-subject comparisons in partially repeated measurements designs is not unequal scoring tendencies but lack of power. The analysis of covariance can be used to enhance the power of statistical tests in such designs, and the technique of using it will be explained in this section. However, the analysis of covariance is only one method for increasing statistical power. As will be explained in Section 8.3, it is probably better to incorporate the covariate as a blocking factor in the experiment and proceed with the usual ANOVA. In other words, if there are 40 subjects, we might use the same covariate to assign subjects to levels of a blocking factor: The 8 subjects with the highest covariates would go in one level, the next 8 would go in another level, and so on. A difficulty with such blocking is that the "covariates" for all experimental subjects must be known before experimentation begins. This constraint does not apply for the analysis of covariance.

Because covariance analysis is questionable in many situations and because there are often superior alternatives to it, the typical experimenter does not find too much use for it. Therefore, the discussion of it here will not be extensive.

8.2.2 How Analysis of Covariance Works

This section explains the use of covariance analysis for a single-factor design and random subject assignment, and then describes other situations in which it may be useful. The score model for this design, ignoring the covariate, is

$$X_{ae} = \mathbf{m} + \mathbf{a}_a + \mathbf{e}_{e(a)}$$

As explained in Section 4.2.3, $\mathbf{e}_{e(a)}$ is simply a residual to explain the discrepancy between an individual's score, X_{ae}, and the expected score for his treatment, $\mathbf{m} + \mathbf{a}_a$. The F-test for the effect of A is based on $\text{MS}_2 = \text{MS}_E$, which estimates σ_E^2. The larger the $\mathbf{e}_{e(a)}$'s, the larger will be MS_E, and the smaller will be the power of the F-test. Term $\mathbf{e}_{e(a)}$ includes a systematic component associated with the individual. We have seen that, by the use of repeated measurements designs, this systematic component can be extracted and attributed to factor S, leaving a smaller σ_E^2 and resulting in more powerful F-tests. The analysis of covariance works in a similar way. Part of $\mathbf{e}_{e(a)}$ is extracted and attributed to the covariate, leaving a smaller error term and resulting in a more powerful test for the effect of A.

We have already noted that, in order for the covariate to serve its purpose, it must correlate with the criterion. The analysis of covariance is simplest if we assume, as is usual, that the relationship between the criterion and the covariate is linear. We also assume that the slope of this linear relationship and the correlation coefficient ρ (rho) is the same within each treatment group.

Simple hypothetical data for such a design are given in Table 8.2-1. The data are plotted in Figure 8.2-1. The scores are also plotted along a vertical line to the right, which shows how the data would appear if there were no covariate Y. We can already intuitively sense that greater discriminability exists among the treatment groups on the basis of the bivariate plot of X and Y with the regression lines than on the basis of score differences along X alone. Such a distinction will also appear in the F-tests, as we shall see.

Figure 8.2-2 is analogous to Figure 8.2-1, but is theoretically oriented. It represents the situation that exists when there are an infinite number of subjects for each a_a. The ellipses are contours of equal density for data points. Assuming that subjects are randomly assigned to treatments and that the covariate measures are not influenced by the treatments, all treatment groups have the same population mean on the covariate Y. This mean is symbolized

Table 8.2-1 Hypothetical data for design A with covariate Y

a_1		a_2		a_3	
X	Y	X	Y	X	Y
2	1	5	2	8	3
1	7	8	6	9	7
4	9	7	12	14	11
7	14	10	16	13	18

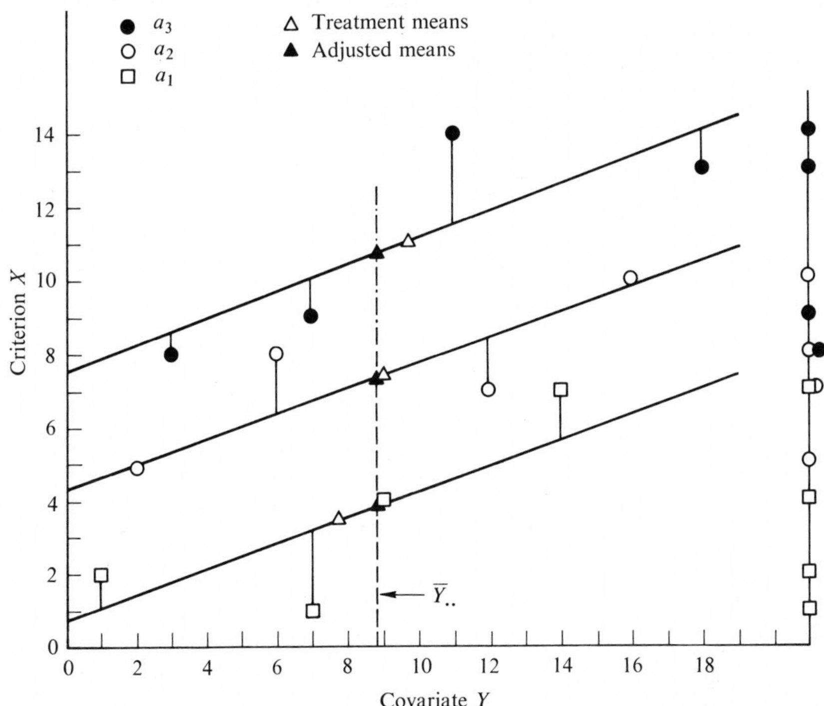

Figure 8.2-1. Plot of the hypothetical data of Table 8.2-1. The straight lines with common slope are those that best fit the data for the three treatments. The data points are also plotted along the single vertical line on the right.

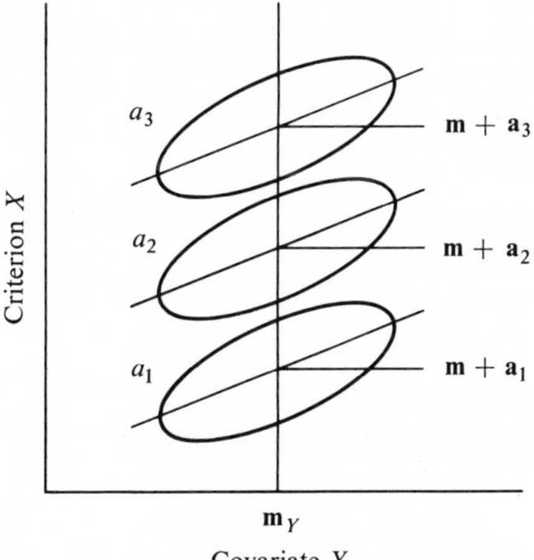

Figure 8.2-2. Illustration of the hypothetical relationship between criterion X and covariate Y for three treatments. The boundary line for each ellipse is a locus of equal probability density for a bivariate probability distribution. For each treatment the best-fitting line relating X to Y is shown.

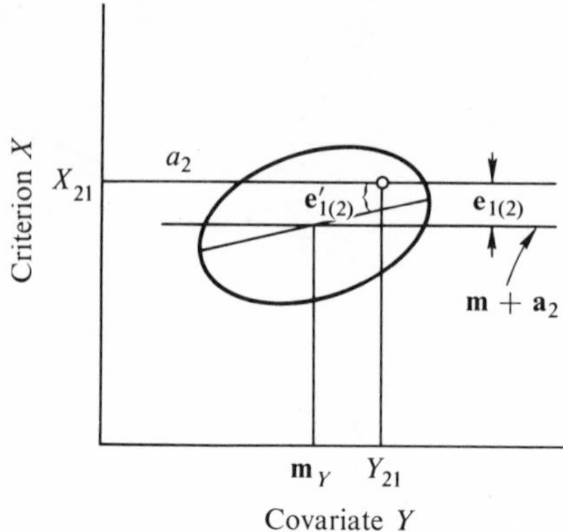

Figure 8.2-3. Without a covariate, the random error component of X_{21} is $\mathbf{e}_{1(2)}$. When a covariate Y_{21} is available, part of $\mathbf{e}_{1(2)}$ is predictable, so the random error is the smaller quantity $\mathbf{e}'_{1(2)}$.

by $\mathbf{m}_Y$. (Mean $\mathbf{m}_Y$ would not generally be the same for treatment groups if intact groups were used or if the treatments affected the covariate measures.) According to our assumptions, the effect of factor A, if any, would be simply to alter the means of the different a_a groups along the criterion. The slope of the elliptical contour and the slope of the (presumed) linear regression of X on Y would remain the same.

Figure 8.2-3 explains why the error terms are generally smaller when a covariate that correlates with the score is available. Without a covariate, the random component $\mathbf{e}_{1(2)} = X_{21} - \mathbf{m} - \mathbf{a}_2$. When the covariate Y_{21} is available, however, part of $\mathbf{e}_{1(2)}$ is predictable, leaving only a generally smaller $\mathbf{e}'_{1(2)}$ as a random component. Because of the smaller error, the power of statistical tests is greater.

8.2.3 Numerical Example

We can study the analysis of covariance by using Table 8.2-1 and Figure 8.2-1. Our analysis yields the following products:

1. Test of a significant difference among the treatment groups.
2. Test for equality of the slope of the regression lines for the separate treatments.
3. Estimates of the treatment means adjusted to account for covariate differences among treatments.

Testing the significance of factor *A*. We begin, as with ANOVA, by computing various sums, and from these we compute SS elements. In this example, however, we have three SS elements for each one we would have for the ANOVA. We have the element for the X's, an analogous element for the

Y's, and a third analogous element on the product of comparable X's and Y's. To distinguish these elements, we append (X), (Y), or (XY) to the SS element subscripts. We also have three SS's for each one we would have for ANOVA, and we distinguish them similarly. Table 8.2-2 illustrates the computations required to find the SS's. The final step is to obtain the adjusted SS's, $SS'_{E(X)}$ and $SS'_{A(X)}$. These values have been corrected by taking account of the covariates.

Table 8.2-3 compares the F-test for factor A obtained from an ANOVA of X alone with that obtained from the analysis of covariance. The latter analysis finds A significant at a smaller significance level than the former. It would be possible for A to be nonsignificant with an ANOVA, but significant with an analysis of covariance. The df for A is the same for both analyses, but the df for E is one less for covariance analysis than for variance analysis, because one df was used up in estimating the slope of the regression line.

Testing the slopes. The best estimate of this slope is:

$$\begin{aligned}\hat{\beta} &= SS_{E(XY)}/SS_{E(Y)} \\ &= 112.5/325.5 = .346\end{aligned}$$

However, the slopes may differ for different treatment groups. We can estimate these separate slopes and test for significant differences among them. The slope β_a for a particular treatment a is estimated as follows:

$$\hat{\beta}_a = SS_{E(XY)_a}/SS_{E(Y)_a},$$

where subscript a means the calculations are limited to treatment a. The SS elements required to compute these estimates are:

$SS_{E(Y)_a}$:

$$\begin{aligned}SS_{E(Y)_1} &= SS_{ae(Y)_1} - SS_{a(Y)_1} \\ &= 327 - [31^2/4] \\ &= 327 - 240.25 = 86.75\end{aligned}$$

$$\begin{aligned}SS_{E(Y)_2} &= SS_{ae(Y)_2} - SS_{a(Y)_2} \\ &= 440 - [36^2/4] \\ &= 440 - 324 = 116\end{aligned}$$

$$\begin{aligned}SS_{E(Y)_3} &= SS_{ae(Y)_3} - SS_{a(Y)_3} \\ &= 503 - [39^2/4] \\ &= 503 - 380.25 = 122.75\end{aligned}$$

$SS_{E(XY)_a}$:

$$\begin{aligned}SS_{E(XY)_1} &= SS_{ae(XY)_1} - SS_{a(XY)_1} \\ &= 143 - [14 \cdot 31/4] \\ &= 143 - 108.5 = 34.5\end{aligned}$$

Table 8.2-2 Analysis of covariance: numerical example

Data and basic sums:

	X_{1e}	Y_{1e}	X_{2e}	Y_{2e}	X_{3e}	Y_{3e}	Totals X	Totals Y
	2	1	5	2	8	3		
	1	7	8	6	9	7		
	4	9	7	12	14	11		
	7	14	10	16	13	18		
$\sum(\ )$	14	31	30	36	44	39	88	106
$\sum(\ )^2$	70	327	238	440	510	503	818	1270
$\sum XY$	143		302		475		920	

SS elements:

SS_{ae}:

$SS_{ae(X)} = \sum X_{ae}^2 = 818$ $\qquad SS_{ae(XY)} = \sum X_{ae}Y_{ae} = 920$

$SS_{ae(Y)} = \sum Y_{ae}^2 = 1270$

SS_a:

$$SS_{a(X)} = \sum X_{a.}^2/E = (14^2 + 30^2 + 44^2)/4 = 3032/4 = 758$$

$$SS_{a(XY)} = \sum X_{a.}Y_{a.}/E = (14\cdot 31 + 30\cdot 36 + 44\cdot 39)/4 = 3230/4 = 807.5$$

$$SS_{a(Y)} = \sum Y_{a.}^2/E = (31^2 + 36^2 + 39^2)/4 = 3778/4 = 944.5$$

SS_1:

$$SS_{1(X)} = X_{..}^2/AE = 88^2/12 = 7744/12 = 645.33$$

$$SS_{1(XY)} = X_{..}Y_{..}/AE = 88\cdot 106/12 = 777.33$$

$$SS_{1(Y)} = Y_{..}^2/AE = 106^2/12 = 11{,}236/12 = 936.33$$

SS's:

SS_{tot}:

$$SS_{tot(X)} = SS_{ae(X)} - SS_{1(X)} = 818 - 645.33 = 172.67$$

$$SS_{tot(XY)} = SS_{ae(XY)} - SS_{1(XY)} = 920 - 777.33 = 142.67$$

$$SS_{tot(Y)} = SS_{ae(Y)} - SS_{1(Y)} = 1270 - 936.33 = 333.67$$

SS_A:

$$SS_{A(X)} = SS_{a(X)} - SS_{1(X)} = 758 - 645.33 = 112.67$$

$$SS_{A(XY)} = SS_{a(XY)} - SS_{1(XY)} = 807.5 - 777.33 = 30.17$$

$$SS_{A(Y)} = SS_{a(Y)} - SS_{1(Y)} = 944.5 - 936.33 = 8.17$$

SS_E:

$$SS_{E(X)} = SS_{ae(X)} - SS_{a(X)} = 818 - 758 = 60$$

$$SS_{E(Y)} = SS_{ae(Y)} - SS_{a(Y)} = 1270 - 944.5 = 325.5$$

$$SS_{E(XY)} = SS_{ae(XY)} - SS_{a(XY)} = 920 - 807.5 = 112.5$$

Table 8.2-2 (continued)

Adjusted SS's:

$$SS'_{tot(X)} = SS_{tot(X)} - (SS^2_{tot(XY)}/SS_{tot(Y)}) = 172.67 - (142.67^2/333.67) = 172.67 - 61.00 = 111.67$$

$$SS'_{E(X)} = SS_{E(X)} - (SS^2_{E(XY)}/SS_{E(Y)}) = 60 - (112.5^2/325.5) = 60 - 38.88 = 21.12$$

$$SS'_{A(X)} = SS'_{tot(X)} - SS'_{E(X)} = 111.67 - 21.12 = 90.55$$

$$SS_{E(XY)_2} = SS_{ae(XY)_2} - SS_{a(XY)_2} = 302 - [30\cdot 36/4] = 302 - 270 = 32$$

$$SS_{E(XY)_3} = 475 - [44\cdot 39/4] = 475 - 429 = 46$$

From these quantities we compute

$$\hat{\beta}_1 = SS_{E(XY)_1}/SS_{E(Y)_1} = 34.5/86.75 = .398$$

$$\hat{\beta}_2 = SS_{E(XY)_2}/SS_{E(Y)_2} = 32/116 = .276$$

$$\hat{\beta}_3 = SS_{E(XY)_3}/SS_{E(Y)_3} = 46/122.75 = .375$$

To determine whether the differences are statistically significant, we compute the F-ratio

Table 8.2-3 Tables for analysis of variance (*top*) and analysis of covariance (*bottom*)

Source	SS	df	MS	F
A	112.67	2	56.34	8.45*
E	60	9	6.67	

* $p < .01$.

Source	SS	df	MS	F
A	90.55	2	45.28	17.15**
E	21.12	8	2.64	

** $p < .001$.

$$F = \frac{SS_{\beta}/(A-1)}{SS_{\beta E}/A(E-1)}$$

where

$$\begin{aligned} SS_{\beta E} &= SS_{E(X)} - \sum \hat{\beta}_a SS_{E(XY)_a} \\ &= 60 - [.398 \cdot 34.5 + .276 \cdot 32 + .375 \cdot 46] \\ &= 60 - 39.81 = 20.19 \\ SS_{\beta} &= \sum \hat{\beta}_a SS_{E(XY)_a} - \hat{\beta} SS_{E(XY)} \\ &= 39.81 - .346 \cdot 112.5 = .89 \end{aligned}$$

The H_N is that the β_a's do not differ, and we find that

$$F = \frac{.89/2}{20.19/6} = .13,$$

which clearly is nonsignificant. We conclude, therefore, that the different treatments have a common regression coefficient β relating X to Y. Although the foregoing F-test has been presented subsequent to the test for the significance of factor A, the validity of the latter test depends on acceptance of H_N that the β_a's are equal.

Adjusting the treatment means. Besides yielding generally smaller error terms, analysis of covariance allows more accurate estimates of the treatment means, and hence of the $\mathbf{a}_a$ parameters. As we have noted, for an infinite population of randomized subjects the distribution of the covariate is the same for each treatment, which implies, among other things, that the mean of the covariates is the same. This does not imply, however, that the mean covariate values for the sample of subjects in a particular experiment will be the same across treatments. On the contrary, these means will in general differ, and these differences are the basis for adjustment of the treatment score means.

We assume here that the same β applies to all treatments. Then the formula for the best estimate of the regression line relating X and Y for a treatment a is

$$X_a = \bar{X}_{a.} + \hat{\beta}(Y - \bar{Y}_{a.})$$

The regression lines in Figure 8.2-1 were determined from this formula. When Y equals $\bar{Y}_{a.}$ (the mean covariate for a treatment), X equals $\bar{X}_{a.}$, the mean score for that treatment. Hence, the regression line for a treatment passes through the point $(\bar{X}_{a.}, \bar{Y}_{a.})$, determined by the X and Y means for the treatment.

Without the covariate, the best estimate for the expected treatment mean is simply $\bar{X}_{a.}$. The adjusted estimate is that value on the treatment regression line corresponding to $\bar{Y}_{..}$, the mean of all covariates in the experiment. We find this value of X by replacing Y in the preceding equation with $\bar{Y}_{..}$, which yields the adjusted estimate

$$\bar{X}'_{a.} = \bar{X}_{a.} + \hat{\beta}(\bar{Y}_{..} - \bar{Y}_{a.})$$

The adjusted estimates are shown in Figure 8.2-1; they are simply those points on the regression lines directly above $Y = \bar{Y}_{..}$, and hence they all fall on the same horizontal line. For positive $\hat{\beta}$, when the mean covariate $\bar{Y}_{a.}$ for a group falls above $\bar{Y}_{..}$, $\bar{X}_{a.}$ is adjusted downward, and when $\bar{Y}_{a.}$ is below $\bar{Y}_{..}$, the adjustment is upward. The reasoning is that if $\bar{Y}_{a.}$ is higher than $\bar{Y}_{..}$, it is so by chance, and because of the relationship between Y and X, $\bar{X}_{a.}$ is also high. In the average replication, both $\bar{Y}_{a.}$ and $\bar{X}_{a.}$ would be lower.

8.2.4 Extensions of Covariance Analysis

Analysis of covariance can be applied more generally than to the one-factor Class I design of the preceding example. The details are not provided here, but the possibilities are indicated to help the student decide whether he should investigate the topic more thoroughly elsewhere in order to achieve his experimental aims. When the main purpose of covariance analysis is to increase statistical power, as in the preceding example, it can only be effective for tests on between-subject effects, not on within-subject effects. Hence, it would not be useful for Class IV designs, where all effects tested are within-subject effects. In a Class III design such as $S(A) \times B$, a covariate could enhance the power for testing the between-subject effect A, but not for testing the within-subject effects of B and AB. Analysis of covariance can increase the power for all tests in Class I and Class II designs. It can be applied to Latin square designs as well as to compact designs, though, again, it would not enhance the power for testing the within-subject effects of repeated measurements Latin square designs.

When the purpose of analysis of covariance is to increase statistical power, each subject will have only one associated covariate measure, even in a Class III design having repeated measures. At times covariance analysis is done on a repeated measurements design in which there is a covariate for each score—that is, there are several covariates for each subject. In such a situation the covariate must be influenced by the treatment (or else there would be no point in having separate covariates for the different treatments). The purpose of the covariance analysis then is primarily to adjust the treatment means via the covariate, so that comparisons among the treatments relate to covariate-adjusted values. As was pointed out earlier, the results of such a covariance analysis must be interpreted with much care and caution.

Although in the example, the regression of X on Y within a treatment was assumed to be linear, any relationship may, in principle, form the basis of a covariance analysis. It should be apparent that a reduction in the size of the effective error and adjustments of the treatment means could occur for nonlinear regressions, although the mathematics would be even more complicated than with the linear analysis. Actually, nonlinear analyses of covariance in psychology are rare.

There is no requirement that an analysis of covariance utilize only one covariate Y. An experimenter may have, in addition, a second covariate Z; that is, each subject might have two other measures besides his score X—a covariate Y and a covariate Z. It is possible to use even more covariates. Whatever one covariate can contribute to a design, two or more covariates may be able to contribute more of, but at an increase in computational effort. Also, in general, the law of diminishing returns applies to additional covariates, so that the marginal advantage of using more than one or two would normally be slight, at best.

8.3 BLOCKING AS AN ALTERNATIVE TO COVARIANCE ANALYSIS

8.3.1 Comparative Merits

Covariance analysis is often considered together with *blocking* (or *stratification*, as it is often called) of subjects, as alternative methods for increasing the power of a design without using repeated measurements. The same variable might be used either as a covariate or as a blocking factor. For example, intelligence scores of subjects could be used as a covariate Y_{ae} in the completely randomized design A, or the subjects could be blocked into levels of factor B, intelligence, and an $A \times B$ analysis of variance performed. Though both approaches accomplish the same end of increasing design power, there are considerations favoring one or the other that the experimenter should keep in mind.

The following considerations favor covariance analysis. First, covariance analysis can result in more powerful tests. In part, this occurs because the exact covariate measure is used in covariate analysis, whereas there is a loss of information when a group of somewhat disparate measures are blocked and treated as if they were equal—that is, available information about the differences in the blocking variable among subjects in the same block is not utilized. This advantage, however, would normally be moderate for most experiments. (In some situations, blocking can have a small advantage over covariate analysis.)

Second, covariance analysis allows greater flexibility in the time and manner of obtaining the covariate measure. When blocking is used, the blocking measure must be available on all subjects before experimentation commences so that an equal number of subjects in each block can be randomized into each treatment. Otherwise, the experimenter simply randomizes all subjects across the treatments, and there would seldom be an equal number of subjects from each block (as determined after randomization) in each treatment. ANOVA requires this equality, however, at least if computations are to be fairly simple.

With covariance analysis, there is more flexibility. The experimenter can have all the covariate measures before experimentation begins; or he can collect each covariate as a pretest as each subject arrives for experimentation; or he can even go to the school records as an afterthought, when experimentation is completed, to get an intelligence or achievement score to use as a covariate. An experimenter can even have subjects return after experimentation to collect the covariates, although there can be some danger here from holding a false assumption that the treatments will not affect the covariates.

The following considerations favor blocking. First, by using blocking an experimenter can avoid making the assumptions required for covariance analysis and thereby avoid the distortions that might enter into the analysis if the assumptions of covariance analysis are faulty. Although ANOVA in general is replete with assumptions having various degrees of dubiousness, it is preferable to avoid compounding the uncertainties of the analysis by conditioning the conclusions on still additional questionable assumptions. For example, the usual form of covariance analysis assumes that the regression of the score on the covariate is linear and has the same slope (the same β) for each treatment. Although these assumptions can be tested, if they are shown to fail it is difficult to proceed with a proper analysis. Also, the power in testing these assumptions may be small.

Second, the analysis of the data is simpler when blocking is used. The experimenter simply performs the usual ANOVA, treating the blocking factor no differently from any other. The analysis of covariance is computationally more complicated.

Although multiple covariates can be used, it is not feasible to use multiple blocking factors, as was explained in Section 3.8.2. Of course, with multiple covariates the risk of proceeding with faulty assumptions is increased. If there are two or more potential covariates, the experimenter may use one of them as a blocking variable and use the other(s) for covariance analysis.

In conclusion, if there is one correlative measure for each subject, it seems simpler and safer to use this measure as a blocking factor rather than as a covariate, if possible. The slight loss of power that results is more than compensated for by the relative simplicity of the assumptions and computations required. When the correlative measures cannot be obtained in time for planning a stratified design, analysis of covariance is a reasonable alternative. It should also be remembered that repeated measurements experimentation can be expected to increase test power for within-subject tests over the power for a completely randomized design more than either blocking or covariance analysis. Repeated measurements experiments, however, have their own drawbacks, as we have noted.

8.3.2 How Many Blocks Should Be Used?

One subject per condition. Given that an experimenter wishes to use a blocking factor, he must still specify how many levels (or strata) this factor should have. Let A be the treatment factor that he sets out to investigate, and let B be the blocking factor. The minimum number of subjects per block would be A, which would permit only one subject per cell. If there were a total of N subjects in the experiment, there would be N/A levels. (For example, if there were $A = 4$ treatments and $N = 40$ subjects, an experimenter could have 10 levels of factor B with 4 subjects per level and one subject per cell.)

A special case of such a randomized blocks design is the *matched-pair design.* Subjects in each adjacent pair along the concomitant variable are randomly assigned to two treatment groups, one subject to each group. The difference score, $X_{2b} - X_{1b}$, is computed for each pair, and a t-test is performed on the differences to test the H_N that $\mathbf{m}_2 - \mathbf{m}_1 = 0$. The test is equivalent to the F-test that would be performed by conceiving the design as an $A \times B$ randomized blocks design with $A = 2$ and B equal to the number of subject-pairs.

The score model and analysis for the randomized blocks design with one subject per condition is analogous to the model and analysis for design $S \times A$, with B equivalent to S. In essence, in design $S \times A$ each "block" s_s has a "group" of "similar" subjects (namely, the same subject, which is nothing but matching carried to the maximum possible limit). The fact that B is taken to be a fixed factor whereas S is a random factor does not affect the MS_2 for testing the effect of A, because when there is only one subject per condition, there can be no separate estimate of error, so MS_2 is the interaction MS in both cases. When the randomized blocks design has more than one subject per condition, however, a separate MS_E can be calculated, and this is used as MS_2 to test factor A, assuming that B is taken to be a fixed factor.

More than one subject per condition. Although randomized blocks designs with one subject per condition are described in textbooks and used by experimenters, as a general rule it is preferable to use at least two subjects per condition. As previously mentioned, when only one subject is used per condition, there can be no separate MS for error. In effect, the significance test for factor A assumes the absence of an AB interaction. When two or more subjects are used per condition, the experimenter can use MS_E to test for the presence of this interaction and can also use MS_E to test the main effects. The existence of an AB interaction means that the effect of factor A differs for different levels of the concomitant variable. For example, two treatments may make a great difference for subjects with low IQ's, but very little difference for those with high IQ's. It would be misleading to interpret the results of an experiment without regard to such an interaction, if one existed.

Granted, then, that the existence of two or more subjects per condition is desirable, exactly how many strata should be used? There is usually a choice ranging from a minimum of two strata (one level amounts to no blocking at all) to the maximum number of levels possible that will still allow two subjects per condition. For example, if there are approximately 50 subjects and $A = 5$, an experimenter has the following possibilities:

Number of strata B	Number of treatments $A = 5$	Subjects per condition E	Total subjects $B \cdot A \cdot E = N$
2	5	5	50
3	5	3	45
4	5	2	40
5	5	2	50

The experimenter might at first think that the more strata the better, for the greater the number of strata the more homogeneous (similar) will be the subjects within each strata, and hence the smaller will be the residual error terms. This reasoning is sound except that it ignores the decrease in df for error as the number of strata increases. Since $\text{df}_E = AB(E - 1)$ in the preceding example, when $B = 2$ strata, $\text{df}_E = 40$, and when $B = 5$ strata, $\text{df}_E = 25$, though the total number of subjects remains the same. Although test power increases when residual error decreases, test power decreases when df_2 decreases, assuming everything else to be constant. Because of this relationship, increasing the number of strata beyond some point can yield diminishing returns.

The optimal number of strata depends on the strength of correlation ρ between the concomitant variable and the criterion for a treatment group. If the correlation is very low (say, zero), stratification is a futile exercise. Subjects so stratified are not more homogeneous on the criterion, and thus residual error is not reduced. Stratification does, however, diminish df_E, and hence test power is smaller than it would be for the completely randomized design. The higher the ρ, the greater the number of strata for which the experimenter can justify sacrificing error df. Feldt has determined the optimal number of levels as a function of ρ and N for $A = 2$ and $A = 5$. His results are given in Table 8.3-1. There is no integral number E for many entries that will give exactly the N tabled, so for these entries an experiment would utilize

Table 8.3-1 Optimal number of strata, B

ρ	A	N					
		20	30	50	70	100	150
.2	2	2	3	4	5	7	9
	5	1	2	2	3	4	6
.4	2	3	4	6	9	13	17
	5	2	3	4	5	7	10
.6	2	4	6	9	13	17	25**
	5	2	3	5	7	9	14
.8	2	5*	7*	12*	17	23	25**
	5	2*	3*	5*	7*	10	15*

SOURCE: From L. S. Feldt, "A comparison of the precision of three experimental designs employing a concomitant variable," in *Psychometrika* 23: 335–353 (1958).
* Limit due to requirement that $E \geq 2$.
** Slight improvement possible with more than 25 levels.

slightly more or fewer than N subjects. Feldt concluded that, for $\rho < .2$, neither the covariance analysis nor blocking allow much improvement in power over the completely randomized design. Correlations above .6 will seldom occur. Therefore, although an experimenter normally does not know the ρ that applies to his experiment, he might assume a value of approximately .4 and stratify accordingly.

It is apparent that specification of the number of strata in an exact, scientific way is generally impossible. Nonetheless, the suggestions that have been mentioned can serve as guidelines:

1. Beware of excessive impoverishment of df_E.
2. Use more strata, the larger ρ is believed to be.
3. A very weakly correlated concomitant variable is not worth using.

8.4 POWER CONSIDERATIONS

8.4.1 In General

An elementary review of the concept of test power was given in Section 2.5. The power of a statistical test is the probability that H_N will be rejected, given that H_A is true. An experiment aimed at establishing H_A is a futile exercise if test power is low, because even if H_A is true, the experimental conclusion is likely to be that it is false.

Ideally, test power should be included among the specifications for a design, at least for those H_N's of greatest importance. Many completed experiments probably would have been aborted as useless before execution, or redesigned for greater power, if test power had been known beforehand. In practice, however, test power is not frequently computed or considered. One reason is that power calculations require that an experimenter have at least an estimate of the error variance (or, more generally, of EMS_2). This information is obtained from the ANOVA, but by this time the experiment has been completed. The estimate must be available before the experiment is run. Although previous experimentation (perhaps a pilot study) may provide an estimate, a reliable estimate is usually lacking, and discrepancies between the estimate of EMS_2 and the true value can seriously distort the estimate of power.

Another difficulty is that power depends on the discrepancy between H_N and H_A. Hence, to compute power, the experimenter must quantitatively specify some H_A. For example, relative to the H_N that $\sigma_A^2 = 0$, for H_A he has to specify σ_A^2 in order to compute a power. For fixed A, this typically means specifying each $\mathbf{a}_a$. Normally, however, an experimenter has no specific H_A in mind. He waits for the data to inform him about the $\mathbf{a}_a$'s.

In specifying H_A, the experimenter does not state his best guesses of the $\mathbf{a}_a$'s. He instead states the smallest values of $\mathbf{a}_a$'s (whether positive or negative) that would be of scientific or practical interest; more precisely, he states the smallest σ_A^2 that would be of interest. If the power is adequate for such $\mathbf{a}_a$'s, the power would be even higher for a stronger effect of factor A. Such judgments about minimal $\mathbf{a}_a$'s are usually highly subjective and arbitrary.

Considering these difficulties, it is understandable that power calculations are not a routine ingredient in experimental design. Nonetheless, the experimenter shows an appreciation of the importance of power when he chooses a design that yields a smaller error variance than a Class I design. In addition, even though a power calculation is ideally used as a design aid, when lack of the required estimate of error variance prevents such use, a power calculation can also be useful in analysis, after the required estimate becomes available. When an effect in which an experimenter had confidence fails to reach significance, he may find it worthwhile to determine whether the power for this effect was reasonably high. If the power was low, the effect may have been lost to a type II error. If the power was high, the experimenter must reconsider his initial belief.

8.4.2 Fixed Effects

We now turn to the specific details of power calculations. These details vary, depending primarily on whether the effect is fixed or random. An effect is fixed if all the bold-faced letters in its model term refer to fixed factors. An effect is random if any of these letters refer to random factors. For example, the effect AB is fixed only if both factors A and B are fixed. We begin by considering power calculations for fixed effects only.

When the test whose power we wish to find is an F-test and the effect we are interested in is fixed, the power is the area above the critical F for a noncentral F-distribution. Such a quantity is not easy to compute, so special charts have been derived to aid in power determinations. In designing an experiment, we are not interested so much in computing a power for a design already specified as in specifying a design so that some desired level of power is reached. In this respect, the Feldt-Mahmoud charts given in Appendix B.1 are helpful. By using these charts we can quickly estimate how many subjects E would be needed per treatment for the one-factor completely randomized design A, in order to attain a given power. Each chart concerns only one value for A, the number of treatments. Only three levels of power (.5, .7, and .9) and only two significance levels (.05 and .01) are graphed. To determine the number of subjects needed per treatment, we enter the abscissa with a parameter ϕ', defined as

$$\phi' = \sqrt{\frac{\sum \mathbf{a}_a^2/A}{\sigma^2}},$$

where σ^2 is the EMS for MS_2 (that is, σ_E^2). We trace up from ϕ' to the appropriate power graph and then over to the abscissa to find E. For example, suppose $A = 4$, and we wish to find E for $\mathbf{a}_1 = -1.5$, $\mathbf{a}_2 = -1$, $\mathbf{a}_3 = 1$, $\mathbf{a}_4 = 1.5$, a power of .9, and $\alpha = .05$. Though σ^2 is not known, data from similar experiments allow us to estimate σ_E^2 as 13. Then

$$\phi' = \sqrt{\frac{[(1.5)^2 + (-1)^2 + 1^2 + 1.5^2]/4}{13}}$$

$$= \sqrt{\frac{6.5}{52}} = \sqrt{.125} = .35$$

We locate .35 on the abscissa of Figure B.1-3 using the upper row for $\alpha = .05$, and using the line for power = .9, we estimate E as 30. The total N would be $30 \cdot A = 120$ subjects. If power = .9 for $\alpha = .05$ is the minimum acceptable power, either we must be able to run 120 subjects or we must use a different design.

The Feldt-Mahmoud charts are strictly valid only for the one-factor completely randomized design. However, they may also be used for two-factor designs with only slight error when $\mathrm{df}_2 \geq 20$. Then the value of "E" taken from the chart is actually the number of subjects at each level of factor A, that is, $B \cdot E$. We may likewise determine the value of "E" to attain a given power for factor B, a value that actually would equal $A \cdot E$. The number of subjects E per condition may not correspond for the two factors, so the larger value must be used to attain the power required for both factors.

Also useful are the Pearson-Hartley charts for the power of the F-test (Appendix B.2). To use these charts, first compute ϕ, which, for a one-factor design, is defined as

$$\phi = \sqrt{\frac{\frac{E}{A}\sum \mathbf{a}_a^2}{\sigma^2}}$$

Note that $\phi = \sqrt{E}\phi'$ for this design. Locate ϕ along the abscissa of the chart for the appropriate df_1. Then, using the graph line for the appropriate df_2, read the power off the ordinate. Suppose, for example, that $E = 30$, $A = 4$, $\sum \mathbf{a}_a^2 = 6.5$, and $\sigma_E^2 = 13$. Then

$$\phi = \sqrt{\frac{30 \cdot 6.5}{4 \cdot 13}} = \sqrt{3.75} = 1.94$$

For a one-factor design, $df_2 = (30 - 1)4 = 116$. Although there is no line for $df_2 = 116$, there are lines for $df_2 = 60$ and $df_2 = \infty$, so we can determine that the power lies very close to .9. Since the same data were used for this example as for the preceding one, the result should be exactly .9, the power we assumed in the preceding example. Because of rounding errors and difficulties of making precise readings from the graphs, especially where interpolation is required, only a close, not an exact, correspondence between the findings can be expected.

In using the Feldt-Mahmoud charts we assume a power and determine the number of subjects N. With the Pearson-Hartley charts, we assume N and calculate the power. If our purpose is to find the N required for a given power, the Feldt-Mahmoud charts can be very handy. However, the range of application of these charts is less extensive than for the Pearson-Hartley charts, which can also be used to find the required N, although with somewhat greater effort. We simply start with some reasonable number N and find the corresponding power. If this power is too low, we try a larger value of N. If this value gives a power that is needlessly large, we try a smaller value of N. In this manner we can eventually find the required N for a given power.

The Pearson-Hartley charts apply to designs with any number of factors, but the formula for ϕ varies depending on the design. The following procedure is used to find ϕ for a specific case:

1. Write the EMS for MS_1.
2. Express the $k_\tau\sigma_\tau^2$ component for the effect under consideration via parameters instead of via σ_τ^2.
3. In $k_\tau\sigma_\tau^2$ isolate the coefficient of $\sum(\ \ )^2$, the sum of the parameters squared for the effect we are interested in.
4. Add one to the denominator of this coefficient.
5. Call the coefficient thus adjusted υ (upsilon).

Then

$$\phi = \sqrt{\frac{\upsilon\sum(\ \ )^2}{\sigma^2}}$$

Again, σ^2 is EMS_2, so it may or may not equal σ_E^2. Recall that the meaning of the σ_τ^2's in terms of model parameters was discussed in Section 6.3. The charts being discussed can be used only if the τ of σ_τ^2 involves only fixed factors.

As an example, consider design $A \times B$ with A and B fixed. In order to find the power for a specific σ_B^2, recall that $\text{EMS}_B = \sigma_E^2 + AE\sigma_B^2$ (Section 6.5.2). According to Section 6.3.3, $\sigma_B^2 = \sum \mathbf{b}_b^2/(B-1)$. The coefficient of $\sum \mathbf{b}_b^2$ is therefore $AE/(B-1)$, so $\upsilon = AE/B$, and

$$\phi = \sqrt{\frac{\frac{AE}{B}\sum \mathbf{b}_b^2}{\sigma^2}}$$

To find the ϕ for the interaction effect AB of the same design, note that $\text{EMS}_{AB} = \sigma_E^2 + E\sigma_{AB}^2$, so the coefficient of $\sum \mathbf{ab}_{ab}^2$ is $E/(A-1)(B-1)$, and hence

$$\phi = \sqrt{\frac{\frac{E}{(A-1)(B-1)+1}\sum \mathbf{ab}_{ab}^2}{\sigma^2}}$$

For design $B(A)$, with both factors fixed, $\text{EMS}_B = \sigma_E^2 + E\sigma_B^2$, where $\sigma_B^2 = \sum_{ab} \mathbf{b}_{b(a)}^2/A(B-1)$, so

$$\phi = \sqrt{\frac{\frac{E}{A(B-1)+1}\sum \mathbf{b}_{b(a)}^2}{\sigma^2}}$$

Note that for each case the numerator of υ is the number of design scores to which a particular parameter applies, and the denominator is df + 1 for the effect under consideration.

8.4.3 Random Effects

When the effect is random, the distribution of the F-ratio over replications is neither the central nor the noncentral F-distribution. The distribution can, however, be expressed in terms of the central F-distribution. For the one-factor design A, A random, the distribution of the F-ratio under the H_N that $\sigma_A^2 = 0$ is $F(\text{df}_1, \text{df}_2)$. Under H_A that $\sigma_A^2 > 0$, the distribution is a constant times $F(\text{df}_1, \text{df}_2)$, the constant being $1 + E\sigma_A^2/\sigma_E^2$. The power of the test is the probability that the F-ratio exceeds the critical F, $F_{100(1-\alpha)}(\text{df}_1, \text{df}_2)$, which is equal to the probability that the central $F(\text{df}_1, \text{df}_2)$ exceeds $F_{100(1-\alpha)}(\text{df}_1, \text{df}_2)/(1 + E\sigma_A^2/\sigma_E^2)$.

Suppose, for example, that $E = 6$, $\text{df}_1 = 3$, $\text{df}_2 = 20$, and $\alpha = .05$. Then $F_{100(1-\alpha)}(\text{df}_1, \text{df}_2) = F_{95}(3, 20) = 3.10$. Under H_N that $\sigma_A^2 = 0$, $F(3, 20)$ exceeds 3.10 with a probability of $\alpha = .05$. The probability of rejecting H_N is likewise $\alpha = .05$. If σ_A^2 exceeds 0 (say, $\sigma_A^2/\sigma_E^2 = 2$), the power is the probability that $F(3, 20)$ exceeds $F_{95}(3, 20)/(1 + 6\cdot 2) = 3.10/13 = .238$. To find the percentile score P for which $F_P(3, 20) = .238$, we require more complete tables of F-distributions than those found in the usual textbook, including this one. We would need a table for $F(3, 20)$ having many closely spaced entries, similar to Table A.3 for the normal distribution. Then we would look for a value of F of .238 and see to which percentile P it corresponds. Although the table given in Graybill[1] is not as complete as Table A.3, from it we see that $F_{10}(3, 20) = .193$ and $F_{25}(3, 20) = .406$. We conclude that the percentile score for $F(3, 20) = .238$ is a little over 10. The power, being the area above .238, would then be a little less than .90.

The power for other random effects is found similarly, except that the constant that was $1 + E\sigma_A^2/\sigma_E^2$ for the one-factor design changes. In general, when a random effect is tested by $F = \text{MS}_1/\text{MS}_2$, then when H_N is true, the F-ratio is distributed over replications as $F(\text{df}_1, \text{df}_2)$; when H_A is true, the F-ratio is distributed as

$$\frac{\text{EMS}_1}{\text{EMS}_2} F(\text{df}_1, \text{df}_2)$$

For the one-factor design just considered, $\text{EMS}_1 = \sigma_E^2 + E\sigma_A^2$, and $\text{EMS}_2 = \sigma_E^2$; hence the constant is

$$\frac{\text{EMS}_1}{\text{EMS}_2} = \frac{\sigma_E^2 + E\sigma_A^2}{\sigma_E^2} = 1 + E\frac{\sigma_A^2}{\sigma_E^2},$$

which accords with the assertion made earlier. Now consider design $A \times B$, with both factors random. A test of H_N that $\sigma_A^2 = 0$ uses the F-ratio $\text{MS}_A/\text{MS}_{AB}$. Since $\text{EMS}_A = \sigma_E^2 + E\sigma_{AB}^2 + BE\sigma_A^2$, and $\text{EMS}_{AB} = \sigma_E^2 + E\sigma_{AB}^2$, the constant is

$$\frac{\sigma_E^2 + E\sigma_{AB}^2 + BE\sigma_A^2}{\sigma_E^2 + E\sigma_{AB}^2} = 1 + \frac{BE\sigma_A^2}{\sigma_E^2 + E\sigma_{AB}^2}$$

8.5 GROUP VERSUS INDIVIDUAL RUNNING OF EXPERIMENTS

Some experiments must be run on only a single subject at a time, perhaps because the experimental equipment can service only a single subject, or perhaps because the experiment requires that the experimenter pay considerable individual attention to the subject. In other experiments, subjects may be run at the same time in groups. For example, the experimenter can read instructions to the group as a whole, if all subjects are to receive the same instructions. Then subjects can be presented with the same stimuli (perhaps visual stimuli large enough to be seen by all, or auditory stimuli that can be heard by all), and then each marks his response on his own score sheet.

[1] F. A. Graybill, *An Introduction to Linear Statistical Models*, Vol. 1 (New York: McGraw-Hill, 1961), pp. 425–443.

Many experiments in social psychology—those concerning problem-solving by "brainstorming," for example—are designed to study interactions of persons in groups. In such experiments, of course, subjects cannot be run individually. This section is not concerned with such experiments in social psychology, but with experiments run on groups of subjects that could logically be run one subject at a time, and whose results would have been the same (it is assumed) had subjects been run singly. The term "group experiments," as used in this section, refers to these latter experiments.

The great advantage of such group experimentation is a considerable saving in time for whoever runs the experiment. There is a similar saving in the number of hours that some room must be reserved for the experiment; this saving can be important when such space is at a premium. For example, suppose that all 10 subjects in each cell of an $A \times B$ design with $A = B = 2$ can be run in a group. If the experiment takes one hour to run, the experimenter who uses groups needs to devote only 4 hours to data collection, instead of the 40 hours required to run subjects individually.

Such savings come at a price, however. There are dangers inherent in group experimentation. These dangers are enumerated in the following paragraphs and some techniques are given for minimizing them, where possible.

(1) Copying. No one can doubt that, in classroom examinations, some students copy answers from those seated nearby. The same temptation and opportunity exist in some experiments, particularly when the responses are perceived to indicate the subject's level of achievement or ability. It matters not that a subject's fate in no way hinges upon his performance in an experiment: He does not want the experimenter to associate his name with inferior performance. The temptation to copy is less when the responses indicate the subject's "opinions," for then there are no "right" or "wrong" answers. Naturally, the subject's experimental scores are assumed to be based on his own responses—not those of his neighbors—and the experimenter should try to make sure that this assumption holds true. (For example, he could use individual cubicles or other visual screening or exercise close supervision. As in a classroom examination, the possibility for close supervision diminishes as group size increases.)

(2) Other interpersonal communication. Apart from the copying of responses, a group-run experiment may evoke various forms of interpersonal communication that may affect subjects' scores. For example, if the experiment uses emotionally laden stimuli such as humorous cartoons or sexually explicit pictures, subjects may emit various gestures or verbalizations indicating amusement, surprise, embarrassment, or puzzlement. Such signals may affect the scores and conclusions in unknown ways. Normally, however, this possibility is ignored in describing experimental results, and the reader of a report must assume that there were no interpersonal communications or that such communications, if existent, had no effect on the scores. It is the experimenter's responsibility, then, to assure as best he can that the reader's assumption is valid.

The danger is not limited to experiments using emotionally laden stimuli. Just as students in a classroom may "cut-up" to enliven a class they find less than completely absorbing, so may subjects seek to add interest to a somewhat tedious task by initiating various forms of verbal and nonverbal communication.

This transformation of a situation in which individuals are intended to be working separately into a group interaction situation is more likely to occur when all subjects are responding to the same stimuli at the same time (for example, when subjects are giving judgments on tones heard by all simultaneously). If subjects are working at their own pace in separate test booklets, interpersonal communication is less meaningful and less rewarding.

When such interpersonal communication is a danger, the experimenter can, in his preliminary instructions, request that subjects work quietly and individually, and avoid making any comments or other extraneous responses to the stimuli. In addition, it may be necessary to provide a full or partial visual block among subjects with the use of screens or separate experimental cubicles.

(3) Conscientious performance. The smaller the number of subjects an experimenter must administer at once, the more likely are the subjects to perform conscientiously and in accordance with instructions. One reason is that it is easier to instruct one or a few subjects than a roomful. When a subject is given individual attention, and when his understanding of the instructions is directly queried, he will be more attentive to instructions and more willing to ask questions when he is uncertain. Furthermore, the subject who is given individual attention will be more likely to feel that his performance in the experiment is important and that his adequacy as a subject is under the personal scrutiny of the experimenter. In contrast, the subject in a large group will usually be satisfied simply to complete the task in a less than completely careful manner. The larger the group, the more care the experimenter should exert to make sure that the instructions are absolutely clear to all, and that each subject is assured of the importance of his own performance to the success of the experiment.

(4) Randomization of sequences. When subjects are run in groups and shown the same stimuli, all subjects naturally must have the same sequence of stimuli (that is, the stimuli must be ordered the same for all subjects in a group). Normally, were the same subjects to be run individually, they would receive different sequences as a precaution against the conclusions being dependent on a particular arbitrary sequence. If the subjects for each treatment are run in several groups, then each group can use a different sequence, thus allowing some protection against this danger.

(5) Other group-specific effects. Innumerable possible influences can affect a particular experimental group but not others, at least not to the same degree. In Section 3.2.9 such influences were called phantom factors. Some examples are the particular manner in which an experimenter presents instructions; the time of day; the effect of a particular subject who arrives late, thus disrupting the session; interpersonal communications that occur in one session but not another; the weather; anecdotes about the experiment told to the present subjects by the earlier participants; and recent news announcements. These and many other possible influences on the scores often vary from group to group. Such differences among groups may have little or no effect on the scores for most experiments. Nonetheless, the experimenter should keep such possible influences in mind when planning his experiment, and try to guard against possible distortions in the conclusions due to them.

Such influences operate whether the experiment is run on individuals or on groups, but the danger that such influences will distort the conclusions is less when subjects are run individually because these influences are then randomly spread across the different treatments. The danger is greatest when all subjects in each treatment are run together in a single group. Then influences specific to a given group are completely confounded with the effect of that group's treatment. When the data are analyzed, effects on the scores due to the phantom factors are mistaken for treatment effects.

If, for the sake of economy in running subjects, group sessions are desirable, the experimenter should at least avoid having only one session for each treatment, if possible. Subjects having the same treatment should be run in several different sessions so that any possible group effects will tend to average out. A particularly desirable arrangement is to include a group factor in the experiment. For example, design $A \times B$ would be converted into design $G(A \times B)$ and the scores would be analyzed accordingly. If there were 12 subjects per ab_{ab} treatment, they might be run in 3 groups of 4 subjects each. It may be difficult to assure, however, that all groups will be of the same size. Even if subjects are scheduled that way, one or more often fail to appear. An experimenter can schedule an excess of subjects to compensate for possible "no-shows," and then discard the scores of excess subjects (randomly selected), but this practice is wasteful when available subject-hours are in short supply. Even though an excess number of subjects appear, they should all participate in the experiment to avoid any disgruntlement among either those excused or those not excused, and to provide extra data in case any subject's responses have to be discarded (see Section 9.1). When groups of different sizes are used, the data are analyzed as if subjects were run individually—the analysis involves no group factor.

It is sometimes possible to run all the subjects of an experiment in one session. Protection is then afforded against group effects; because all treatments are subjected to the same group-specific effects, these effects are controlled insofar as comparisons among treatments are concerned. Such experimentation does, of course, require a large group, which has its drawbacks.

The one-session experiment is likely to be used with an intact group, such as a class. The class is a convenient captive group of subjects, and the experimenter thus avoids the difficulties of recruiting subjects and arranging for their appearance and for the use of an experimental room, in addition to the many additional experimenter-hours that would be required if the subjects were run in small groups or as individuals.

CHAPTER NINE

Miscellaneous Analysis Considerations

9.1 MISSING SCORES

9.1.1 Introduction

It is not uncommon for an experimenter to find, after running his subjects, that one or more of the scores called for by the design are unavailable, possibly because some subjects failed to show up as scheduled, or because some subjects simply omitted some of the responses required or failed to follow the instructions, or because the equipment misbehaved during the session. It is best to exercise whatever care is possible before and during the experimental sessions to check out the equipment and to see that subjects understand the requirements of the experiment. In spite of such precautions, however, some scores will often be missing. Since the methods described thus far in this book assume that all scores are available, missing scores require special consideration.

Four general kinds of actions are possible. The validity of all the methods described here is based on an assumption that missing scores are unrelated to the particular conditions from which they are missing. To provide a contrary example, if a particular treatment exposed subjects to electric shock of such intensity that some immediately got up and abandoned the experiment, then the missing scores could be attributed to the condition, and the techniques described here should not be used except with special explanation.

The four kinds of action are as follows. First, an experimenter can run additional subjects to obtain the required scores. Second, he can delete some scores to bring about the balance required, though he will have fewer scores than originally planned. Third, he can insert score estimates in place of the missing scores, and then analyze the scores in almost the same way as if all scores were authentic. Finally, the experimenter can analyze the data that include the missing scores, using different and more complicated procedures than those that have been described.

When subjects are run individually and each experimental session is relatively short, the no-show subjects present little problem if additional subjects are available in the subject pool. The experimenter assumes that some subjects will not appear at their scheduled times and plans accordingly for extra subjects. However, if the semester ends or if no more subjects are available from the same source, extra sessions may not be possible. Also, when an experiment requires the participation of a subject over a longer period, such easy substitutions may not be possible. For example, suppose the students in a grade are divided into two classes and taught the same material in two different ways. Midway through the course some students are taken ill and some others leave school. It is not possible to start over with replacements.

9.1.2 Deletion and Insertion

When bona fide scores cannot be obtained to fill all score positions, an experimenter can resort to deletion or insertion. A simple example of deletion is as follows. An experimenter planned to perform an $A \times B$ experiment with two levels for each factor, using 10 subjects in each cell, or 40 altogether. His subjects, university-level psychology students, were required to serve in that capacity for 5 hours during the semester. At the end of the semester only 39 subjects had been run, and it was too late to obtain another subject. Examination of the data revealed that one subject misunderstood instructions, so his score was eliminated from the analysis. The 2 missing scores were from different cells, ab_{12} and ab_{21}. One score was randomly chosen from each of the other 2 cells and discarded from the analysis, leaving a balanced design with 9 subjects per cell instead of the 10 originally planned. Since the original specification of 10 subjects per cell probably had no exact scientific basis, there was no need for concern because the design ended up with only 9 subjects per cell. Of course, statistical tests using only 9 subjects per cell have somewhat less power and the estimates have somewhat less precision than tests that use 10 subjects per cell.

If the df for error is small to begin with, an experimenter usually hesitates to reduce it further by deleting scores, because he thereby also reduces df_E, but if only one or a few deletions are required to achieve balance, this technique is acceptable. An alternative technique for gaining balance is to insert an estimate for each missing score. Then the scores are analyzed as if none were missing, but df_E (or some other df) is reduced by one for each score estimated.

In a Class I or Class II design with at least several scores per condition, the mean of a condition's score can be inserted for a missing score in that condition. The analysis then proceeds as usual except that df_E is reduced by one for each score so estimated. A somewhat superior but more complicated alternative is discussed in Section 9.1.3.

When the design calls for only one score per condition, a missing score implies no score for a condition. Therefore, that condition's mean cannot be computed from available scores. As a specific example, assume an $A \times B$ Class II randomized blocks design with only one score per condition. There are A treatments and B blocks. The following formula may be used to estimate a missing score X_{ab}, where all summations are over the available scores:

$$\text{Estimated } X_{ab} = \frac{B\sum_a X_{ab} + A\sum_b X_{ab} - \sum_{ab} X_{ab}}{(A - 1)(B - 1)}$$

Suppose the following scores were available, with only X_{23} missing:

	a_1	a_2	a_3
b_1	2	4	0
b_2	3	6	–
b_3	3	5	4
b_4	0	3	2

An X_{23} for insertion can be estimated to be

$$\frac{4 \cdot 9 + 3 \cdot 6 - 32}{2 \cdot 3} = \frac{36 + 18 - 32}{6} = \frac{22}{6} = 3.67$$

If more than one score is missing, insert a reasonable guess for all but one of these, based on an inspection of the data. Then consider such guessed scores to be available and estimate the "one" missing score using the procedure just described. Insert this estimate for the missing score, delete a guessed score, and find an estimate for it using the preceding formula, considering all other design scores to be "available," whether they be guesses, observed scores, or estimates. Proceed in this manner until a calculated estimate has replaced each guess. Then start over and re-estimate each missing score, but use the scores inserted instead of the guesses. Continue to cycle through the missing scores in this manner until the current cycle produces estimates close to those of the preceding cycle.

There is no MS_E for the design under discussion, nor any df_E. The test of H_N that $\sigma_A^2 = 0$ uses MS_{AB} as MS_2, and df_{AB} must be diminished by one for each missing score that is replaced by an estimate. In the example with $A = 3$ and $B = 4$, normally $\text{df}_{AB} = (A - 1)(B - 1) = 6$, but when there is one insertion for a missing score use $\text{df}_{AB} = 6 - 1 = 5$.

The procedure and formula just described can be applied generally when there is one score per condition. For example, it would apply to a Class I completely randomized design with one score per condition, though such designs rarely occur. It can apply to a design with more factors. For example, suppose that the randomized blocks design were $A \times B \times C$, C being another treatment factor. Then instead of $\sum_a$ we would have $\sum_{ac}$, instead of $\sum_{ab}$ we would have $\sum_{abc}$, and instead of A we would have AC. If a block is nested, estimation proceeds as if the design involved only that one nest level. For example, if the design were $A \times B(C)$, the preceding estimation formula would be used with B equaling the number of blocks in the level c_c of the missing score, and with all summations also confined to that level.

In a repeated measurements experiment, if a subject is missing it means that more than one score is missing. In a Class III design, the ideal solution would be to run another subject. Otherwise, the experimenter could resort to deletion or to computational techniques that are outside the scope of this book. In a Class IV design, all subjects receive all treatments; if a subject is missing, it simply means that the number of subjects is reduced by one, but there is no need to estimate any missing scores.

Should only one score of a subject be missing, then it might be better to estimate that missing score rather than to discard all scores for the subject. When S is a nested factor, estimation involves only scores within the nest level(s) containing the missing score. For example, if the design is $S(A) \times B$, estimation involves only scores within the level a_a containing the missing score. The estimation formula given previously for the randomized blocks design would then apply with only a change of symbolism:

$$\text{Estimated } X_{abs} = \frac{S_a \sum_b X_{abs} + B \sum_s X_{abs} - \sum_{bs} X_{abs}}{(B - 1)(S_a - 1)}$$

The subscript on S_a is required only if the number of subjects differs among a_a levels. If several scores are missing within the same nest level(s), the iterative technique with guessed scores must be used. If the missing scores are at different nest levels, only a single application of the formula is required for each. The reduction in df for $S(A) \times B$ would be in df_{BS}. In general, the reduction is in the df for the model term having all design subscripts (term $\mathbf{bs}_{bs(a)}$ in the present example).

Although deletion and insertion have been presented as alternative approaches to be used when data are missing, in practice they are likely to be combined. For example, suppose that, in a Class I $A \times B$ design planned for 10 subjects per cell, 2 scores are missing for ab_{11} and one subject each is missing for ab_{12} and ab_{22}. Then one randomly chosen score can be deleted from ab_{21} and one estimated score can be inserted for ab_{11}, balancing the design with 9 scores per condition. Such an approach avoids the need to delete 3 more scores to balance the design with 8 scores per cell, and also avoids the need to use more than one estimated score.

9.1.3 Unweighted-Means Analysis

The unweighted-means analysis can be applied to Class I or II designs with at least several scores available per condition, when the number of scores differs across conditions. The calculations are applied directly to the available scores; no insertion or deletion is required. It is assumed that the design originally called for an equal number of scores per condition, but that due to various circumstances unrelated to the treatments per se, some scores could not be obtained. The unweighted-means analysis is more complex than the method that combines deletion or insertion with the usual analysis of variance, but it is a superior method otherwise.

Recall that the formulas described in Chapter 5 for finding sums of squares required computations of the scores directly; condition means were required in Chapter 4 for estimating model terms, but they are not normally required in computing SS's. With the unweighted-means analysis, condition means are computed as the first step in finding the SS's. As a specific example, suppose the design is $A \times B$ of Class I. Although the specifications originally called for E subjects per cell with one score each, there turned out to be a variable number E_{ab} of scores per cell.

The first step, then, is to compute the mean score for each cell, $\bar{X}_{ab.} = \sum_e X_{abe}/E_{ab}$. All subsequent means required are computed from these cell means, not from the scores directly.

The next step is to compute estimates of the model parameters, in this example of the $\mathbf{a}_a$'s, $\mathbf{b}_b$'s, and $\mathbf{ab}_{ab}$'s. These estimates are found from formulas analogous to those presented in Chapter 4, except that all averaging is over the cell means—never over the scores directly. For example, we estimate $\mathbf{a}_a$ to be

$$\hat{\mathbf{a}}_a = \bar{X}_{a..} - \bar{X}_{...},$$

just as before, except that here we must compute $\bar{X}_{a..}$ as $\sum_b \bar{X}_{ab.}/B$, and $\bar{X}_{...}$ as $\sum_{ab} \bar{X}_{ab.}/AB$ or $\sum_a \bar{X}_{a..}/A$ or $\sum_b \bar{X}_{.b.}/B$. In each case the quantity summed is based on cell means. When E is the same for all cells, we could also compute $\bar{X}_{a..}$, for example, as $\sum_{be} X_{abe}/BE$ (that is, directly from the scores), but such formulations, requiring division by a presumed constant E, are not meaningful in the present context. The estimates so found for the unweighted-means analysis are not least-squares estimates, as they are when E is the same across cells.

The name of the procedure, "unweighted-means analysis," derives from the fact that the various cell means used in computing, say, $\bar{X}_{a..}$, or $\bar{X}_{...}$, enter into the result with equal weight; that is, a cell mean of 5, based on 6 scores, is given no more weight than a mean of 7, based on 3 scores. The unweighted mean of these 2 cells is computed simply as $(5 + 7)/2 = 6$.

Recall from Section 5.3 that we can compute an SS for a term from the parameter estimates. We simply sum the squares of each parameter and multiply by the number of scores to which a single parameter applies. For example, for $A \times B$ we have $SS_A = BE\sum \hat{\mathbf{a}}_a^2$, $SS_B = AE\sum \hat{\mathbf{b}}_b^2$, and $SS_{AB} = E\sum \widehat{\mathbf{ab}}_{ab}^2$. Analogous formulas are used to compute the SS's in the unweighted-means analysis, the only differences being that the parameter estimates are based on the unweighted means, as explained previously, and that wherever E appears as a multiplier we replace it with $\tilde{E}$, the harmonic mean of the E_{ab}'s:

$$\tilde{E} = \frac{AB}{\sum_{ab}(1/E_{ab})}$$

To find SS_E we can estimate each error parameter $\hat{\mathbf{e}}_{e(ab)}$ as $X_{abe} - \bar{X}_{ab.}$ and compute $SS_E = \sum \hat{\mathbf{e}}_{e(ab)}^2$. Equivalently, we can find the SS for a cell by the formula $SS_{E(ab)} = \sum_e X_{abe}^2 - (\sum_e X_{abe})^2/E_{ab}$, and $SS_E = \sum_{ab} SS_{E(ab)}$. We find df_E by the formula $S_{\text{tot}} - A \cdot B$, where S_{tot} is the total number of subjects actually appearing in the design, not the number called for by the design. The other df's are the same as for the balanced design.

The formulas used in an unweighted-means analysis of Class I and II designs with more factors generalize readily from those presented for $A \times B$. The unweighted-means analysis, as presented, requires that at least one score be available for each condition. The analysis has also been used when a design condition has no available scores. Then a cell mean must be estimated from data of other conditions. Such estimations are, needless to say, problematical, and are beyond the scope of this book.[1]

9.2 FAILURE OF ASSUMPTIONS

9.2.1 The Problem of Failed Assumptions

Our inferences about effects based on F-ratios depend on the presumed conformance of some F-ratio over replications to the theoretical F-distribution. This conformance requires that various assumptions about the model terms be true. Should these assumptions fail, singly or jointly, the conformance would fail. Such assumptions have been mentioned here and there throughout the text. Among the assumptions required are the following:

[1] See W. T. Federer, *Experimental Design: Theory and Application* (New York: Macmillan, 1955), pp. 124–127, 133–134.

1. A score is a point along a continuum of potential scores.
2. The error parameter for each score is a random sample from a normal probability distribution.
3. The error parameters for different positions of a design are statistically independent, that is, uncorrelated.
4. The parameters for an effect of a random factor have a normal distribution and are uncorrelated.
5. The variance of the error parameters is the same for each condition of the design; that is, there is *homogeneity of variance.*
6. A like homogeneity of variance exists for the parameters of a random factor effect.

In addition, we assume that those factors designated as random and fixed operate across replications in accordance with our description of random and fixed factors. A further assumption of some importance is required for designs having both fixed and random factors that has not been mentioned heretofore. Since it requires some explanation, more thorough discussion of it is postponed until Section 9.2.3.

Should these assumptions fail, singly or jointly, the F-ratio would still have some probability distribution over replications, but the distribution would depart more or less from a true F-distribution, depending on how much the scores deviate from the assumptions about them. We call the distribution of the F-ratio in the actual experimental situation the *effective distribution* of the F-ratio to distinguish it from the theoretical F-distribution.

When the assumptions fail, it would not be true, for example, that under H_N the F-ratio would exceed $F_{100(1-\alpha)}(\text{df}_1, \text{df}_2)$ with a probability of α. The true probability that the F-ratio will exceed $F_{100(1-\alpha)}(\text{df}_1, \text{df}_2)$ equals the proportion of the effective distribution that exceeds $F_{100(1-\alpha)}(\text{df}_1, \text{df}_2)$. This proportion may be greater or less than α. When the proportion is greater than α, we say that the F-test is *positively biased.* When the proportion is less than α, we say that the F-test is *negatively biased.*

The problem with positively biased tests is that H_N's are rejected too often; that is, large F-ratios occur with greater probability than our theory states, so we burden and delude ourselves with more false reports of significant effects (type 1 errors) than we would like. The problem with negatively biased tests, on the contrary, is that false H_N's are too frequently accepted; that is, small F-ratios occur with greater probability than our theory states, so we are too often deluded into making type 2 errors (overlooking the real influences that various effects have on the scores).

It might be thought that, since the value of α is rather arbitrarily chosen to begin with, the fact that some other value of α is the effective value applying to some experimental analysis is not of great importance. To a certain extent, this is undoubtedly true, if the discrepancy between the nominal α and the effective α is not large. If the discrepancy is large, however, and varies in unknown and unreported ways from experiment to experiment, considerable support would be given to arguments holding that the analytic techniques described in this and other statistical books are unsound and misleading, and could as well be dispensed with. If, when an experimenter reports that his conclusions are based on an α of .05, the effective α might actually be .005, .02, .09, .38, or some other unknown quantity, the value of F-testing is diminished. For this reason, statistical theorists and practitioners interest themselves in the possible failure of the assumptions underlying F-testing.

Two general approaches are used in dealing with the problem of failed assumptions: empirical and theoretical. The empirical approach consists of examination of the scores of an experiment to determine whether the assumptions made actually hold. Various statistical tests are available to assist the experimenter in deciding whether the assumptions hold. The theoretical approach consists of mathematical investigations designed to assess the extent to which failed assumptions affect the conclusions of an experiment. These mathematical investigations do not dwell on the data of particular experiments, but rather determine how conclusions would be affected if assumptions failed in various possible ways. If an F-test would be only slightly affected by a moderate departure from the assumptions, we say that the F-test is *robust* with respect to that departure.

It can be very awkward for an experimenter to discover that the assumptions on which his planned analysis was to be based do not hold. There he is, all dressed up with his experimental data, and no place to go. Frequently, the assumptions may hold well enough for scores after they have been mathematically transformed, even though they fail when applied to the original scores. Such transformations are discussed in Section 9.3. Otherwise, if the experimenter is to proceed with his planned analysis, he must hope that any failure of his assumptions will not seriously alter his reported results. He must hope, in other words, that his analysis is robust with respect to deviations in his data from the assumptions. To some extent he can rely on theoretical investigations that have demonstrated the existence of robustness. To a large extent, however, he proceeds because he has no satisfactory alternative. It is no wonder that experimeters are not as avid as they might be in searching for failed assumptions in their scores. In fact, many reports of research make no mention of any attention to the matter, though this does not mean that none was paid.

An experimenter cannot possibly check out all the assumptions on which his F-tests depend. It is hardly possible, for example, for an experimenter to verify the assumptions about the distribution of the error parameter over replications on the basis of one replication. However, various checks are possible within the one replication. For example, the previously mentioned assumptions imply that the scores within each condition should be normally distributed with equal variance across conditions. If there are several scores per condition, statistical tests are available to check out normality and homogeneity of variance. To a lesser extent, testing of assumptions is also possible for designs having only one score per condition.

9.2.2 Normality and Homogeneity of Variance

In general an experimenter does not perform statistical tests to determine whether the distributions of scores within a condition conform to the normal distribution. One reason is that it hardly matters whether such conformance occurs unless a discrepancy distorts the conclusions of the analysis. In a theoretical investigation, Norton[2] showed that the distribution of the F-ratio for non-normal score distributions conforms remarkably well to the theoretical F-distribution, so that the effective α in such cases does not depart greatly from the nominal α. Norton studied the effects of both asymmetry (skewing) and kurtosis on the effective α. He found that the greatest discrepancy occurred for a leptokurtic distribution (relatively too dense at the center and the extreme tails), but even then the effective α was only approximately .08 compared with a nominal value of .05, and approximately .03 compared with a nominal level of .01. A highly platykurtic distribution (rectangular shaped) had effective α levels that were very close to the nominal values. Norton also found that the F-test is fairly robust with respect to heterogeneity of variance of normal distributions, and also with respect to simultaneous failures of both normality and heterogeneity of variance, at least for the situations investigated.

There are several well-known tests for heterogeneity of variance. One of them, the Hartley $F_{\max}$ test, is described here. If heterogeneity of variance is extreme enough to cause concern, however, it will undoubtedly be apparent from a simple visual inspection of the scores. In such instances, a nonlinear transformation (Section 9.3) can usually be found that will eliminate the heterogeneity, or at least reduce it to an acceptable level. Frequently, a nonlinear transformation that reduces heterogeneity of variance also reduces deviations from normality.

The Hartley $F_{\max}$ test can be used to test the H_N of homogeneity of variance for balanced Class I and II designs having more than one score per condition. The test statistic required is

$$F_{\max} = \frac{\hat{\sigma}^2_{E\ \max}}{\hat{\sigma}^2_{E\ \min}}$$

To obtain the statistic, the experimenter must compute a separate estimate of σ^2_E for each design condition; $\hat{\sigma}^2_{E\ \max}$ is the largest of such estimates and

[2] The study is summarized in E. F. Lindquist, *Design and Analysis of Experiments in Psychology and Education* (Boston: Houghton Mifflin, 1953).

$\hat{\sigma}^2_{E\ \min}$ is the smallest. An estimate $\hat{\sigma}^2_{E\ \text{cell}}$ for a condition (cell) is simply $\text{SS}_{E\ \text{cell}}/(E - 1)$, where E is the number of scores per condition:

$$\text{SS}_{E\ \text{cell}} = \sum_{\text{cell}} X^2 - \frac{(\sum_{\text{cell}} X)^2}{E}$$

Critical values of the $F_{\max}$ statistic are given in Table A.6. The critical value differs depending on the number of variances, the number of scores per condition, and the significance level. The df that is to be entered in Table A.6 is the df for calculating one cell's variance, that is, $E - 1$. If the $F_{\max}$ statistic for an experiment exceeds the critical $F_{\max}$, then the experimenter rejects the H_N of homogeneity of variance.

The $F_{\max}$ test can also be used with Class III designs. Consider, for example, design $S(A) \times B$. This design has two different MS_2's: MS_S for testing the between-subject effect A, and MS_{BS} for testing the within-subject effects B and AB. Quantity MS_S estimates $\sigma^2_E + B\sigma^2_S$, which is assumed to be homogeneous across the levels of factor A. To test this assumption, compute an MS_S separately for each level of A, and then compute an $F_{\max}$ statistic as the ratio of the largest of these to the smallest. This $F_{\max}$ statistic is compared with a critical value from Table A.6, using A, the number of levels of factor A, as the "number of variances," and $S - 1$ in place of $E - 1$.

Quantity MS_{BS} is an estimate of $\sigma^2_E + \sigma^2_{BS}$, which is also assumed to be homogeneous across the levels of factor A. To test this assumption, compute an MS_{BS} separately for each level of factor A, and then compute an $F_{\max}$ statistic as the ratio of the largest of these to the smallest. To find the critical value of $F_{\max}$ from Table A.6, use A for the "number of variances" and $(S - 1)(B - 1)$ in place of $E - 1$.

These procedures can be generalized to other Class III designs. The "number of variances" equals the number of different MS's from which the largest and smallest were chosen. In place of $E - 1$ use the df associated with an MS in the $F_{\max}$ ratio.

9.2.3 Homogeneity of Variance-Covariance

An additional assumption is required for some designs so that all the F-ratios will be in accord with the theoretical F-distribution. This assumption can conveniently be explained in relation to design $S \times A$, assuming that factor A is fixed and that factor S is random as usual. A score in design $S \times A$ is symbolized by X_{as}. The mean of the scores for some level a of factor A is symbolized by $\bar{X}_{a.}$. Let $\bar{X}_{a'.}$ be the mean of the scores at some other level a'. The difference $\bar{X}_{a.} - \bar{X}_{a'.}$ equals a single quantity for a single replication, but over a series of replications this quantity would vary; hence over replications, $\bar{X}_{a.} - \bar{X}_{a'.}$ would have a variance. The assumption is that the variance is the same for any pair of levels a and a'.

The assumption will be true if the covariances between all pairs of different treatments, a and a', are equal, and the variances of the scores for each treatment are equal. These quantities are often displayed in a table called the *variance-covariance* matrix, which is a square array of numbers. The positive diagonal consists of the variances for the different treatments; the off-diagonal entries are covariances. By definition, the covariance between two treatments equals

$$\frac{\sum_s (X_{as} - \bar{X}_{a.})(X_{a's} - \bar{X}_{a'.})}{S - 1}$$

We distinguish the true variance-covariance matrix, which would obtain if scores were available for the entire (and presumably infinite) population of subjects, from the estimated matrix, which is computed from the scores of subjects used in the experiment.

A true variance-covariance matrix for $A = 4$ can be symbolized as follows:

$$\begin{matrix} \sigma_{11}^2 & \sigma_{12}^2 & \sigma_{13}^2 & \sigma_{14}^2 \\ \sigma_{21}^2 & \sigma_{22}^2 & \sigma_{23}^2 & \sigma_{24}^2 \\ \sigma_{31}^2 & \sigma_{32}^2 & \sigma_{33}^2 & \sigma_{34}^2 \\ \sigma_{41}^2 & \sigma_{42}^2 & \sigma_{43}^2 & \sigma_{44}^2 \end{matrix}$$

The entries along the positive diagonal, that is, σ_{11}^2, σ_{22}^2, σ_{33}^2, σ_{44}^2, are the variances of the scores for a_1, a_2, a_3, and a_4, respectively. The other entries are covariances. By definition, two entries that are symmetrically placed across the positive diagonal are equal, that is, $\sigma_{21}^2 = \sigma_{12}^2$, $\sigma_{42}^2 = \sigma_{24}^2$, and so on.

When there is homogeneity of variances and of covariances, the matrix has the following form:

$$\begin{matrix} \sigma_X^2 & \rho\sigma_X^2 & \rho\sigma_X^2 & \rho\sigma_X^2 \\ \rho\sigma_X^2 & \sigma_X^2 & \rho\sigma_X^2 & \rho\sigma_X^2 \\ \rho\sigma_X^2 & \rho\sigma_X^2 & \sigma_X^2 & \rho\sigma_X^2 \\ \rho\sigma_X^2 & \rho\sigma_X^2 & \rho\sigma_X^2 & \sigma_X^2 \end{matrix}$$

When there is homogeneity of variance and covariance, then the variance of $\bar{X}_{a.} - \bar{X}_{a'.}$ is the same for all pairs of treatments, so if the other assumptions are met, the distribution of the F-ratio for factor A accords with the theoretical F-distribution. Although the assumption about $\bar{X}_{a.} - \bar{X}_{a'.}$ may hold when homogeneity of variance-covariance fails, in practice failure of the latter would generally imply failure of the former.

The constant covariance can be expressed as $\rho\sigma_X^2$, where ρ is the correlation between the scores of two treatments. Since $\rho\sigma_X^2 = \sigma_{aa'}^2$, the covariance between two treatments, it follows that $\rho = \sigma_{aa'}^2/\sigma_X^2$; that is, the covariance divided by the common variance equals the correlation. The covariance is zero only if the correlation is zero. When the covariance is positive the correlation is positive, and when the covariance is negative the correlation is negative.

Normally the correlation (and covariance) is positive, because the subject who scores high for one treatment tends to score high for another treatment, and the subject who scores low in one treatment tends to score low in another. Considered in relation to the score model, the $\mathbf{s}_s$ parameter for a subject is a constant across all treatments, so if the parameter is positive, the subject tends to score high across all treatments, and if it is negative he tends to score low across all treatments.

It would be fanciful to assume that, in general, the true variance-covariance matrix has the ideal pattern that has been described, and due to that discrepancy alone, the effective F-distribution would differ from the theoretical one. Theoretical investigations have shown that when, contrary to our assumption, the variances and covariances are heterogeneous, the effective α is larger than the nominal one; that is, the usual F-test has a positive bias. A statistical test is available to determine whether homogeneity of the variances and covariances obtains, but because the test requires rather complicated calculations, it is not routinely used.[3]

In order to correct for the positive bias that heterogeneity entails, it has been suggested that the experimenter use $F_{100(1-\alpha)}(1, S-1)$ as the critical F instead of the usual $F_{100(1-\alpha)}(A-1, (A-1)(S-1))$. In other words, df_1 is reduced from the usual $A-1$ to 1, and df_2 is reduced from the usual $(A-1)(S-1)$ to $S-1$. The altered F-test is conservative; that is, it normally would have a negative bias. The test is unbiased only when the heterogeneity of covariance is extreme. The conservative test cannot really be recommended as a routine replacement for the usual test, because a negative bias is no more desirable than a positive one. The conservative test might be recommended as a supplement to the usual F-test, however. When the conservative test rejects the H_N, the rejection is justified in spite of any heterogeneity of covariance. However, if the usual test rejects H_N whereas the conservative test does not, it is not clear whether the rejection is justified or whether it is due to heterogeneity of covariance.

The problem of heterogeneity of covariance has been discussed in the context of the repeated measurements design $S \times A$. The problem arises much more generally, however. For example, in design $S(A) \times B$, the covariance matrix must exhibit homogeneity within a particular level a_a, and the matrices must be the same at the different a_a levels. The problem of heterogeneity of covariance arises not out of some peculiarity of repeated measurements designs per se, but out of the combination of fixed and random factors in the same design. Hence, the problem can affect single measurement designs also.

[3] For information on the test, see R. E. Kirk, *Experimental Design: Procedures for the Behavioral Sciences* (Belmont, Calif.: Brooks/Cole, 1968), pp. 139–142.

9.3 TRANSFORMATIONS OF THE SCORES

Before doing an analysis of variance (ANOVA), an experimenter may wish to apply a nonlinear transformation to the scores. In general, there are two reasons for applying such a transformation: (1) to bring the scores into accord with the normality and homogeneity of variance assumptions, and (2) to eliminate interaction terms, so that the scores can be described with a main effects ("additive") model. A linear transformation, $X' = aX + b$ (or $X' = ax$), where a and b are constants, would serve neither purpose. Although such a transformation could make all variances smaller, and hence in a sense less discrepant, the F-ratios and the distortions of them caused by the failure of assumptions would be unchanged.

Any nonlinear transformation may be used, as long as it is monotonic (that is, if $X_1 > X_2$, then for the transformed scores, $X_1' > X_2'$). The transformations most often used are log X, square root of X, and inverse sine of X. If log X is to be used, for example, the experimenter finds $X' = \log X$ for each score X, and then performs the ANOVA on the X' scores. The conclusions reached apply to the X' scores, not to the X scores, which complicates interpretation of the results. For example, if the design is $A \times B$, then the model being tested is $X' = \mathbf{m} + \mathbf{a}_a + \mathbf{b}_b + \mathbf{ab}_{ab} + \mathbf{e}_{e(ab)}$, not the model for X. If the interaction term is not significant, interaction is lacking in the X' scores. Interaction may still occur in the X scores.

Both the log X and $\sqrt{X}$ transformations produce X' scores that are smaller than the corresponding X scores. The larger the X, the greater the reduction. Therefore, these transformations tend to normalize positively skewed distributions. In addition, these transformations reduce the variances of distributions; the reduction is greater for distributions having large variances and means. Therefore, when cells with large means also have larger variances than cells with smaller means, the log X and $\sqrt{X}$ transformations tend to equalize the variances. Such equalization is desirable, since ANOVA assumes that the variance is homogeneous in different cells. Fortunately, experiments having scores that are positively skewed within cells also tend to have cell variances that are positively correlated with cell means, so that a suitable transformation can correct both discrepancies at once. If the scores fail on only one of the assumptions, use of a nonlinear transformation to correct that deviation will yield X' scores failing on the other assumption. For example, suppose that all cells have the same variance regardless of cell mean, but that the distribution in each cell is positively skewed. Then a logarithmic transformation might normalize the distributions, but at the same time yield unequal cell variances.

As we have noted, there are statistical tests for normality and homogeneity of variance. Frequently, however, visual inspection of the scores will be adequate. Arrange the scores in increasing order within each cell. This makes it

Table 9.3-1 Original and transformed scores

	X			$X' = \sqrt{X}$	
	a_1	a_2		a_1	a_2
	1	4		1	2
	1	16		1	4
b_1	9	16	b_1	3	4
	9	36		3	6
	16	49		4	7
	1	4		1	2
	4	9		2	3
b_2	4	16	b_2	2	4
	9	25		3	5
	16	49		4	7

easier to compare the distributions. Observe the range of the scores within different cells, and the approximate means. Also observe whether the cells with larger means also have larger ranges. To check for skewness compare the median score in a cell with the smaller and larger scores. If the median is closer to the smaller scores for all or most of the cells, the distributions are probably positively skewed.

Table 9.3-1 shows a hypothetical set of scores X for design $A \times B$. The scores for the two a_1 cells have relatively small means and score ranges, whereas the scores for the two a_2 cells have larger means and larger ranges. For three of the four cells, the median cell score is much closer to the lowest score than to the highest, and for the other cell the median is approximately midway between the lowest and highest scores. Actually, the F_{max} test (Section 9.2.2) does not reject the H_N that the population variances are equal across the cells for the data shown, because with only 5 scores per cell considerable variation in sample variances is possible, even if the population variances are equal. Nonetheless, the data serve to illustrate how scores may be inspected visually and transformed.

The square roots of the scores are shown to the right in Table 9.3-1. These X' scores meet the requirements of normality and homogeneity better than the original scores. The smallest range is $4 - 1 = 3$ and the largest is $7 - 2 = 5$. It is not the difference between 3 and 5 that is important, but the ratio, $5/3 = 1.67$. The ratio of the largest range to the smallest for the original scores is $45/15 = 3$. Likewise, the cell medians are more centrally located in the transformed scores than in the original. Again, we should think in terms of ratios instead of absolutes, since the whole scale is contracted. In the original data, for example, for cell ab_{22}, the median, 16, differs from the lowest score by 12, and from the largest score by 33, a ratio of $33/12 = 2.75$. In the transformed data, the ratio is $3/2 = 1.5$. With symmetric distributions, the figure should vary around one for different cells, a condition better satisfied with the transformed scores than with the original.

To summarize, for the original scores of Table 9.3-1 there appears to be a positive correlation of cell mean with variance, together with a positive skewing of the within-cell distributions. The square roots of the scores adhere to the homogeneity and normality requirements for ANOVA better than the original scores.

A normal distribution is symmetric, not skewed. Therefore, if the within-cell distributions are skewed, they cannot be normal, but normality requires more than symmetry. Nonetheless, unless there are specific reasons for a more thorough investigation, or unless the data obviously depart from normality, experimenters are usually willing to take symmetry as an indication of normality. The student should feel free to follow the same policy.

Of the several commonly used transformations—and the innumerable possible ones—the one that should be used is the one that best accomplishes the intended purpose. It need not be a commonly used transformation, nor one describable by any well-known mathematical function. The experimenter may contrive his own. He may do analyses with several different transformations, and finally report results for the transformation that seems to work best. Some guidance can be given for choosing among the better-known distributions, depending on the circumstances.

(1) Cell means and variances proportional. The square root transformation is applicable here. The score in this situation may be a frequency, such as number of responses emitted or number of errors made during a trial. If some of the entries are less than 10, either of the following alternative transformations may be more suitable:

$$X' = \sqrt{X} + \sqrt{X + 1}$$

$$X' = \sqrt{X + \frac{1}{2}}$$

(2) Binomial score distribution. This situation may exist when the score is a proportion, such as the proportion of items correctly answered in a recall test. If the distribution is binomial, the cell variance is related to the cell mean by $\sigma^2 = \mathbf{m}(1 - \mathbf{m})$. The appropriate transformation for the proportion (score) is then

$$X' = 2 \arcsin \sqrt{X}$$

The student will recall from trigonometry that $y = \sin \theta$ is a ratio between two sides of a right triangle. The arcsin y, also symbolized by $\sin^{-1} y$ and called the inverse sine, means the angle having y as the sine. In other words, if $y = \sin \theta$, then $\theta = \arcsin y$. In terms of the present notation, $X' =$

2 arcsin $\sqrt{X}$. Direct tables of 2 arcsin $\sqrt{X}$ can be found in Kirk or Winer.[4] If $X = 1$, go to the tables with $X = 1 - (1/2n)$ instead, where n is the number of cases on which the proportion was based. If $X = 0$, go to the tables with $X = 1 + (1/2n)$.

(3) Cell means and standard deviations proportional. The logarithmic transformation is useful here. Log X exists only for positive values of X. As X approaches 0 from above, log X approaches negative infinity. To avoid taking logs of numbers equal to or close to zero, we may instead use the transformation

$$X' = \log (X + 1)$$

The logarithmic transformation is particularly useful for handling skewed distributions, such as frequently occur when the criterion is response time (for example, time required to complete a task or solve a problem). The reciprocal transformation, $X' = 1/X$, has also been useful with response time measures.

A slide rule is a handy device for making the transformations discussed in this section. Although for some fields of science, the slide rule may not be accurate enough, the two to three significant figures available from the slide rule are usually adequate for the behavioral sciences.

Computer programs for ANOVA often allow specification of any of several standard score transformations, and also provide statistical tests for normality and homogeneity of the scores, both before or after transformation. Under such circumstances it is feasible and may very well be worthwhile to carry out a complete analysis with the original scores as well as with one or more different transformations.

9.4 SIGNIFICANCE TESTING OF CONTRASTS

9.4.1 Introduction

Up to now we have considered significance testing for H_N's that assert that all parameters of a particular model term (effect) equal zero. For example, H_N would be of the form $\mathbf{a}_a = 0$ for each of A parameters. Rejection of H_N means only that we conclude that at least two $\mathbf{a}_a$ parameters do not equal zero; however, an experimenter frequently wishes to explore the relationships among the $\mathbf{a}_a$ parameters more fully. For instance, he might want answers to questions such as:

1. Do $\bar{X}_{2.}$ and $\bar{X}_{3.}$ differ significantly? In other words, can the H_N that $\mathbf{a}_2 = \mathbf{a}_3$ be sustained?

[4] Kirk, *op. cit.*, p. 539; B. J. Winer, *Statistical Principles in Experimental Design*, 2nd ed. (New York: McGraw-Hill, 1971), p. 872.

2. Does $\bar{X}_{1.}$ differ significantly from the mean of $\bar{X}_{3.}$ and $\bar{X}_{4.}$? In other words, can the H_N that $\mathbf{a}_1 = (\mathbf{a}_3 + \mathbf{a}_4)/2$ be sustained?

If factor A is quantitative, then each level a_a has an associated quantitative value depending on the location of a_a along some scale (Section 3.2.8). Call this scale x, and let the x value for a_a be symbolized by x_a. Then the experimenter may wish to ask the following question:

3. What is the best guess for the mathematical function relating the population values X for the a_a's to the values x_a?

For example, an experimenter might find that $X = k_0 + k_1 x$, where k_0 and k_1 are constants. If $x = 4$ for a_3, then the associated population value X would be estimated as $X = k_0 + k_1 \cdot 4$.

Consideration of such mathematical functions will be postponed until Section 9.5. Here we consider only such questions as (1) and (2), which require no associated scale x for the a_a's, and hence may be asked in relation to qualitative factors. (They may also be asked in relation to quantitative factors, by simply ignoring x.)

9.4.2 Contrasts Defined

Questions of the sort we wish to ask can be expressed using *contrasts* (also called *comparisons*). A contrast among a set of A means in design A is defined to equal

$$w_1\bar{X}_{1.} + w_2\bar{X}_{2.} + \cdots + w_A\bar{X}_{A.},$$

where the w's are constants such that $\sum_a w_a = 0$. Such a contrast, using treatment means, is an estimate of a corresponding contrast of population means $\mathbf{m} + \mathbf{a}_1$, $\mathbf{m} + \mathbf{a}_2$, and so on. Contrasts may be defined among treatment totals as well as among means. In more complicated designs, to compute contrasts among the treatments for factor A, an experimenter must take the average (or just the sum) across the levels of the other factors.

Two contrasts among the same treatments are said to be *orthogonal* if $\sum_a w_a w'_a = 0$, where w'_a is a coefficient for the second contrast. Suppose that the contrasts had coefficients as follows:

	a_1	a_2	a_3	a_4
w_a	0	1	-1	0
w'_a	1	0	$-\frac{1}{2}$	$-\frac{1}{2}$

The contrasts are not orthogonal, since

$$\sum_a w_a w'_a = 0 \cdot 1 + 1 \cdot 0 + (-1)\left(-\frac{1}{2}\right) + 0\left(-\frac{1}{2}\right) = \frac{1}{2}$$

If two contrasts are orthogonal, their numerical values over an infinity of replications would be uncorrelated. A set of A means has $A - 1$ df, and can have at most $A - 1$ mutually orthogonal contrasts (that is, each contrast is orthogonal to every other one). For example, with four treatment means ($A = 4$), it would be impossible to find more than three mutually orthogonal contrasts. No fourth contrast could be found that would be orthogonal to all three. The three orthogonal contrasts would not be unique, however. Many different sets of three mutually orthogonal contrasts could be found.

Now consider how the preceding questions can be expressed in terms of contrasts. To ask whether $\bar{X}_{2.}$ and $\bar{X}_{3.}$ differ significantly is to ask whether $1(\mathbf{m} + \mathbf{a}_2) - 1(\mathbf{m} + \mathbf{a}_3) = 0$. The expression to the left of the equals sign is a contrast between two population means, where $w_2 = 1$, $w_3 = -1$, and all other w's equal zero. The estimate of this contrast is $1 \cdot \bar{X}_{2.} - 1 \cdot \bar{X}_3$, a contrast between treatment means. Our question, then, is whether the contrast $\bar{X}_{2.} - \bar{X}_{3.}$ differs significantly from zero. Likewise, question (2) asks whether the contrast $\bar{X}_{1.} - (1/2)(\bar{X}_{3.} + \bar{X}_{4.})$ differs significantly from zero. For this contrast, $w_1 = 1$, $w_3 = -1/2$, and $w_4 = -1/2$. Note that the two contrasts just mentioned are not orthogonal—in fact, they are the very ones that were used previously to illustrate lack of orthogonality.

9.4.3 Sums of Squares and *F*-Ratios

We have seen that many of the questions we might wish to ask about significant differences for a set of treatments can be expressed in terms of contrasts. Once a question is posed, however, the test of significance remains to be conducted. Associated with any contrast is a sum of squares having 1 df, by means of which the contrast may be tested. A problem arises, however, when *multiple contrasts* are tested for the same set of means, since the more contrasts that are tested, the greater is the probability of one or more type 1 errors. Discussion of this problem of multiple contrasts is postponed until Section 9.4.4; let us suppose for now that only one contrast is of interest.

The sum of squares for a contrast of treatment means is the same as the sum of squares for a contrast of treatment sums. The formula presented here utilizes treatment sums. Let W be the value of the contrast of treatment sums. Then the sum of squares for the comparison is

$$\mathrm{SS}_W = \frac{W^2}{E \sum w_a^2}$$

If the number of scores differs among treatments, then the sum of squares is defined in terms of treatment means:

$$\mathrm{SS}_W = \frac{(\sum w_a \bar{X}_{a.})^2}{\sum_a (w_a^2 / E_a)}$$

This formula is equivalent to the preceding one when E_a is the constant E. Since SS_W has 1 df, it is also a mean square, and if the normal distribution and sampling assumptions are met, SS_W/MS_E is distributed as F over replications. Hence, a significance test can be performed based on the F-ratio SS_W/MS_E.

For a numerical example, assume that the scores for design A are as follows:

	a_1	a_2	a_3	a_4
	1	10	16	17
	2	14	17	17
	4	15	18	20
	7	18	20	22
$X_{a.}$	14	57	71	76
$\bar{X}_{a.}$	3.50	14.25	17.75	19.00

The analysis of variance table is:

Source	SS	df	MS	F
A	595.25	3	198.42	29.61*
E	80.5	12	6.71	

* $p < .01$.

The ANOVA tells us that the population means for the four treatments are not the same. We wish, however, to have more detail about specific differences among the population means. Suppose we wanted to answer question (1): Do $\bar{X}_{2.}$ and $\bar{X}_{3.}$ differ significantly? The pertinent contrast for treatment totals equals

$$\begin{aligned} W &= w_2 X_{2.} + w_3 X_{3.} \\ &= 1 \cdot X_{2.} + (-1) X_{3.} \\ &= X_{2.} - X_{3.} \\ &= 57 - 71 = -14 \end{aligned}$$

Since $\sum w_a^2 = 1^2 + (-1)^2 = 2$, we have

$$\begin{aligned} SS_W &= \frac{W^2}{E \sum w_a^2} \\ &= \frac{14^2}{4 \cdot 2} = \frac{196}{8} = 24.5 \end{aligned}$$

The F-ratio, then, is

$$F = \frac{24.5}{6.71} = 3.65$$

Since $F_{95}(1, 12) = 4.75$, we conclude that the population means for a_2 and a_3 do not differ.

Let us now consider question (2). Does $\bar{X}_{1.}$ differ significantly from the mean of $\bar{X}_{3.}$ and $\bar{X}_{4.}$? The pertinent contrast for treatment totals equals

$$\begin{aligned} W &= w_1X_{1.} + w_3X_{3.} + w_4X_{4.} \\ &= 1 \cdot X_{1.} + \left(-\frac{1}{2}\right) X_{3.} + \left(-\frac{1}{2}\right) X_{4.} \\ &= X_{1.} - \frac{1}{2}(X_{3.} + X_{4.}) \\ &= 14 - \frac{1}{2}(71 + 76) \\ &= 14 - 73.5 = -59.5 \end{aligned}$$

Since $\sum w_a^2 = 1^2 + (-1/2)^2 + (-1/2)^2 = 1.5$, we have

$$\begin{aligned} \text{SS}_W &= \frac{W^2}{E\sum w_a^2} \\ &= \frac{59.5^2}{4 \cdot 1.5} = \frac{3540.25}{6} = 590.04 \end{aligned}$$

The F-ratio, then, is

$$F = \frac{590.04}{6.71} = 87.9$$

Since the critical value $F_{99.9}(1, 12) = 18.6$, we conclude that the population mean for a_1 differs from the mean of the population means for a_3 and a_4.

9.4.4 The Error Rate Problem

An ever-present danger in statistical testing is that the null hypothesis H_N will be falsely rejected. The probability for such a type 1 error for a single test is α, the significance level for the test. Normally, α is set no higher than .05, so the probability of a type 1 error for a single test is only .05. When an experimenter performs many tests, however, each having a significance level α, the probability of a type 1 error occurring is considerably higher than α. For example, for N independent tests, each having H_N true, the probability of one or more type 1 errors is given by the formula $1 - (1 - \alpha)^N$. When α is .05 and N is 10, for example, the probability of one or more type 1 errors is .40. The problem is that, whereas for one test the probability of a type 1 error may be kept low, in an experiment involving tests of many H_N's, the occurrence of type 1 errors becomes rather likely.

In the present subsection the term "error rate" always refers to type 1 errors. The "error rate problem" refers to the increasing probability of type 1 errors occurring in an experiment as the number of H_N's tested increases. The error rate problem arises from two sources: (1) The greater the number of factors a design has, the greater the number of effects there are for which H_N's must be tested; (2) each contrast tested has an associated H_N, so error rate increases with the number of contrasts tested.

Bases for error rate. Generally, when we have considered type 1 errors, we have been concerned with the probability relative to each test; that is, we have said that the probability is α that a false rejection of H_N will occur for a single test, given that H_N is true. However, other bases for the error probability are possible, and they are germane to the present discussion. First consider the (possibly) multifactor experiment as a whole, and assume that all H_N's are true. Though all H_N's are true, one or more of these H_N's may be rejected; that is, there may be one or more type 1 errors. Now imagine that the experiment is replicated innumerable times. The proportion of replications having one or more type 1 errors is called the *error rate experimentwise*. The average number of type 1 errors per replication is a somewhat larger number, because some replications have more than one type 1 error. This number is called the *error rate per experiment*. If multiple contrasts are tested for a particular effect, the comparable terms are the *error rate effectwise* and the somewhat larger *error rate per effect* (also called the *error rate familywise* and the *error rate per family*). When a one-factor design is used, the error rates based on the experiment and the effect are the same. Various authorities have proposed that the significance level α should equal one or the other of these error rates rather than the error rate per H_N.

Actually, there is very little acceptance of the idea that a multifactor experiment should be the basis of α. The testing of each effect in a multifactor experiment with the usual α (or α's), regardless of the number of effects being tested, is a well-established practice. No concern is shown for what an experiment-based α would be. Given that each effect is tested with the usual α, there is frequently concern that the effect-based error rates may become large when multiple contrasts are tested. Most discussions of the error rate problem concern the effect-based rates, and many of the statistical techniques proposed are designed to allow the experimenter to specify and maintain an effect-based α, which may be kept conservatively low regardless of the number of contrasts tested.

Multifactor experiments. Before proceeding to a consideration of effect-based rates, it would be well to concern ourselves with the appearance of type 1 errors when effects are tested in a multifactor experiment. The greater the number of factors, the greater the number of H_N's that must be tested. In a four-factor Class I design, for example, a total of 14 H_N's are tested, excluding any follow-up tests of contrasts: 4 main effects, 6 two-factor interaction effects, 3 three-factor interaction effects, and one four-factor interaction effect. Even if all H_N's are true, it is not unlikely that one or more significant effects will be found. Suppose that a design has 20 effects to be tested, that each is tested with an α of .05, and that one significant source is found. Since each test has a probability of a type 1 error of 1/20, the significant result may very well be spurious. If only one effect were tested, as with a one-factor design, and the source were significant, more confidence could be had in the conclusion.

The experimenter should avoid thinking, however, that if only one significant effect is found in 20 tests using $\alpha = .05$, the result must be spurious. The result may be valid. Likewise, if two significant effects are found out of 20 tests, the experimenter should avoid thinking that one must be valid and one is spurious. Both may be valid, or both may be spurious.

Although, as previously mentioned, there seems to be very little interest among experimenters in using an experiment-based α, three alternative options are available to the experimenter to keep the number of type 1 errors low when many effects are tested.

1. Do preliminary overall F-testing.

With a Class I design, for example, imagine that all treatments are levels of one factor. The SS's and df's for all effects except error are pooled to get the SS and df for this "one" factor, which is then tested for significance. Unless H_N is rejected, do not go on to test the effects separately. The idea here is that unless an experimenter can demonstrate differences among all the treatments in an experiment, he should not proceed to search for significant effects. When the various effects in a design are tested with different MS_2's, as in Class III designs, an overall test must be performed for each MS_2: All the terms tested by that MS_2 are pooled and thought of as a single factor. This first option is not frequently followed, but it is a relatively simple procedure that could offer some protection against type 1 errors. It would be primarily useful when all the effect H_N's are true, since one false H_N with a large σ_τ^2 can lead to a rejection of the overall H_N, and hence to tests of all the effect H_N's.

2. Use a smaller α.

Instead of basing significance on an α of .05, an experimenter can use a more stringent value, say, .025 or .01. Some journal editors and consultants consider that an α of .05 is too high, in general; this opinion would appear to have merit, at least when many H_N's are tested.

3. Ignore interaction effects involving four or more factors.

When many interaction effects involving four or more factors are to be tested, significant results are awkward for the experimenter; these significant effects are difficult to describe and comprehend and are unlikely to be of any practical or theoretical importance (Section 1.4). A solution is simply to ignore such effects and to concentrate on effects involving fewer factors. When fewer effects are tested, the probability of type 1 errors is smaller. This option is only helpful for designs having more than four factors.

Multiple contrasts. When an experimenter tests many contrasts in addition to effects, the possibilities for type 1 errors increase. This is particularly true when the factor under consideration has many levels so that many contrasts may be of interest. Suppose, for example, that a one-factor design has 8 levels, and the experimenter wants to test all pairs of means for a significant difference. There are 28 such pairs of means, and hence 28 associated contrasts. If all 28 were made with $\alpha = .05$ for each, it is very likely that one or more spurious significant differences would be found. On the other hand, if only a few contrasts were tested, chances are that any significant differences found would be valid.

Various strategies have been devised to assist the experimenter and research consumer in performing and interpreting statistical tests of multiple contrasts. The purpose of these strategies is to exert some control over the extreme variations in the proportion of type 1 errors among significant effects, so that significant effects for different experiments have comparable probabilities of being valid. For an understanding of these strategies, two distinctions are particularly important. One is the distinction between orthogonal and nonorthogonal contrasts, which has already been explained. Another is the distinction between a priori or planned contrasts and a posteriori or post hoc contrasts. Sometimes an experimenter has a small number of specific contrasts in mind before he collects the data. These are called *a priori* or *planned contrasts.* At other times the experimenter intends to test a large number of contrasts to see which turn out to be significant; or, after collecting his data he examines his treatment means and tests various contrasts that seem to be possibly significant. Such contrasts are called *a posteriori* or *post hoc contrasts.* The error rate problem is of particular concern with post hoc contrasts, for two reasons. First, the number of post hoc contrasts may be rather large, whereas the number of planned contrasts tested is generally rather small. Second, post hoc contrasts capitalize on the chance differences that happen to be large for that particular replication due to sampling variability. This is not likely to be true for planned contrasts, since the data are not available when these contrasts are specified.

Likewise, when the only contrasts tested are those from within one orthogonal set, the number of tests is relatively small. For example, with 8 levels there can be at most 7 orthogonal contrasts, even though there are 28 pairs of means that might be tested.

At least four options are available to the experimenter for dealing with the error rate problem he encounters when testing multiple contrasts:

1. He may decline to perform a quantitative test on any contrast and thus form any conclusions about such contrasts subjectively, while leaving the consumer free to agree, to disagree, to form additional impressions, or to do his own testing.

In favor of this option it can be said that for any experiment, a limit will always be reached beyond which subjective evaluation must replace quantitative analysis. Considering multiplicity of approaches and the difficulty in choosing with certitude among them for the testing of multiple contrasts, it is understandable that judgments about the significance of contrasts are often made without objective testing.

2. He may test each contrast at the same significance level α as was used for testing effects.

Although, as we have seen, such additional tests can result in effect-based error rates far in excess of α, the danger is not extensive if only a few contrasts are tested. In order to test for a significant difference between any pair of means in design A, the experimenter would calculate

$$t = \frac{\bar{X}_{a.} - \bar{X}_{a'.}}{\sqrt{\dfrac{2\ \mathrm{MS}_E}{E}}}$$

and (for a two-tailed test) compare this with $t_c = \pm t_{100[1-(\alpha/2)]}(\mathrm{df}_E)$. The test is the same as the test for a difference between means described in Section 2.4.3, except that the estimate of variance MS_E used here is based on all treatments instead of on only the two treatments whose means are compared. Because of this difference we must use df_E. This second option is generally acceptable for testing planned contrasts, which, as a rule, are not numerous. Some authorities suggest that this option should also be limited to orthogonal contrasts, but since there is no consensus on this point, the experimenter should not be bound by it.

3. He may test each contrast at some significance level α' that is smaller than the significance level α used for testing effects.

If the major concern in the testing of multiple contrasts is that too many type 1 errors may occur, the number can be decreased simply by using a lower level α'.

4. He may use a special test devised for multiple contrasts that allows specification of an effect-based error rate.

Option (4) is especially advisable when there are many post hoc contrasts of interest and when the experimenter wishes to perform statistical tests to decide which ones are significant. Various tests have been devised. They are derived from different bases and often lead to different conclusions about the significance of a contrast. It is normally hard to choose between tests on the basis of statistical theory, and well-established conventions do not exist to guide the experimenter in the choice of a test. Some of these tests are discussed in the next section.

9.4.5 Post Hoc Tests for Contrasts

Many different post hoc tests for contrasts have been proposed—so many, in fact, that it is difficult to know which, if any, should be used in a specific case. Only four such tests are presented here, so selection among them is simplified. A more comprehensive treatment of multiple comparisons can be found in Kirk (*op. cit.*, chap. 3) or in Keppel.[5] One of the four tests, the Newman-Keuls test, is only used to determine which pairs of means for some effect differ significantly. All pairs of means are tested. Regardless of whether the experimenter "planned" to test all such pairs before examining the data, the tests are considered post hoc; by definition, "planned contrasts" should be more specific and selective—at least in this author's opinion. Tukey's HSD (honestly significant difference) test (the T-method) can be used more generally to test any contrast, but, like the Newman-Keuls test, it is normally used to test all pairs of means. The Scheffé test (the S-method) can be used to test all contrasts, but it is normally limited to testing more complicated contrasts than differences between pairs of means; the Tukey test is more sensitive for such testing of pairs. Dunnett's test only applies when an experimenter wishes to test each of a set of means against a control group mean, and is recommended over the other tests for this purpose.

Post hoc tests, particularly the T-method and the S-method, tend to be conservative; that is, the power to reject false H_N's is relatively small for moderate departures from H_N. An experimenter may be disturbed by the failure of these tests to reject H_N's. It is possible to reject more H_N's by using larger α's or by using other tests. However, α's larger than .05 are not well accepted, in general. The use of alternative, less conservative procedures may often be more acceptable, while accomplishing the same result. It is undoubtedly bad statistical practice, however, to fiddle with different α's and tests in order to reach the desired conclusions.

Tukey's HSD test. The HSD (honestly significant difference) test, also known as the T-method, allows the experimenter to set the error rate effect-

[5] G. Keppel, *Design and Analysis: A Researcher's Handbook* (Englewood Cliffs, N. J.: Prentice-Hall, 1973), chap. 8.

wise at a known, controlled value α. The test was designed for testing the significance of the difference between each pair of means; though its applicability has been extended to contrasts in general, its primary usage is still for testing pairs of means. The test requires an equal number of scores for each mean and access to a table of the so-called studentized range statistic, $q_{100(1-\alpha)}(A, \text{df}_E)$ (Table A.7), where α is the effectwise error rate and A is the number of means in the set. The difference between two means is significant if it exceeds T_c, where

$$T_c = q_{100(1-\alpha)}(A, \text{df}_E)\sqrt{\frac{\text{MS}_E}{E}}$$

For the numerical example of Section 9.4.3, $\text{MS}_E = 6.71$, $\text{df}_E = 12$, $A = 4$, and $E = 4$. From Table A.8 we find that, for $\alpha = .05$, $q_{95}(4, 12) = 4.20$. Hence,

$$\begin{aligned} T_c &= 4.20\sqrt{\frac{6.71}{4}} \\ &= 4.20\sqrt{1.68} \\ &= 4.20 \cdot 1.30 = 5.46 \end{aligned}$$

In other words, only differences between means exceeding 5.46 are considered to be significant. The means are $\bar{X}_{1.} = 3.5$, $\bar{X}_{2.} = 14.25$, $\bar{X}_{3.} = 17.75$, and $\bar{X}_{4.} = 19$. According to this test, then, treatment a_1 differs from the other three, but none of these three differ from each other.

The preceding discussion assumed a one-factor design. For more complicated designs, the means may be formed by averaging across other factors. In computing T_c, use whatever MS would be used as MS_2 for the effect being tested; for df use the corresponding df_2; and for E use the total number of scores summed to get each mean.

Scheffé's test. Scheffé's test, also known as the *S*-method, allows the experimenter to set the effective error rate at some specific value α, as does the *T*-method. The *T*-method is recommended for pairwise comparisons because it yields a shorter significant difference. For more complicated contrasts, however, the *S*-method is generally more sensitive; also, it does not require an equal number of scores per mean. Again, the test is explained here in terms of a one-factor design, but the test can be extended to more complicated designs in the manner indicated for the *T*-method.

For a contrast to be significant, it must exceed the following quantity:

$$S_c = \sqrt{(A-1)F_{100(1-\alpha)}(A-1, \text{df}_E)}\sqrt{\text{MS}_E \sum \frac{w_a^2}{E_a}},$$

where the w_a's are the coefficients of the contrast being tested, and the E_a's are the number of scores for the various means. Let us see how this test compares with the T-method, using the same data that were used to illustrate the T-method. For any contrast describing simply a difference between two means, the two w_a coefficients are 1 and -1. Since, in our example, E_a is the constant $E = 4$ for each mean, $\sum w_a^2/E_a = (1 + 1)/4 = 1/2$. If we let $\alpha = .05$, then from Table A.5 of the F-distribution we find that $F_{95}(3, 12) = 3.49$.

The critical value, then, for Scheffé's test is

$$\sqrt{3 \cdot 3.49}\,\sqrt{6.71 \cdot .5} = \sqrt{10.47}\,\sqrt{3.355} = \sqrt{35.127} = 5.93$$

The contrast we employ for the difference between a pair of means is simply $\bar{X}_{a.} - \bar{X}_{a'.}$, the difference itself. Hence only differences between means that exceed 5.93 are significant. As with the T-method, we conclude that treatment a_1 differs from the others, but that these others do not differ among themselves. Note, however, that the critical difference for the T-method is only 5.46, whereas for Scheffé's test it is 5.93. In other words, a difference between pairs must be larger to be judged significant by the Scheffé test, which confirms the previous assertion about the greater sensitivity of the T-method for pair differences.

Newman-Keuls test. The Newman-Keuls test can be used only for testing all pairs of means for an effect. Like the T-method, it uses the studentized range table. Also, the critical difference required to declare the largest and smallest means significantly different is precisely the same for the two tests. But whereas the T-method has only one critical difference, which applies to all pairs of means, the Newman-Keuls test has a variety of critical differences.

The means are arranged in order, according to the numerical value of each. Then a separate critical value applies to each adjacent pair of means, to each adjacent set of three, to each adjacent set of four, and so on. If the difference between the largest and smallest means in the adjacent set does not exceed the associated critical difference, then no pair of means in the set differs significantly.

Let us apply the test to the same numerical example we used previously. The means, in order of magnitude, are:

$\bar{X}_{1.}$	$\bar{X}_{2.}$	$\bar{X}_{3.}$	$\bar{X}_{4.}$
3.50	14.25	17.75	19.00

The critical difference for a set of n adjacent means is symbolized by NK_n:

$$NK_n = q_{100(1-\alpha)}(n, \text{df}_E)\sqrt{\frac{\text{MS}_E}{E}}$$

The quantity $\sqrt{\text{MS}_E/E} = \sqrt{6.71/4} = 1.30$ is the same regardless of n. The values of NK_n, then, differ only according to $q_{100(1-\alpha)}(n, \text{df}_E)$. We find these values from Table A.7 for $\alpha = .05$ to be:

$$q_{95}(2, 12) = 3.08$$

$$q_{95}(3, 12) = 3.77$$

$$q_{95}(4, 12) = 4.20$$

Hence, we have

$$NK_2 = 3.08 \cdot 1.30 = 4.00$$

$$NK_3 = 3.77 \cdot 1.30 = 4.90$$

$$NK_4 = 4.20 \cdot 1.30 = 5.46$$

Begin by comparing the smallest mean with the largest. A total of four means is encompassed by this range, so the critical difference is $NK_4 = 5.46$. Since $\bar{X}_{4.} - \bar{X}_{1.} = 15.5$ exceeds 5.46, the difference is significant. Now compare the next largest mean with $\bar{X}_{1.}$. Since $\bar{X}_{3.} - \bar{X}_{1.} = 14.25$ exceeds $NK_3 = 4.90$, the difference is significant. Note that $\bar{X}_{2.} - \bar{X}_{1.} = 10.75$ exceeds $NK_2 = 4.00$, so this difference is significant also.

Now test the second smallest mean $\bar{X}_{2.}$. First compare it with the largest mean, the next largest, and so on, always using the NK_n whose subscript equals the number of means included in the range being tested. This number always equals 2 (for the two means being tested) plus the number of intermediate means. Since $\bar{X}_{4.} - \bar{X}_{2.} = 4.75$ does not exceed $NK_3 = 4.90$, the difference is not significant. Whenever a nonsignificant difference is found, perform no more tests on the smaller mean (in this example $\bar{X}_{2.}$). If $\bar{X}_{2.}$ does not differ significantly from $\bar{X}_{4.}$, assume that it cannot differ significantly from means smaller than $\bar{X}_{4.}$, that is, from means between $\bar{X}_{2.}$ and $\bar{X}_{4.}$. Actually, if an experimenter did not stop, but tested the intermediate means, he could come to the anomalous conclusion that, for example, $\bar{X}_{2.}$ does not differ significantly from $\bar{X}_{4.}$, but does differ from the smaller value $\bar{X}_{3.}$, since the latter difference is tested with a smaller critical NK.

Having completed the tests on $\bar{X}_{2.}$, we proceed to tests on $\bar{X}_{3.}$. For our example, we need only compare $\bar{X}_{4.} - \bar{X}_{3.} = 1.25$ with $NK_2 = 4.00$, and we conclude that no significant difference exists. In summary, $\bar{X}_{1.}$ differs significantly from the other means, which do not differ significantly from each other.

It so happens that we reached the same conclusions with the S- and T-methods, but this is not a necessary result. Because in the Newman-Keuls test, the critical differences between means diminish as the number of steps between them decreases, this test yields, in general, more significant differences than the others. However, α in the Newman-Keuls test does not refer to an error rate on any basis that has been discussed; it is neither an effect-based nor a contrast-based value.

Although in our example, the means were ordered $\bar{X}_{1.}$, $\bar{X}_{2.}$, $\bar{X}_{3.}$, $\bar{X}_{4.}$, in the Newman-Keuls test the ordering is based on the means, not on treatment number. Hence, in general, some scrambled ordering of the means, such as $\bar{X}_{3.}$, $\bar{X}_{2.}$, $\bar{X}_{1.}$, $\bar{X}_{4.}$, would occur, and testing would begin by comparing $\bar{X}_{3.}$ with $\bar{X}_{4.}$, then with $\bar{X}_{1.}$, and so on.

Dunnett's test. Dunnett's test is designed for a situation in which one treatment, the "control" group, is to be tested for a significant difference against all other means for some effect, but these other means are not to be tested against each other. The error rate effectwise is α.

For a treatment group to differ significantly from the control group, the mean score difference must exceed

$$D_c = t'_c \sqrt{\frac{2\text{MS}_E}{E}},$$

where t'_c is a value that is found in Table A.8 and that depends on α, df_E, and whether the test is one-tailed or two-tailed. Using the numerical example of this section, imagine that treatment a_1 is the control group, and that each of the other treatments might deviate in either direction so that a two-tailed test is used. When there are four treatment means, $\text{df}_E = 12$, and $\alpha = .05$. Just as we did when using the t-test, when using a two-tailed test we look up $t'_{100[1-(\alpha/2)]}$, but if a one-tailed test is used we look up $t'_{100(1-\alpha)}$. Here, $t'_c = t'_{100[1-(.05/2)]} = t_{97.5} = 2.68$. Hence,

$$\begin{aligned} D_c &= 2.68 \sqrt{\frac{2 \cdot 6.71}{4}} \\ &= 2.68\sqrt{3.355} \\ &= 2.68 \cdot 1.83 = 4.90 \end{aligned}$$

Since $\bar{X}_{1.}$ differs from all the other means by more than 4.90, each of these means differs significantly from $\bar{X}_{1.}$. No statement is made here concerning significant differences among the other means, since Dunnett's test makes no allowance for testing such differences.

9.5 TREND ANALYSIS

9.5.1 Introduction

In Section 3.2 it was mentioned that a factor can be qualitative or quantitative. The levels of a qualitative factor differ, but are not related according to any particular mathematical scheme. The levels of a quantitative factor, on the other hand, can be described by the quantitative value that each level has along some scale other than the criterion. The scale might be physical, such as room temperature or size of stimulus measured in square centimeters or subjective, such as the pitch of a sine wave tone measured in mels.

A quantitative factor can be analyzed in the same way as a qualitative factor, by simply ignoring the scale values of the levels and considering the levels to be only different from each other, as we do with qualitative factors. The analysis in this text has proceeded in such a manner. The ANOVA procedures described in Chapters 5 and 6 apply whether the factors are qualitative or quantitative; quantitative factors are analyzed as if they were qualitative. If a design has quantitative factors, however, additional analytical possibilities exist. In particular, it is possible to examine the form of the mathematical function relating the mean cell scores to the factor scale values.

Consider, for example, the one-factor design A. The scores are symbolized by X_{ae}, and the scale values associated with the a_a's are measured on a variable we call x. The particular x values used in the experiment are symbolized by $x_1, x_2, \ldots, x_A$. From the experimental scores, we have a set of cell means and their associated scale values, as follows:

$\bar{X}_{a.}$	$\bar{X}_{1.}$	$\bar{X}_{2.}$	$\bar{X}_{3.}$	$\cdots$	$\bar{X}_{A}$
x_a	x_1	x_2	x_3	$\cdots$	x_A

We want to find a mathematical function relating the X's and the x's. The function might be simply a straight line function, such as $X = k_0 + k_1x$, where k_0 and k_1 are constants. The function might be logarithmic, for example, $X = \log(x/x_0)$, where x_0 is a constant. It might be a power function, like $X = kx^n$, where k and n are constants. A priori, any conceivable function of one variable x might apply. We would not expect the true function underlying the data to fit the data exactly, if for no other reason than that the cell means include components based on error parameters. Ideally, we would like to find a function that fits the cell means well enough so that any discrepancies can be attributed to the **e**'s.

There are considerable advantages to knowing the mathematical function relating X and x. From ANOVA of Chapters 5 and 6, we can say that true differences exist among the population means for the different cells. But the scientific achievement is greater if we can express the differences as a mathematical function of the x's. For example, consider the difference in importance, for the science of physics, between an experiment that simply concludes that the pressure exerted by a gas at constant volume differs according to the temperature, and an experiment that shows that an inverse function exists. In science there is a large and important step between showing that one variable affects another and discovering mathematical relationships between the variables. One reason is that the mathematical relationship enables the scientist to assert, with some trepidation, that certain values of X would hold for values of x other than those used in the experiment. He simply takes whatever x he desires and uses the function to compute the corresponding X. If x is somewhere within the range of the experimental x's, the process is called *interpolation*. If x is above or below the extreme values used in the experiment, the process is called *extrapolation*. Although a scientist can err by using either interpolation or extrapolation, interpolation is generally safer than extrapolation. Besides allowing such generalizations, functional relationships allow scientists to make complex inferences among a set of variables, including those used in different experiments.

In spite of the advantages of using mathematical functions relating the dependent and independent variables, trend analysis is not used in most psychological experiments, partly because many of the factors used in experiments are qualitative, not quantitative. In addition, psychologists typically use only a few levels of a factor, whereas many levels are advisable to ascertain the shape of a function. Undoubtedly, psychological experimenters would take a greater interest in trend analysis if the simple, basic functional relations of the sort found in expositions of fundamental physics or chemistry existed for psychological data, but they seldom do. Although experimenters often use simple functions to describe the trends found, if more data had been collected so that the errors would cancel out more thoroughly and the cell means would be more precise, it would become apparent that the simple functions used are only crude approximations of the true functions underlying data.

9.5.2 Polynomial Functions

Actually, when psychologists perform trend analyses, they seldom attempt to compare the fit of a variety of functions, which, a priori, might be adequate. Instead, they normally attempt to fit the data with a *polynomial function*, which has the following form:

$$X = k_0 + k_1x + k_2x^2 + k_3x^3 + \cdots + k_{n-1}x^{n-1}$$

The $n - 1$ may be any integer, so the number of terms in a polynomial function can vary. The value $n - 1$ is called the *order* of the polynomial function. A set of n cell means can always be fit perfectly, with the proper choice for the k coefficients, by a polynomial function of order $n - 1$. Therefore, there is never any question about whether a polynomial function can fit a set of cell means. An experimenter need only use a polynomial function of sufficiently high order—and that order may not be very high in most psychology experiments, since relatively few cell means are available to be fit. Even if the cell means actually reflect some other function, such as the logarithmic or the inverse, a polynomial function can fit the data. The question is not, then, whether the polynomial function can fit the data, but why the experimenter should want to achieve this fit, and what order of function should be used.

We have already considered why an experimenter would want to fit a function in general. But why should a polynomial be used? Because with a polynomial, success is assured and the calculations are relatively easy.

The question remains, what order polynomial should be used? Although a perfect fit can be obtained to n means with a polynomial of order $n - 1$, such a fit serves no purpose. The cell means include error components, so the means would vary from replication to replication. Nothing is gained from reporting a function that describes the random fluctuations of a particular replication. The experimenter desires instead the true function, which would hold for every replication, realizing that the cell means for a particular replication will fluctuate around this function due to sampling error.

Normally, the approach taken is to use the polynomial function of smallest order that will fit the cell means, leaving only variations between the means and the fitted function that can be attributed to sampling variability. Oftentimes the polynomial function of order 1 suffices. This function is simply $X = k_0 + k_1x$, a *linear function*. Constants k_0 and k_1 are chosen so that the function lies as close as possible to the cell means, based on the criterion of least squares.[6] Then the deviations between the function and the means are examined. The SS for the quantitative factor is divided into a component accountable to the linear function derived and a component not accountable, that is, the remainder. Then the remainder is F-tested to see whether to accept the null hypothesis that the remainder can be attributed to random error. If the null hypothesis is accepted, the linear function is considered to be adequate to represent the cell means. If the null hypothesis is rejected, the procedure is to find the best-fitting function of order 2: $X = k_0 + k_1x + k_2x^2$. This is called a *quadratic function*. Next the SS is found that is associated with the quadratic. This SS is larger than the linear SS, so the remainder is correspondingly smaller. Again an F-test is done on the residual SS, and if the residual is attributable to error (if the H_N is accepted), the quadratic is taken to represent the trend adequately. Otherwise, the best fitting *cubic function*, $X = k_0 + k_1X + k_2x^2 + k_3x^3$, is found. It is unusual for psychological data to require anything beyond the cubic function.

Graphed, the linear function appears as a straight line. A quadratic function is curved, but the curvature is either everywhere upward or everywhere downward. The segments of a cubic function are curved upward and downward. Figure 9.5-1 illustrates a linear, a quadratic, and a cubic equation.

The quadratic function illustrated in the figure is *convex downward;* that is,

[6] The function is placed so that it minimizes the sum of the squared discrepancies between the scores and the function.

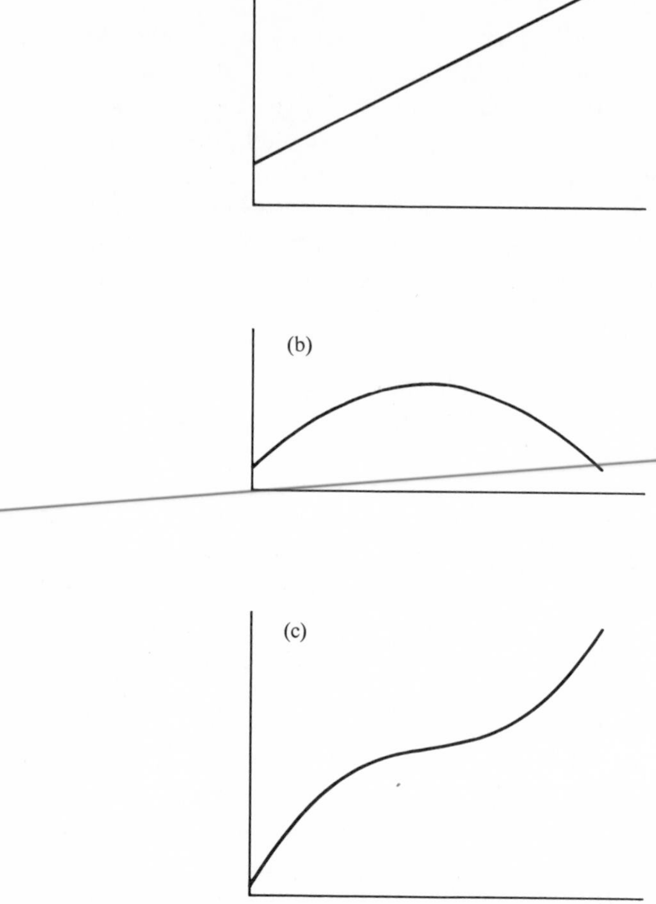

Figure 9.5-1. Thre polynomial functions: (a) a linear function; (b) a quadratic function; (c) a cubic function.

its curvature is like a bowl facing downward. A quadratic function may also be convex upward—like a bowl right-side up. The cubic equation illustrated has an initial segment convex downward followed by a segment convex upward. The order of the convexity may be opposite in other cubic equations. Also, the function illustrated is monotonic; that is, an increase in x always implies an increase in X. A cubic function may, however, have a peak and valley, rather like a letter S on its side.

The testing sequence that has been described here is sometimes altered when the experimenter has a priori hypotheses concerning which polynomial components should appear. Such a priori hypotheses are tested individually without first testing them pooled with other components.

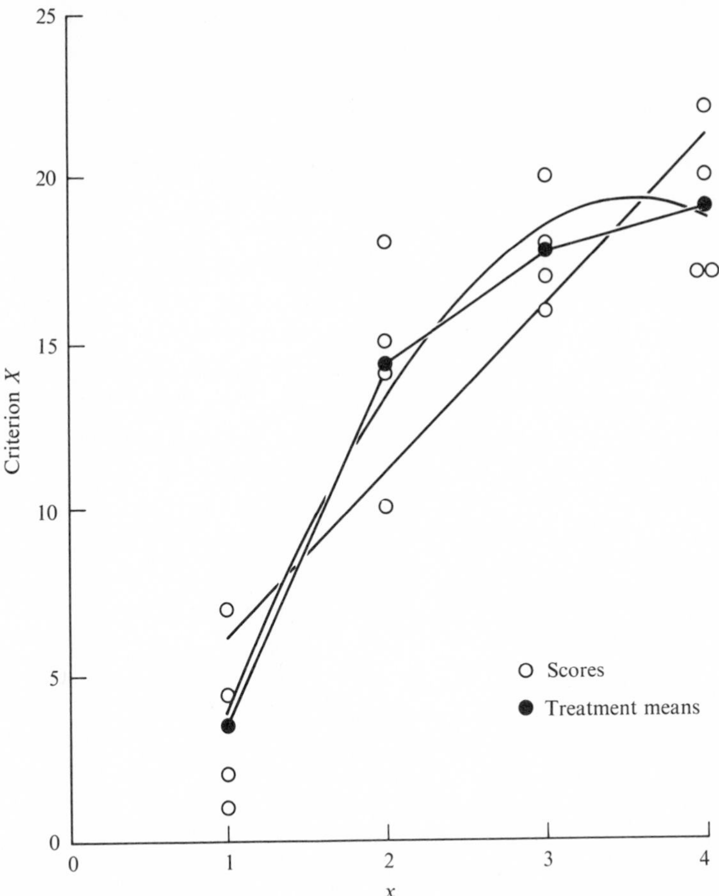

Figure 9.5-2. Plot of the scores, mean cell scores (connected by straight-line segments), the best-fitting linear function, and the best-fitting quadratic function, for the numerical example.

9.5.3 Numerical Example: One Quantitative Factor

Significance testing. In this subsection, polynomial trend analysis is applied to the data for design *A* presented in Section 9.4.3. Polynomial trend analysis is simplified if the spacing between the successive values of the quantitative factor is the same for each adjacent pair of levels. It is usually convenient to use equal spacing, and the techniques demonstrated here assume that equal spacing was used. Myers[7] describes the alterations required when equal spacing is not used.

The data are plotted in Figure 9.5-2. Treatment a_1 has an associated x_1 of 1, treatment a_2 has an x_2 of 2, and so on. The figure shows the best-fitting

[7] J. L. Myers, *Fundamentals of Experimental Design* (Boston: Allyn and Bacon, 1966), chap. 14.

linear function and the best-fitting quadratic function. The methods used to derive these functions are given later in this subsection. For now let us concern ourselves with the statistical testing required to determine what order of polynomial should be fit to the cell means.

The analysis of variance table (Section 9.4.3) showed that $SS_A = 595.25$, that $MS_E = 6.71$, and that the A effect is significant at the .01 significance level. We proceed to partition SS_A into linear and supralinear components. The required calculations will be shown momentarily. For now we must simply accept that the linear part, $SS_{lin} = 500.00$, and $SS_{sup\ lin} = SS_A - SS_{lin} = 95.25$. We can do separate F-tests on the linear and supralinear parts, in each case using MS_E as MS_2. The df for SS_{lin} is 1, and the df for $SS_{sup\ lin}$ is $A - 2 = 2$. Hence, the F-statistics are:

$$\text{Linear:} \qquad F = \frac{500.00}{6.71} = 74.52$$

$$\text{Supralinear:} \qquad F = \frac{95.25/2}{6.71} = 7.10$$

Since $F_{99}(1, 12) = 9.33$ and $F_{99}(2, 12) = 6.93$, both the linear and supralinear components are significant.

To say that the supralinear SS is significant is to say that more than error contributes to it: specifically, other higher-order polynomial components. In other words, we conclude that a straight-line function is not an adequate description of the trend. The next step is to decompose $SS_{sup\ lin}$ into quadratic and supraquadratic components. The calculational details are postponed momentarily, but once we compute SS_{quad} as 90.25, we find $SS_{sup\ quad}$ as $SS_{sup\ lin} - SS_{quad} = 95.25 - 90.25 = 5.00$. Again, we do separate F-tests on these components. The df for SS_{quad} is 1, and the df for $SS_{sup\ quad}$ is $A - 3 = 1$. Hence, the F-statistics are:

$$\text{Quadratic:} \qquad F = \frac{90.25}{6.71} = 13.45$$

$$\text{Supraquadratic:} \qquad F = \frac{5.00}{6.71} = .745$$

Since $F_{99}(1, 12) = 9.33$, the quadratic component is significant at the .01 level. The supraquadratic component is not significant.

We conclude that a polynomial function of order 2—a quadratic function—is the proper one to use to describe our data. To the extent that the best-fitting quadratic function fails to fit the treatment means perfectly, the discrepancy can be attributed to error that causes the treatment means to depart from their true values.

Computing the polynomial SS's. The method for computing the SS for each polynomial order (SS_{lin}, SS_{quad}, and SS_{cub}) will now be presented. These

Table 9.5-1 Coefficients of orthogonal polynomials for $A = 4$

	w_1	w_2	w_3	w_4	$\sum w_a^2$
Linear	-3	-1	1	3	20
Quadratic	1	-1	-1	1	4
Cubic	-1	3	-3	1	20

computations are particularly simple when the spacing of the treatments is equal on x, as in the present example. For each polynomial SS, for a specified number of treatments A, there is a set of integral numbers called *coefficients of orthogonal polynomials.* These coefficients, symbolized by w's, are given in Table A.9. For $A = 4$, the coefficients are as shown in Table 9.5-1.
To compute a polynomial SS, proceed as follows:

1. Compute $W_{\text{poly}} = \sum w_a X_{a.}$.
2. Compute $\text{SS}_{\text{poly}} = W^2_{\text{poly}}/E\sum w_a^2$.

The $X_{a.}$'s, sums of the scores for the different treatments, are:

$X_{1.}$	$X_{2.}$	$X_{3.}$	$X_{4.}$
14	57	71	76

The $\sum w_a^2$ quantities, together with the w_a's, are given in Table 9.5-1. We have, then,

$$\begin{aligned} W_{\text{lin}} &= (-3)(14) + (-1)(57) + (1)(71) + (3)(76) \\ &= -42 - 57 + 71 + 228 = 200 \\ \text{SS}_{\text{lin}} &= 200^2/4\cdot 20 = 40{,}000/80 = 500 \end{aligned}$$

$$\begin{aligned} W_{\text{quad}} &= (1)(14) + (-1)(57) + (-1)(71) + (1)(76) \\ &= 14 - 57 - 71 + 76 = -38 \\ \text{SS}_{\text{quad}} &= (-38)^2/4\cdot 4 = 1444/16 = 90.25 \end{aligned}$$

$$\begin{aligned} W_{\text{cub}} &= (-1)(14) + (3)(57) + (-3)(71) + (1)(76) \\ &= -14 + 171 - 213 + 76 = 20 \\ \text{SS}_{\text{cub}} &= 20^2/4\cdot 20 = 400/80 = 5 \end{aligned}$$

Note that W_{lin}, W_{quad}, and W_{cub} are mutually orthogonal contrasts. As pointed out in the preceding section, the SS for an effect can be subdivided into SS's associated with orthogonal contrasts on the effect, and that each such component SS has 1 df. The breakdown of SS_A into polynomial components illustrates this phenomenon. The significance tests for the SS components are simply tests of contrasts, as described in Section 9.4.3. A special feature of the polynomial contrast tests is the testing of the supra-SS's, which are, in effect, SS's formed by pooling SS's for separate contrasts.

Finding the equation. In finding the polynomial equation that is appropriate for the data, instead of working with the form

$$X = k_0 + k_1x + k_2x^2 + k_3x^3 + \cdots + k_{n-1}x^{n-1},$$

it is convenient to use the form

$$X = \bar{X} + k_1'z_1 + k_2'z_2 + k_3'z_3 + \cdots + k_{n-1}z_{n-1}$$

where

$$k_1' = \frac{W_{\text{lin}}}{E\sum w_{\text{lin}}^2} \quad \text{and} \quad z_1 = \lambda_1(x - \bar{x})$$

$$k_2' = \frac{W_{\text{quad}}}{E\sum w_{\text{quad}}^2} \quad \text{and} \quad z_2 = \lambda_2\left[(x - \bar{x})^2 - \frac{A^2 - 1}{12}\right]$$

$$k_3' = \frac{W_{\text{cub}}}{E\sum w_{\text{cub}}^2} \quad \text{and} \quad z_3 = \lambda_3\left[(x - \bar{x})^3 - (x - \bar{x})\frac{3A^2 - 7}{20}\right]$$

Since psychologists seldom find significant SS's higher than the cubic, the preceding formulas should suffice.

Let us see how they would apply to our numerical example. The λ's (Greek lambdas), together with the w's, are given in Table A.9. When $A = 4$, $\lambda_1 = 2$, $\lambda_2 = 1$, and $\lambda_3 = 10/3$. Then

$$k_1' = \frac{200}{80} = 2.5$$

$$k_2' = \frac{-38}{16} = -2.375$$

$$k_3' = \frac{20}{80} = .25$$

$$z_1 = 2(x - \bar{x})$$

$$z_2 = (x - \bar{x})^2 - \frac{4^2 - 1}{12} = (x - \bar{x})^2 - 1.25$$

$$z_3 = \frac{10}{3}\left[(x - \bar{x})^3 - (x - \bar{x})\frac{3\cdot 4^2 - 7}{20}\right] = \frac{10}{3}[(x - \bar{x})^3 - 2.05(x - \bar{x})]$$

If only the linear SS were significant, the formula for fitting the data would be

$$X = \bar{X} + k_1'z_1 = 5x + 1.12$$

In the example, both the linear and the quadratic trend components were significant, so the equation for fitting the data is

$$X = \bar{X} + k_1'z_1 + k_2'z_2 = -10.75 + 16.88x - 2.38x^2$$

The linear and quadratic equations are plotted in Figure 9.5-2. The cubic equation that fits the four treatment means perfectly is

$$X = \bar{X} + k_1'z_1 + k_2'z_2 + k_3'z_3 = -19.5 + 30.8x - 8.63x^2 + .83x^3$$

Although an experimenter may wish to find the polynomial function in the usual form (that is, $X = k_0 + k_1x + k_2x^2$), for computing values of X for different x's it is simpler to use the form $X = \bar{X} + k_1'z_1 + k_2'z_2$.

Differences between means. The preceding section focused on significant differences among treatment means. The same data that are used in this section were used to test the various contrasts among means. For several tests, it was found that, whereas $\bar{X}_{1.}$ differed significantly from the other means, these latter did not differ significantly among themselves. Although such tests apply whether the factor is quantitative or qualitative, the results must be interpreted in a different light when the factor is quantitative. It does not make sense to conclude that $\bar{X}_{2.}$, $\bar{X}_{3.}$, and $\bar{X}_{4.}$ are not significantly different, and also that a best-fitting function shows them to be different. When reason and a trend analysis of the data suggest that the true cell means vary with x, an experimenter normally concludes that true differences exist, and uses the fitted curve to estimate them. If statistical tests that apply to qualitative factors imply that no true differences exist, it only means that the differences are small relative to the sampling variability.

9.5.4 Trend Analysis in Other Designs

The preceding discussion of trend analysis concerned only the Class I, single measurement design A. To perform a polynomial trend analysis of the repeated measurements design $S \times A$, we can proceed in much the same way as with design A. In the formula for W, substitute S, the number of subjects, for E. In design $S \times A$, MS_{AS} is used as MS_2 for testing the main effect A, and it is likewise used for MS_2 in trend analysis.

For designs with many factors, a polynomial trend analysis can be done for any one quantitative factor of the design by summing across all scores at each level of this factor. In place of E in the formula for W, use the number of scores occurring for each level of the quantitative factor. For MS_2 in the tests, use the same MS_2 as is used to test the main effect of the quantitative factor.

If the quantitative factor does not interact with other design factors, then the same polynomial function applies across all levels of these other factors, except possibly for a constant. For example, suppose that we have performed a polynomial trend analysis for factor A of design $A \times B$ by using sums of $B \cdot E$ scores at each level of A. We found that the trend could be described by a linear (straight-line) function, with no significant supralinear components.

The main effect for factor B was significant, but there was no AB interaction. Then at each level of factor B, there is a best linear function describing the trend for factor A. These linear functions are parallel.

When the quantitative factor interacts with another factor, the trend lines are not parallel at the different levels of the other factor, and these trend lines need not have the same shape. Winer (*op. cit.*) gives a more detailed description of polynomial trend analysis involving more than one factor.

APPENDIX A

Statistical Tables

Table A.1 Random digits

Line \ Col.	(1)	(2)	(3)	(4)	(5)	(6)	(7)	(8)	(9)	(10)	(11)	(12)	(13)	(14)
1	10480	15011	01536	02011	81647	91646	69179	14194	62590	36207	20969	99570	91291	90700
2	22368	46573	25595	85393	30995	89198	27982	53402	93965	34095	52666	19174	39615	99505
3	24130	48360	22527	97265	76393	64809	15179	24830	49340	32081	30680	19655	63348	58629
4	42167	93093	06243	61680	07856	16376	39440	53537	71341	57004	00849	74917	97758	16379
5	37570	39975	81837	16656	06121	91782	60468	81305	49684	60672	14110	06927	01263	54613
6	77921	06907	11008	42751	27756	53498	18602	70659	90655	15053	21916	81825	44394	42880
7	99562	72905	56420	69994	98872	31016	71194	18738	44013	48840	63213	21069	10634	12952
8	96301	91977	05463	07972	18876	20922	94595	56869	69014	60045	18425	84903	42508	32307
9	89579	14342	63661	10281	17453	18103	57740	84378	25331	12566	58678	44947	05585	56941
10	85475	36857	53342	53988	53060	59533	38867	62300	08158	17983	16439	11458	18593	64952
11	28918	69578	88231	33276	70997	79936	56865	05859	90106	31595	01547	85590	91610	78188
12	63553	40961	48235	03427	49626	69445	18663	72695	52180	20847	12234	90511	33703	90322
13	09429	93969	52636	92737	88974	33488	36320	17617	30015	08272	84115	27156	30613	74952
14	10365	61129	87529	85689	48237	52267	67689	93394	01511	26358	85104	20285	29975	89868
15	07119	97336	71048	08178	77233	13916	47564	81056	97735	85977	29372	74461	28551	90707
16	51085	12765	51821	51259	77452	16308	60756	92144	49442	53900	70960	63990	75601	40719
17	02368	21382	52404	60268	89368	19885	55322	44819	01188	65255	64835	44919	05944	55157
18	01011	54092	33362	94904	31273	04146	18594	29852	71585	85030	51132	01915	92747	64951
19	52162	53916	46369	58586	23216	14513	83149	98736	23495	64350	94738	17752	35156	35749
20	07056	97628	33787	09998	42698	06691	76988	13602	51851	46104	88916	19509	25625	58104
21	48663	91245	85828	14346	09172	30168	90229	04734	59193	22178	30421	61666	99904	32812
22	54164	58492	22421	74103	47070	25306	76468	26384	58151	06646	21524	15227	96909	44592
23	32639	32363	05597	24200	13363	38005	94342	28728	35806	06912	17012	64161	18296	22851
24	29334	27001	87637	87308	58731	00256	45834	15398	46557	41135	10367	07684	36188	18510
25	02488	33062	28834	07351	19731	92420	60952	61280	50001	67658	32586	86679	50720	94953
26	81525	72295	04839	96423	24878	82651	66566	14778	76797	14780	13300	87074	79666	95725
27	29676	20591	68086	26432	46901	20849	89768	81536	86645	12659	92259	57102	80428	25280
28	00742	57392	39064	66432	84673	40027	32832	61362	98947	96067	64760	64584	96096	98253
29	05366	04213	25669	26422	44407	44048	37937	63904	45766	66134	75470	66520	34693	90449
30	91921	26418	64117	94305	26766	25940	39972	22209	71500	64568	91402	42416	07844	69618

31	[illegible]	[illegible]	[illegible]	[illegible]	[illegible]	[illegible]	[illegible]	[illegible]	[illegible]	[illegible]	[illegible]	[illegible]	[illegible]	[illegible]
32	00725	69884	62797	56170	86324	88072	76222	36086	84637	93161	76038	65855	77919	88006
33	69011	65795	95876	55293	18988	27354	26575	08625	40801	59920	29841	80150	12777	48501
34	25976	57948	29888	88604	67917	48708	18912	82271	65424	69774	33611	54262	85963	03547
35	09763	83473	73577	12908	30883	18317	28290	35797	05998	41688	34952	37888	38917	88050
36	91567	42595	27958	30134	04024	86385	29880	99730	55536	84855	29080	09250	79656	73211
37	17955	56349	90999	49127	20044	59931	06115	20542	18059	02008	73708	83517	36103	42791
38	46503	18584	18845	49618	02304	51038	20655	58727	28168	15475	56942	53389	20562	87338
39	92157	89634	94824	78171	84610	82834	09922	25417	44137	48413	25555	21246	35509	20468
40	14577	62765	35605	81263	39667	47358	56873	56307	61607	49518	89656	20103	77490	18062
41	98427	07523	33362	64270	01638	92477	66969	98420	04880	45585	46565	04102	46880	45709
42	34914	63976	88720	82765	34476	17032	87589	40836	32427	70002	70663	88863	77775	69348
43	70060	28277	39475	46473	23219	53416	94970	25832	69975	94884	19661	72828	00102	66794
44	53976	54914	06990	67245	68350	82948	11398	42878	80287	88267	47363	46634	06541	97809
45	76072	29515	40980	07391	58745	25774	22987	80059	39911	96189	41151	14222	60697	59583
46	90725	52210	83974	29992	65831	38857	50490	83765	55657	14361	31720	57375	56228	41546
47	64364	67412	33339	31926	14883	24413	59744	92351	97473	89286	35931	04110	23726	51900
48	08962	00358	31662	25388	61642	34072	81249	35648	56891	69352	48373	45578	78547	81788
49	95012	68379	93526	70765	10592	04542	76463	54328	02349	17247	28865	14777	62730	92277
50	15664	10493	20492	38391	91132	21999	59516	81652	27195	48223	46751	22923	32261	85653
51	16408	81899	04153	53381	79401	21438	83035	92350	36693	31238	59649	91754	72772	02338
52	18629	81953	05520	91962	04739	13092	97662	24822	94730	06496	35090	04822	86774	98289
53	73115	35101	47498	87637	99016	71060	88824	71013	18735	20286	23153	72924	35165	43040
54	57491	16703	23167	49323	45021	33132	12544	41035	80780	45393	44812	12515	98931	91202
55	30405	83946	23792	14422	15059	45799	22716	19792	09983	74353	68668	30429	70735	25499
56	16631	35006	85900	98275	32388	52390	16815	69298	82732	38480	73817	32523	41961	44437
57	96773	20206	42559	78985	05300	22164	24369	54224	35083	19687	11052	91491	60383	19746
58	38935	64202	14349	82674	66523	44133	00697	35552	35970	19124	63318	29686	03387	59846
59	31624	76384	17403	53363	44167	64486	64758	75366	76554	31601	12614	33072	60332	92325
60	78919	19474	23632	27889	47914	02584	37680	20801	72152	39339	34806	08930	85001	87820

SOURCE: Abridged from "Table of 105,000 Random Decimal Digits," Statement 4914, Bureau of Transport Economics and Statistics, Interstate Commerce Commission, 1949.

Table A.2 Squares and square roots

N	N^2	$\sqrt{N}$	$\sqrt{10N}$	N	N^2	$\sqrt{N}$	$\sqrt{10N}$
1	1	1.00 000	3.16 228	51	2 601	7.14 143	22.58 32
2	4	1.41 421	4.47 214	52	2 704	7.21 110	22.80 35
3	9	1.73 205	5.47 723	53	2 809	7.28 011	23.02 17
4	16	2.00 000	6.32 456	54	2 916	7.34 847	23.23 79
5	25	2.23 607	7.07 107	55	3 025	7.41 620	23.45 21
6	36	2.44 949	7.74 597	56	3 136	7.48 331	23.66 43
7	49	2.64 575	8.36 660	57	3 249	7.54 983	23.87 47
8	64	2.82 843	8.94 427	58	3 364	7.61 577	24.08 32
9	81	3.00 000	9.48 683	59	3 481	7.68 115	24.28 99
10	100	3.16 228	10.00 00	60	3 600	7.74 597	24.49 49
11	121	3.31 662	10.48 81	61	3 721	7.81 025	24.69 82
12	144	3.46 410	10.95 45	62	3 844	7.87 401	24.89 98
13	169	3.60 555	11.40 18	63	3 969	7.93 725	25.09 98
14	196	3.74 166	11.83 22	64	4 096	8.00 000	25.29 82
15	225	3.87 298	12.24 74	65	4 225	8.06 226	25.49 51
16	256	4.00 000	12.64 91	66	4 356	8.12 404	25.69 05
17	289	4.12 311	13.03 84	67	4 489	8.18 535	25.88 44
18	324	4.24 264	13.41 64	68	4 624	8.24 621	26.07 68
19	361	4.35 890	13.78 40	69	4 761	8.30 662	26.26 79
20	400	4.47 214	14.14 21	70	4 900	8.36 660	26.45 75
21	441	4.58 258	14.49 14	71	5 041	8.42 615	26.64 58
22	484	4.69 042	14.83 24	72	5 184	8.48 528	26.83 28
23	529	4.79 583	15.16 58	73	5 329	8.54 400	27.01 85
24	576	4.89 898	15.49 19	74	5 476	8.60 233	27.20 29
25	625	5.00 000	15.81 14	75	5 625	8.66 025	27.38 61
26	676	5.09 902	16.12 45	76	5 776	8.71 780	27.56 81
27	729	5.19 615	16.43 17	77	5 929	8.77 496	27.74 89
28	784	5.29 150	16.73 32	78	6 084	8.83 176	27.92 85
29	841	5.38 516	17.02 94	79	6 241	8.88 819	28.10 69
30	900	5.47 723	17.32 05	80	6 400	8.94 427	28.28 43
31	961	5.56 776	17.60 68	81	6 561	9.00 000	28.46 05
32	1 024	5.65 685	17.88 85	82	6 724	9.05 539	28.63 56
33	1 089	5.74 456	18.16 59	83	6 889	9.11 043	28.80 97
34	1 156	5.83 095	18.43 91	84	7 056	9.16 515	28.98 28
35	1 225	5.91 608	18.70 83	85	7 225	9.21 954	29.15 48
36	1 296	6.00 000	18.97 37	86	7 396	9.27 362	29.32 58
37	1 369	6.08 276	19.23 54	87	7 569	9.32 738	29.49 58
38	1 444	6.16 441	19.49 36	88	7 744	9.38 083	29.66 48
39	1 521	6.24 500	19.74 84	89	7 921	9.43 398	29.83 29
40	1 600	6.32 456	20.00 00	90	8 100	9.48 683	30.00 00
41	1 681	6.40 312	20.24 85	91	8 281	9.53 939	30.16 62
42	1 764	6.48 074	20.49 39	92	8 464	9.59 166	30.33 15
43	1 849	6.55 744	20.73 64	93	8 649	9.64 365	30.49 59
44	1 936	6.63 325	20.97 62	94	8 836	9.69 536	30.65 94
45	2 025	6.70 820	21.21 32	95	9 025	9.74 679	30.82 21
46	2 116	6.78 233	21.44 76	96	9 216	9.79 796	30.98 39
47	2 209	6.85 565	21.67 95	97	9 409	9.84 886	31.14 48
48	2 304	6.92 820	21.90 89	98	9 604	9.89 949	31.30 50
49	2 401	7.00 000	22.13 59	99	9 801	9.94 987	31.46 43
50	2 500	7.07 107	22.36 07	100	10 000	10.00 000	31.62 28
N	N^2	$\sqrt{N}$	$\sqrt{10N}$	N	N^2	$\sqrt{N}$	$\sqrt{10N}$

Table A.3 Percentiles of the standard normal distribution

P (%)	z	P (%)	z	P (%)	z
50	0.00	75	0.67	95.0	1.645
51	0.03	76	0.71	95.5	1.695
52	0.05	77	0.74	96.0	1.751
53	0.08	78	0.77	96.5	1.812
54	0.10	79	0.81	97.0	1.881
55	0.13	80	0.84	97.5	1.960
56	0.15	81	0.88	98.0	2.054
57	0.18	82	0.92	98.5	2.170
58	0.20	83	0.95	99.0	2.326
59	0 23	84	0 99	99.5	2.576
60	0.25	85	1.04	99.6	2.652
61	0.28	86	1.08	99.7	2.748
62	0.30	87	1.13	99.8	2.878
63	0.33	88	1.17	99.9	3.090
64	0.36	89	1.23		
65	0.39	90	1.28	99.95	3.291
66	0.41	91	1.34	99.995	3.891
67	0.44	92	1.41		
68	0.47	93	1.48	99.9995	4.417
69	0.50	94	1.55		
				99.99995	5.327
70	0.52				
71	0.55				
72	0.58				
73	0.61				
74	0.64				

SOURCE: This table is abridged from Table 9 in *Biometrika Tables for Statisticians*, 3rd ed., vol. 1 (New York: Cambridge University Press, 1966), edited by E. S. Pearson and H. O. Hartley, by permission of the *Biometrika* Trustees.

P = percentile score.
z = percentile.

Table A.4 Percentiles of the *t*-distribution

	P (%)					
df	90	95	97.5	99	99.5	99.95
1	3.078	6.314	12.706	31.821	63.657	636.619
2	1.886	2.920	4.303	6.965	9.925	31.598
3	1.638	2.353	3.182	4.541	5.841	12.924
4	1.533	2.132	2.776	3.747	4.604	8.610
5	1.476	2.015	2.571	3.365	4.032	6.869
6	1.440	1.943	2.447	3.143	3.707	5.959
7	1.415	1.895	2.365	2.998	3.499	5.408
8	1.397	1.860	2.306	2.896	3.355	5.041
9	1.383	1.833	2.262	2.821	3.250	4.781
10	1.372	1.812	2.228	2.764	3.169	4.587
11	1.363	1.796	2.201	2.718	3.106	4.437
12	1.356	1.782	2.179	2.681	3.055	4.318
13	1.350	1.771	2.160	2.650	3.012	4.221
14	1.345	1.761	2.145	2.624	2.977	4.140
15	1.341	1.753	2.131	2.602	2.947	4.073
16	1.337	1.746	2.120	2.583	2.921	4.015
17	1.333	1.740	2.110	2.567	2.898	3.965
18	1.330	1.734	2.101	2.552	2.878	3.922
19	1.328	1.729	2.093	2.539	2.861	3.883
20	1.325	1.725	2.086	2.528	2.845	3.850
21	1.323	1.721	2.080	2.518	2.831	3.819
22	1.321	1.717	2.074	2.508	2.819	3.792
23	1.319	1.714	2.069	2.500	2.807	3.767
24	1.318	1.711	2.064	2.492	2.797	3.745
25	1.316	1.708	2.060	2.485	2.787	3.725
26	1.315	1.706	2.056	2.479	2.779	3.707
27	1.314	1.703	2.052	2.473	2.771	3.690
28	1.313	1.701	2.048	2.467	2.763	3.674
29	1.311	1.699	2.045	2.462	2.756	3.659
30	1.310	1.697	2.042	2.457	2.750	3.646
40	1.303	1.684	2.021	2.423	2.704	3.551
60	1.296	1.671	2.000	2.390	2.660	3.460
120	1.289	1.658	1.980	2.358	2.617	3.373
∞	1.282	1.645	1.960	2.326	2.576	3.291

SOURCE: This table is taken from Table 3 of Fisher and Yates, *Statistical Tables for Biological, Agricultural and Medical Research,* 1973, published by Longman Group Ltd., London (previously published in 1963 by Oliver and Boyd Ltd., Edinburgh), and reprinted by permission of the authors and publishers.

Table A.5 Percentiles of the *F*-distribution

df_2	P (%)	df_1: 1	2	3	4	5	6	7	8	9	10	12	15	20	24	30	40	60	∞
1	75	5.83	7.50	8.20	8.58	8.82	8.98	9.10	9.19	9.26	9.32	9.41	9.49	9.58	9.63	9.67	9.71	9.76	9.85
	90	39.9	49.5	53.6	55.8	57.2	58.2	58.9	59.4	59.9	60.2	60.7	61.2	61.7	62.0	62.3	62.5	62.8	63.3
	95	161	200	216	225	230	234	237	239	241	242	244	246	248	249	250	251	252	254
2	75	2.57	3.00	3.15	3.23	3.28	3.31	3.34	3.35	3.37	3.38	3.39	3.41	3.43	3.43	3.44	3.45	3.46	3.48
	90	8.53	9.00	9.16	9.24	9.29	9.33	9.35	9.37	9.38	9.39	9.41	9.42	9.44	9.45	9.46	9.47	9.47	9.49
	95	18.5	19.0	19.2	19.2	19.3	19.3	19.4	19.4	19.4	19.4	19.4	19.4	19.4	19.5	19.5	19.5	19.5	19.5
	99	98.5	99.0	99.2	99.2	99.3	99.3	99.4	99.4	99.4	99.4	99.4	99.4	99.4	99.5	99.5	99.5	99.5	99.5
3	75	2.02	2.28	2.36	2.39	2.41	2.42	2.43	2.44	2.44	2.44	2.45	2.46	2.46	2.46	2.47	2.47	2.47	2.47
	90	5.54	5.46	5.39	5.34	5.31	5.28	5.27	5.25	5.24	5.23	5.22	5.20	5.18	5.18	5.17	5.16	5.15	5.13
	95	10.1	9.55	9.28	9.12	9.01	8.94	8.89	8.85	8.81	8.79	8.74	8.70	8.66	8.64	8.62	8.59	8.57	8.53
	99	34.1	30.8	29.5	28.7	28.2	27.9	27.7	27.5	27.4	27.2	27.0	26.9	26.7	26.6	26.5	26.4	26.3	26.1
	99.9	167	148	141	137	135	133	132	131	130	129	128	127	126	126	125	125	124	124
4	75	1.81	2.00	2.05	2.06	2.07	2.08	2.08	2.08	2.08	2.08	2.08	2.08	2.08	2.08	2.08	2.08	2.08	2.08
	90	4.54	4.32	4.19	4.11	4.05	4.01	3.98	3.95	3.94	3.92	3.90	3.87	3.84	3.83	3.82	3.80	3.79	3.76
	95	7.71	6.94	6.59	6.39	6.26	6.16	6.09	6.04	6.00	5.96	5.91	5.86	5.80	5.77	5.75	5.72	5.69	5.63
	99	21.2	18.0	16.7	16.0	15.5	15.2	15.0	14.8	14.7	14.6	14.4	14.2	14.0	13.9	13.8	13.8	13.6	13.5
	99.9	74.1	61.2	56.2	53.4	51.7	50.5	49.7	49.0	48.5	48.0	47.4	46.8	46.1	45.8	45.4	45.1	44.8	44.0
5	75	1.69	1.85	1.88	1.89	1.89	1.89	1.89	1.89	1.89	1.89	1.89	1.89	1.88	1.88	1.88	1.88	1.87	1.87
	90	4.06	3.78	3.62	3.52	3.45	3.40	3.37	3.34	3.32	3.30	3.27	3.24	3.21	3.19	3.17	3.16	3.14	3.10
	95	6.61	5.79	5.41	5.19	5.05	4.95	4.88	4.82	4.77	4.74	4.68	4.62	4.56	4.53	4.50	4.46	4.43	4.36
	99	16.3	13.3	12.1	11.4	11.0	10.7	10.5	10.3	10.2	10.0	9.89	9.72	9.55	9.47	9.38	9.29	9.20	9.02
	99.9	47.2	37.1	33.2	31.1	29.8	28.8	28.2	27.6	27.2	26.9	26.4	25.9	25.4	25.1	24.9	24.6	24.3	23.8

(*continued*)

Table A.5 (*continued*)

df_2	P (%)	df_1: 1	2	3	4	5	6	7	8	9	10	12	15	20	24	30	40	60	∞
6	75	1.62	1.76	1.78	1.79	1.79	1.78	1.78	1.78	1.77	1.77	1.77	1.76	1.76	1.75	1.75	1.75	1.74	1.74
	90	3.78	3.46	3.29	3.18	3.11	3.05	3.01	2.98	2.96	2.94	2.90	2.87	2.84	2.82	2.80	2.78	2.76	2.72
	95	5.99	5.14	4.76	4.53	4.39	4.28	4.21	4.15	4.10	4.06	4.00	3.94	3.87	3.84	3.81	3.77	3.74	3.67
	99	13.8	10.9	9.78	9.15	8.75	8.47	8.26	8.10	7.98	7.87	7.72	7.56	7.40	7.31	7.23	7.14	7.06	6.88
	99.9	35.5	27.0	23.7	21.9	20.8	20.0	19.5	19.0	18.7	18.4	18.0	17.6	17.1	16.9	16.7	16.4	16.2	15.8
7	75	1.57	1.70	1.72	1.72	1.71	1.71	1.70	1.70	1.69	1.69	1.68	1.68	1.67	1.67	1.66	1.66	1.65	1.65
	90	3.59	3.26	3.07	2.96	2.88	2.83	2.78	2.75	2.72	2.70	2.67	2.63	2.59	2.58	2.56	2.54	2.51	2.47
	95	5.59	4.74	4.35	4.12	3.97	3.87	3.79	3.73	3.68	3.64	3.57	3.51	3.44	3.41	3.38	3.34	3.30	3.23
	99	12.2	9.55	8.45	7.85	7.46	7.19	6.99	6.84	6.72	6.62	6.47	6.31	6.16	6.07	5.99	5.91	5.82	5.65
	99.9	29.2	21.7	18.8	17.2	16.2	15.5	15.0	14.6	14.3	14.1	13.7	13.3	12.9	12.7	12.5	12.3	12.1	11.7
8	75	1.54	1.66	1.67	1.66	1.66	1.65	1.64	1.64	1.63	1.63	1.62	1.62	1.61	1.60	1.60	1.59	1.59	1.58
	90	3.46	3.11	2.92	2.81	2.73	2.67	2.62	2.59	2.56	2.54	2.50	2.46	2.42	2.40	2.38	2.36	2.34	2.29
	95	5.32	4.46	4.07	3.84	3.69	3.58	3.50	3.44	3.39	3.35	3.28	3.22	3.15	3.12	3.08	3.04	3.01	2.93
	99	11.3	8.65	7.59	7.01	6.63	6.37	6.18	6.03	5.91	5.81	5.67	5.52	5.36	5.28	5.20	5.12	5.03	4.86
	99.9	25.4	18.5	15.8	14.4	13.5	12.9	12.4	12.0	11.8	11.5	11.2	10.8	10.5	10.3	10.1	9.92	9.73	9.33
9	75	1.51	1.62	1.63	1.63	1.62	1.61	1.60	1.60	1.59	1.59	1.58	1.57	1.56	1.56	1.55	1.54	1.54	1.53
	90	3.36	3.01	2.81	2.69	2.61	2.55	2.51	2.47	2.44	2.42	2.38	2.34	2.30	2.28	2.25	2.23	2.21	2.16
	95	5.12	4.26	3.86	3.63	3.48	3.37	3.29	3.23	3.18	3.14	3.07	3.01	2.94	2.90	2.86	2.83	2.79	2.71
	99	10.6	8.02	6.99	6.42	6.06	5.80	5.61	5.47	5.35	5.26	5.11	4.96	4.81	4.73	4.65	4.57	4.48	4.31
	99.9	22.9	16.4	13.9	12.6	11.7	11.1	10.7	10.4	10.1	9.89	9.57	9.24	8.90	8.72	8.55	8.37	8.19	7.81
10	75	1.49	1.60	1.60	1.59	1.59	1.58	1.57	1.56	1.56	1.55	1.54	1.53	1.52	1.52	1.51	1.51	1.50	1.48
	90	3.29	2.92	2.73	2.61	2.52	2.46	2.41	2.38	2.35	2.32	2.28	2.24	2.20	2.18	2.16	2.13	2.11	2.06
	95	4.96	4.10	3.71	3.48	3.33	3.22	3.14	3.07	3.02	2.98	2.91	2.85	2.77	2.74	2.70	2.66	2.62	2.54
	99	10.0	7.56	6.55	5.99	5.64	5.39	5.20	5.06	4.94	4.85	4.71	4.56	4.41	4.33	4.25	4.17	4.08	3.91
	99.9	21.0	14.9	12.6	11.3	10.5	9.92	9.52	9.20	8.96	8.75	8.45	8.13	7.80	7.64	7.47	7.30	7.12	6.76

	[illegible]	[illegible]	[illegible]	[illegible]	[illegible]	[illegible]	[illegible]	[illegible]	[illegible]	[illegible]	[illegible]	[illegible]	[illegible]	[illegible]	[illegible]	[illegible]	[illegible]	[illegible]	[illegible]
	90	3.23	2.86	2.66	2.54	2.45	2.39	2.34	2.30	2.27	2.25	2.21	2.17	2.12	2.10	2.08	2.05	2.03	1.97
11	95	4.84	3.98	3.59	3.36	3.20	3.09	3.01	2.95	2.90	2.85	2.79	2.72	2.65	2.61	2.57	2.53	2.49	2.40
	99	9.65	7.21	6.22	5.67	5.32	5.07	4.89	4.74	4.63	4.54	4.40	4.25	4.10	4.02	3.94	3.86	3.78	3.60
	99.9	19.7	13.8	11.6	10.4	9.58	9.05	8.66	8.35	8.12	7.92	7.63	7.32	7.01	6.85	6.68	6.52	6.35	6.00
	75	1.46	1.56	1.56	1.55	1.54	1.53	1.52	1.51	1.51	1.50	1.49	1.48	1.47	1.46	1.45	1.45	1.44	1.42
	90	3.18	2.81	2.61	2.48	2.39	2.33	2.28	2.24	2.21	2.19	2.15	2.10	2.06	2.04	2.01	1.99	1.96	1.90
12	95	4.75	3.89	3.49	3.26	3.11	3.00	2.91	2.85	2.80	2.75	2.69	2.62	2.54	2.51	2.47	2.43	2.38	2.30
	99	9.33	6.93	5.95	5.41	5.06	4.82	4.64	4.50	4.39	4.30	4.16	4.01	3.86	3.78	3.70	3.62	3.54	3.36
	99.9	18.6	13.0	10.8	9.63	8.89	8.38	8.00	7.71	7.48	7.29	7.00	6.71	6.40	6.25	6.09	5.93	5.76	5.42
	75	1.45	1.55	1.55	1.53	1.52	1.51	1.50	1.49	1.49	1.48	1.47	1.46	1.45	1.44	1.43	1.42	1.42	1.40
	90	3.14	2.76	2.56	2.43	2.35	2.28	2.23	2.20	2.16	2.14	2.10	2.05	2.01	1.98	1.96	1.93	1.90	1.85
13	95	4.67	3.81	3.41	3.18	3.03	2.92	2.83	2.77	2.71	2.67	2.60	2.53	2.46	2.42	2.38	2.34	2.30	2.21
	99	9.07	6.70	5.74	5.21	4.86	4.62	4.44	4.30	4.19	4.10	3.96	3.82	3.66	3.59	3.51	3.43	3.34	3.17
	99.9	17.8	12.3	10.2	9.07	8.35	7.86	7.49	7.21	6.98	6.80	6.52	6.23	5.93	5.78	5.63	5.47	5.30	4.97
	75	1.44	1.53	1.53	1.52	1.51	1.50	1.49	1.48	1.47	1.46	1.45	1.44	1.43	1.42	1.41	1.41	1.40	1.38
	90	3.10	2.73	2.52	2.39	2.31	2.24	2.19	2.15	2.12	2.10	2.05	2.01	1.96	1.94	1.91	1.89	1.86	1.80
14	95	4.60	3.74	3.34	3.11	2.96	2.85	2.76	2.70	2.65	2.60	2.53	2.46	2.39	2.35	2.31	2.27	2.22	2.13
	99	8.86	6.51	5.56	5.04	4.69	4.46	4.28	4.14	4.03	3.94	3.80	3.66	3.51	3.43	3.35	3.27	3.18	3.00
	99.9	17.1	11.8	9.73	8.62	7.92	7.43	7.08	6.80	6.58	6.40	6.13	5.85	5.56	5.41	5.25	5.10	4.94	4.60
	75	1.43	1.52	1.52	1.51	1.49	1.48	1.47	1.46	1.46	1.45	1.44	1.43	1.41	1.41	1.40	1.39	1.38	1.36
	90	3.07	2.70	2.49	2.36	2.27	2.21	2.16	2.12	2.09	2.06	2.02	1.97	1.92	1.90	1.87	1.85	1.82	1.76
15	95	4.54	3.68	3.29	3.06	2.90	2.79	2.71	2.64	2.59	2.54	2.48	2.40	2.33	2.29	2.25	2.20	2.16	2.07
	99	8.68	6.36	5.42	4.89	4.56	4.32	4.14	4.00	3.89	3.80	3.67	3.52	3.37	3.29	3.21	3.13	3.05	2.87
	99.9	16.6	11.3	9.34	8.25	7.57	7.09	6.74	6.47	6.26	6.08	5.81	5.54	5.25	5.10	4.95	4.80	4.64	4.31

(*continued*)

Table A.5 (*continued*)

df_2	P (%)	df_1: 1	2	3	4	5	6	7	8	9	10	12	15	20	24	30	40	60	∞
16	75	1.42	1.51	1.51	1.50	1.48	1.47	1.46	1.45	1.44	1.44	1.43	1.41	1.40	1.39	1.38	1.37	1.36	1.34
	90	3.05	2.67	2.46	2.33	2.24	2.18	2.13	2.09	2.06	2.03	1.99	1.94	1.89	1.87	1.84	1.81	1.78	1.72
	95	4.49	3.63	3.24	3.01	2.85	2.74	2.66	2.59	2.54	2.49	2.42	2.35	2.28	2.24	2.19	2.15	2.11	2.01
	99	8.53	6.23	5.29	4.77	4.44	4.20	4.03	3.89	3.78	3.69	3.55	3.41	3.26	3.18	3.10	3.02	2.93	2.75
	99.9	16.1	11.0	9.00	7.94	7.27	6.81	6.46	6.19	5.98	5.81	5.55	5.27	4.99	4.85	4.70	4.54	4.39	4.06
17	75	1.42	1.51	1.50	1.49	1.47	1.46	1.45	1.44	1.43	1.43	1.41	1.40	1.39	1.38	1.37	1.36	1.35	1.33
	90	3.03	2.64	2.44	2.31	2.22	2.15	2.10	2.06	2.03	2.00	1.96	1.91	1.86	1.84	1.81	1.78	1.75	1.69
	95	4.45	3.59	3.20	2.96	2.81	2.70	2.61	2.55	2.49	2.45	2.38	2.31	2.23	2.19	2.15	2.10	2.06	1.96
	99	8.40	6.11	5.18	4.67	4.34	4.10	3.93	3.79	3.68	3.59	3.46	3.31	3.16	3.08	3.00	2.92	2.83	2.65
	99.9	15.7	10.7	8.73	7.68	7.02	6.56	6.22	5.96	5.75	5.58	5.32	5.05	4.78	4.63	4.48	4.33	4.18	3.85
18	75	1.41	1.50	1.49	1.48	1.46	1.45	1.44	1.43	1.42	1.42	1.40	1.39	1.38	1.37	1.36	1.35	1.34	1.32
	90	3.01	2.62	2.42	2.29	2.20	2.13	2.08	2.04	2.00	1.98	1.93	1.89	1.84	1.81	1.78	1.75	1.72	1.66
	95	4.41	3.55	3.16	2.93	2.77	2.66	2.58	2.51	2.46	2.41	2.34	2.27	2.19	2.15	2.11	2.06	2.02	1.92
	99	8.29	6.01	5.09	4.58	4.25	4.01	3.84	3.71	3.60	3.51	3.37	3.23	3.08	3.00	2.92	2.84	2.75	2.57
	99.9	15.4	10.4	8.49	7.46	6.81	6.35	6.02	5.76	5.56	5.39	5.13	4.87	4.59	4.45	4.30	4.15	4.00	3.67
19	75	1.41	1.49	1.49	1.47	1.46	1.44	1.43	1.42	1.41	1.41	1.40	1.38	1.37	1.36	1.35	1.34	1.33	1.30
	90	2.99	2.61	2.40	2.27	2.18	2.11	2.06	2.02	1.98	1.96	1.91	1.86	1.81	1.79	1.76	1.73	1.70	1.63
	95	4.38	3.52	3.13	2.90	2.74	2.63	2.54	2.48	2.42	2.38	2.31	2.23	2.16	2.11	2.07	2.03	1.98	1.88
	99	8.18	5.93	5.01	4.50	4.17	3.94	3.77	3.63	3.52	3.43	3.30	3.15	3.00	2.92	2.84	2.76	2.67	2.49
	99.9	15.1	10.2	8.28	7.26	6.62	6.18	5.85	5.59	5.39	5.22	4.97	4.70	4.43	4.29	4.14	3.99	3.84	3.51
20	75	1.40	1.49	1.48	1.47	1.45	1.44	1.43	1.42	1.41	1.40	1.39	1.37	1.36	1.35	1.34	1.33	1.32	1.29
	90	2.97	2.59	2.38	2.25	2.16	2.09	2.04	2.00	1.96	1.94	1.89	1.84	1.79	1.77	1.74	1.71	1.68	1.61
	95	4.35	3.49	3.10	2.87	2.71	2.60	2.51	2.45	2.39	2.35	2.28	2.20	2.12	2.08	2.04	1.99	1.95	1.84
	99	8.10	5.85	4.94	4.43	4.10	3.87	3.70	3.56	3.46	3.37	3.23	3.09	2.94	2.86	2.78	2.69	2.61	2.42
	99.9	[illegible]	[illegible]	[illegible]	[illegible]	[illegible]	[illegible]	[illegible]	[illegible]	[illegible]	[illegible]	[illegible]	[illegible]	[illegible]	[illegible]	[illegible]	[illegible]	[illegible]	[illegible]

22	75	1.40	1.48	1.47	1.45	1.44	1.42	1.41	1.40	1.39	1.39	1.37	1.36	1.34	1.33	1.32	1.31	1.30	1.28
	90	2.95	2.56	2.35	2.22	2.13	2.06	2.01	1.97	1.93	1.90	1.86	1.81	1.76	1.73	1.70	1.67	1.64	1.57
	95	4.30	3.44	3.05	2.82	2.66	2.55	2.46	2.40	2.34	2.30	2.23	2.15	2.07	2.03	1.98	1.94	1.89	1.78
	99	7.95	5.72	4.82	4.31	3.99	3.76	3.59	3.45	3.35	3.26	3.12	2.98	2.83	2.75	2.67	2.58	2.50	2.31
	99.9	14.4	9.61	7.80	6.81	6.19	5.76	5.44	5.19	4.99	4.83	4.58	4.33	4.06	3.92	3.78	3.63	3.48	3.15
24	75	1.39	1.47	1.46	1.44	1.43	1.41	1.40	1.39	1.38	1.38	1.36	1.35	1.33	1.32	1.31	1.30	1.29	1.26
	90	2.93	2.54	2.33	2.19	2.10	2.04	1.98	1.94	1.91	1.88	1.83	1.78	1.73	1.70	1.67	1.64	1.61	1.53
	95	4.26	3.40	3.01	2.78	2.62	2.51	2.42	2.36	2.30	2.25	2.18	2.11	2.03	1.98	1.94	1.89	1.84	1.73
	99	7.82	5.61	4.72	4.22	3.90	3.67	3.50	3.36	3.26	3.17	3.03	2.89	2.74	2.66	2.58	2.49	2.40	2.21
	99.9	14.0	9.34	7.55	6.59	5.98	5.55	5.23	4.99	4.80	4.64	4.39	4.14	3.87	3.74	3.59	3.45	3.29	2.97
26	75	1.38	1.46	1.45	1.44	1.42	1.41	1.39	1 38	1.37	1.37	1.35	1.34	1.32	1.31	1.30	1.29	1.28	1.25
	90	2.91	2.52	2.31	2.17	2.08	2.01	1.96	1.92	1.88	1.86	1.81	1.76	1.71	1.68	1.65	1.61	1.58	1.50
	95	4.23	3.37	2.98	2.74	2.59	2.47	2.39	2.32	2.27	2.22	2.15	2.07	1.99	1.95	1.90	1.85	1.80	1.69
	99	7.72	5.53	4.64	4.14	3.82	3.59	3.42	3.29	3.18	3.09	2.96	2.81	2.66	2.58	2.50	2.42	2.33	2.13
	99.9	13.7	9.12	7.36	6.41	5.80	5.38	5.07	4.83	4.64	4.48	4.24	3.99	3.72	3.59	3.44	3.30	3.15	2.82
28	75	1.38	1.46	1.45	1.43	1.41	1.40	1.39	1.38	1.37	1.36	1.34	1.33	1.31	1.30	1.29	1.28	1.27	1.24
	90	2.89	2.50	2.29	2.16	2.06	2.00	1.94	1.90	1.87	1.84	1.79	1.74	1.69	1.66	1.63	1.59	1.56	1.48
	95	4.20	3.34	2.95	2.71	2.56	2.45	2.36	2.29	2.24	2.19	2.12	2.04	1.96	1.91	1.87	1.82	1.77	1.65
	99	7.64	5.45	4.57	4.07	3.75	3.53	3.36	3.23	3.12	3.03	2.90	2.75	2.60	2.52	2.44	2.35	2.26	2.06
	99.9	13.5	8.93	7.19	6.25	5.66	5.24	4.93	4.69	4.50	4.35	4.11	3.86	3.60	3.46	3.32	3.18	3.02	2.69
30	75	1.38	1.45	1.44	1.42	1.41	1.39	1.38	1.37	1.36	1.35	1.34	1.32	1.30	1.29	1.28	1.27	1.26	1.23
	90	2.88	2.49	2.28	2.14	2.05	1.98	1.93	1.88	1.85	1.82	1.77	1.72	1.67	1.64	1.61	1.57	1.54	1.46
	95	4.17	3.32	2.92	2.69	2.53	2.42	2.33	2.27	2.21	2.16	2.09	2.01	1.93	1.89	1.84	1.79	1.74	1.62
	99	7.56	5.39	4.51	4.02	3.70	3.47	3.30	3.17	3.07	2.98	2.84	2.70	2.55	2.47	2.39	2.30	2.21	2.01
	99.9	13.3	8.77	7.05	6.12	5.53	5.12	4.82	4.58	4.39	4.24	4.00	3.75	3.49	3.36	3.22	3.07	2.92	2.59

(continued)

Table A.5 (*continued*)

df_2	P (%)	df_1: 1	2	3	4	5	6	7	8	9	10	12	15	20	24	30	40	60	∞
40	75	1.36	1.44	1.42	1.40	1.39	1.37	1.36	1.35	1.34	1.33	1.31	1.30	1.28	1.26	1.25	1.24	1.22	1.19
	90	2.84	2.44	2.23	2.09	2.00	1.93	1.87	1.83	1.79	1.76	1.71	1.66	1.61	1.57	1.54	1.51	1.47	1.38
	95	4.08	3.23	2.84	2.61	2.45	2.34	2.25	2.18	2.12	2.08	2.00	1.92	1.84	1.79	1.74	1.69	1.64	1.51
	99	7.31	5.18	4.31	3.83	3.51	3.29	3.12	2.99	2.89	2.80	2.66	2.52	2.37	2.29	2.20	2.11	2.02	1.80
	99.9	12.6	8.25	6.60	5.70	5.13	4.73	4.44	4.21	4.02	3.87	3.64	3.40	3.15	3.01	2.87	2.73	2.57	2.23
60	75	1.35	1.42	1.41	1.38	1.37	1.35	1.33	1.32	1.31	1.30	1.29	1.27	1.25	1.24	1.22	1.21	1.19	1.15
	90	2.79	2.39	2.18	2.04	1.95	1.87	1.82	1.77	1.74	1.71	1.66	1.60	1.54	1.51	1.48	1.44	1.40	1.29
	95	4.00	3.15	2.76	2.53	2.37	2.25	2.17	2.10	2.04	1.99	1.92	1.84	1.75	1.70	1.65	1.59	1.53	1.39
	99	7.08	4.98	4.13	3.65	3.34	3.12	2.95	2.82	2.72	2.63	2.50	2.35	2.20	2.12	2.03	1.94	1.84	1.60
	99.9	12.0	7.76	6.17	5.31	4.76	4.37	4.09	3.87	3.69	3.54	3.31	3.08	2.83	2.69	2.55	2.41	2.25	1.89
120	75	1.34	1.40	1.39	1.37	1.35	1.33	1.31	1.30	1.29	1.28	1.26	1.24	1.22	1.21	1.19	1.18	1.16	1.10
	90	2.75	2.35	2.13	1.99	1.90	1.82	1.77	1.72	1.68	1.65	1.60	1.55	1.48	1.45	1.41	1.37	1.32	1.19
	95	3.92	3.07	2.68	2.45	2.29	2.17	2.09	2.02	1.96	1.91	1.83	1.75	1.66	1.61	1.55	1.50	1.43	1.25
	99	6.85	4.79	3.95	3.48	3.17	2.96	2.79	2.66	2.56	2.47	2.34	2.19	2.03	1.95	1.86	1.76	1.66	1.38
	99.9	11.4	7.32	5.79	4.95	4.42	4.04	3.77	3.55	3.38	3.24	3.02	2.78	2.53	2.40	2.26	2.11	1.95	1.54
∞	75	1.32	1.39	1.37	1.35	1.33	1.31	1.29	1.28	1.27	1.25	1.24	1.22	1.19	1.18	1.16	1.14	1.12	1.00
	90	2.71	2.30	2.08	1.94	1.85	1.77	1.72	1.67	1.63	1.60	1.55	1.49	1.42	1.38	1.34	1.30	1.24	1.00
	95	3.84	3.00	2.60	2.37	2.21	2.10	2.01	1.94	1.88	1.83	1.75	1.67	1.57	1.52	1.46	1.39	1.32	1.00
	99	6.63	4.61	3.78	3.32	3.02	2.80	2.64	2.51	2.41	2.32	2.18	2.04	1.88	1.79	1.70	1.59	1.47	1.00
	99.9	10.8	6.91	5.42	4.62	4.10	3.74	3.47	3.27	3.10	2.96	2.74	2.51	2.27	2.13	1.99	1.84	1.66	1.00

SOURCE: This table is abridged from Table 18 in *Biometrika Tables for Statisticians*, 3rd ed., vol. 1 (New York: Cambridge University Press, 1966), edited by E. S. Pearson and H. O. Hartley, by permission of the *Biometrika* Trustees.

Table A.6 Percentiles of the F_{max} distribution

$E-1$	$P(\%)$	Number of variances										
		2	3	4	5	6	7	8	9	10	11	12
4	95	9.60	15.5	20.6	25.2	29.5	33.6	37.5	41.4	44.6	48.0	51.4
	99	23.2	37.	49.	59.	69.	79.	89.	97.	106.	113.	120.
5	95	7.15	10.8	13.7	16.3	18.7	20.8	22.9	24.7	26.5	28.2	29.9
	99	14.9	22.	28.	33.	38.	42.	46.	50.	54.	57.	60.
6	95	5.82	8.38	10.4	12.1	13.7	15.0	16.3	17.5	18.6	19.7	20.7
	99	11.1	15.5	19.1	22.	25.	27.	30.	32.	34.	36.	37.
7	95	4.99	6.94	8.44	9.70	10.8	11.8	12.7	13.5	14.3	15.1	15.8
	99	8.89	12.1	14.5	16.5	18.4	20.	22.	23.	24.	26.	27.
8	95	4.43	6.00	7.18	8.12	9.03	9.78	10.5	11.1	11.7	12.2	12.7
	99	7.50	9.9	11.7	13.2	14.5	15.8	16.9	17.9	18.9	19.8	21.
9	95	4.03	5.34	6.31	7.11	7.80	8.41	8.95	9.45	9.91	10.3	10.7
	99	6.54	8.5	9.9	11.1	12.1	13.1	13.9	14.7	15.3	16.0	16.6
10	95	3.72	4.85	5.67	6.34	6.92	7.42	7.87	8.28	8.66	9.01	9.34
	99	5.85	7.4	8.6	9.6	10.4	11.1	11.8	12.4	12.9	13.4	13.9
12	95	3.28	4.16	4.79	5.30	5.72	6.09	6.42	6.72	7.00	7.25	7.48
	99	4.91	6.1	6.9	7.6	8.2	8.7	9.1	9.5	9.9	10.2	10.6
15	95	2.86	3.54	4.01	4.37	4.68	4.95	5.19	5.40	5.59	5.77	5.93
	99	4.07	4.9	5.5	6.0	6.4	6.7	7.1	7.3	7.5	7.8	8.0
20	95	2.46	2.95	3.29	3.54	3.76	3.94	4.10	4.24	4.37	4.49	4.59
	99	3.32	3.8	4.3	4.6	4.9	5.1	5.3	5.5	5.6	5.8	5.9
30	95	2.07	2.40	2.61	2.78	2.91	3.02	3.12	3.21	3.29	3.36	3.39
	99	2.63	3.0	3.3	3.4	3.6	3.7	3.8	3.9	4.0	4.1	4.2
60	95	1.67	1.85	1.96	2.04	2.11	2.17	2.22	2.26	2.30	2.33	2.36
	99	1.96	2.2	2.3	2.4	2.4	2.5	2.5	2.6	2.6	2.7	2.7
∞	95	1.00	1.00	1.00	1.00	1.00	1.00	1.00	1.00	1.00	1.00	1.00
	99	1.00	1.00	1.00	1.00	1.00	1.00	1.00	1.00	1.00	1.00	1.00

SOURCE: This table is abridged from Table 31 in *Biometrika Tables for Statisticians*, 3rd ed., vol. 1 (New York: Cambridge University Press, 1966), edited by E. S. Pearson and H. O. Hartley, by permission of the *Biometrika* Trustees.

Table A.7 Percentiles of the studentized range distribution

		Number of means in range (including end means)									
df_2	*P* (%)	2	3	4	5	6	7	8	9	10	11
5	95	3.64	4.60	5.22	5.67	6.03	6.33	6.58	6.80	6.99	7.17
	99	5.70	6.98	7.80	8.42	8.91	9.32	9.67	9.97	10.24	10.48
6	95	3.46	4.34	4.90	5.30	5.63	5.90	6.12	6.32	6.49	6.65
	99	5.24	6.33	7.03	7.56	7.97	8.32	8.61	8.87	9.10	9.30
7	95	3.34	4.16	4.68	5.06	5.36	5.61	5.82	6.00	6.16	6.30
	99	4.95	5.92	6.54	7.01	7.37	7.68	7.94	8.17	8.37	8.55
8	95	3.26	4.04	4.53	4.89	5.17	5.40	5.60	5.77	5.92	6.05
	99	4.75	5.64	6.20	6.62	6.96	7.24	7.47	7.68	7.86	8.03
9	95	3.20	3.95	4.41	4.76	5.02	5.24	5.43	5.59	5.74	5.87
	99	4.60	5.43	5.96	6.35	6.66	6.91	7.13	7.33	7.49	7.65
10	95	3.15	3.88	4.33	4.65	4.91	5.12	5.30	5.46	5.60	5.72
	99	4.48	5.27	5.77	6.14	6.43	6.67	6.87	7.05	7.21	7.36
11	95	3.11	3.82	4.26	4.57	4.82	5.03	5.20	5.35	5.49	5.61
	99	4.39	5.15	5.62	5.97	6.25	6.48	6.67	6.84	6.99	7.13
12	95	3.08	3.77	4.20	4.51	4.75	4.95	5.12	5.27	5.39	5.51
	99	4.32	5.05	5.50	5.84	6.10	6.32	6.51	6.67	6.81	6.94
13	95	3.06	3.73	4.15	4.45	4.69	4.88	5.05	5.19	5.32	5.43
	99	4.26	4.96	5.40	5.73	5.98	6.19	6.37	6.53	6.67	6.79
14	95	3.03	3.70	4.11	4.41	4.64	4.83	4.99	5.13	5.25	5.36
	99	4.21	4.89	5.32	5.63	5.88	6.08	6.26	6.41	6.54	6.66
15	95	3.01	3.67	4.08	4.37	4.59	4.78	4.94	5.08	5.20	5.31
	99	4.17	4.84	5.25	5.56	5.80	5.99	6.16	6.31	6.44	6.55
16	95	3.00	3.65	4.05	4.33	4.56	4.74	4.90	5.03	5.15	5.26
	99	4.13	4.79	5.19	5.49	5.72	5.92	6.08	6.22	6.35	6.46
17	95	2.98	3.63	4.02	4.30	4.52	4.70	4.86	4.99	5.11	5.21
	99	4.10	4.74	5.14	5.43	5.66	5.85	6.01	6.15	6.27	6.38
18	95	2.97	3.61	4.00	4.28	4.49	4.67	4.82	4.96	5.07	5.17
	99	4.07	4.70	5.09	5.38	5.60	5.79	5.94	6.08	6.20	6.31
19	95	2.96	3.59	3.98	4.25	4.47	4.65	4.79	4.92	5.04	5.14
	99	4.05	4.67	5.05	5.33	5.55	5.73	5.89	6.02	6.14	6.25
20	95	2.95	3.58	3.96	4.23	4.45	4.62	4.77	4.90	5.01	5.11
	99	4.02	4.64	5.02	5.29	5.51	5.69	5.84	5.97	6.09	6.19
24	95	2.92	3.53	3.90	4.17	4.37	4.54	4.68	4.81	4.92	5.01
	99	3.96	4.55	4.91	5.17	5.37	5.54	5.69	5.81	5.92	6.02
30	95	2.89	3.49	3.85	4.10	4.30	4.46	4.60	4.72	4.82	4.92
	99	3.89	4.45	4.80	5.05	5.24	5.40	5.54	5.65	5.76	5.85
40	95	2.86	3.44	3.79	4.04	4.23	4.39	4.52	4.63	4.73	4.82
	99	3.82	4.37	4.70	4.93	5.11	5.26	5.39	5.50	5.60	5.69
60	95	2.83	3.40	3.74	3.98	4.16	4.31	4.44	4.55	4.65	4.73
	99	3.76	4.28	4.59	4.82	4.99	5.13	5.25	5.36	5.45	5.53
120	95	2.80	3.36	3.68	3.92	4.10	4.24	4.36	4.47	4.56	4.64
	99	3.70	4.20	4.50	4.71	4.87	5.01	5.12	5.21	5.30	5.37
∞	95	2.77	3.31	3.63	3.86	4.03	4.17	4.29	4.39	4.47	4.55
	99	3.64	4.12	4.40	4.60	4.76	4.88	4.99	5.08	5.16	5.23

Number of means in range (including end means)										
12	13	14	15	16	17	18	19	20	*P* (%)	df_2
7.32	7.47	7.60	7.72	7.83	7.93	8.03	8.12	8.21	95	5
10.70	10.89	11.08	11.24	11.40	11.55	11.68	11.81	11.93	99	
6.79	6.92	7.03	7.14	7.24	7.34	7.43	7.51	7.59	95	6
9.48	9.65	9.81	9.95	10.08	10.21	10.32	10.43	10.54	99	
6.43	6.55	6.66	6.76	6.85	6.94	7.02	7.10	7.17	95	7
8.71	8.86	9.00	9.12	9.24	9.35	9.46	9.55	9.65	99	
6.18	6.29	6.39	6.48	6.57	6.65	6.73	6.80	6.87	95	8
8.18	8.31	8.44	8.55	8.66	8.76	8.85	8.94	9.03	99	
5.98	6.09	6.19	6.28	6.36	6.44	6.51	6.58	6.64	95	9
7.78	7.91	8.03	8.13	8.23	8.33	8.41	8.49	8.57	99	
5.83	5.93	6.03	6.11	6.19	6.27	6.34	6.40	6.47	95	10
7.49	7.60	7.71	7.81	7.91	7.99	8.08	8.15	8.23	99	
5.71	5.81	5.90	5.98	6.06	6.13	6.20	6.27	6.33	95	11
7.25	7.36	7.46	7.56	7.65	7.73	7.81	7.88	7.95	99	
5.61	5.71	5.80	5.88	5.95	6.02	6.09	6.15	6.21	95	12
7.06	7.17	7.26	7.36	7.44	7.52	7.59	7.66	7.73	99	
5.53	5.63	5.71	5.79	5.86	5.93	5.99	6.05	6.11	95	13
6.90	7.01	7.10	7.19	7.27	7.35	7.42	7.48	7.55	99	
5.46	5.55	5.64	5.71	5.79	5.85	5.91	5.97	6.03	95	14
6.77	6.87	6.96	7.05	7.13	7.20	7.27	7.33	7.39	99	
5.40	5.49	5.57	5.65	5.72	5.78	5.85	5.90	5.96	95	15
6.66	6.76	6.84	6.93	7.00	7.07	7.14	7.20	7.26	99	
5.35	5.44	5.52	5.59	5.66	5.73	5.79	5.84	5.90	95	16
6.56	6.66	6.74	6.82	6.90	6.97	7.03	7.09	7.15	99	
5.31	5.39	5.47	5.54	5.61	5.67	5.73	5.79	5.84	95	17
6.48	6.57	6.66	6.73	6.81	6.87	6.94	7.00	7.05	99	
5.27	5.35	5.43	5.50	5.57	5.63	5.69	5.74	5.79	95	18
6.41	6.50	6.58	6.65	6.73	6.79	6.85	6.91	6.97	99	
5.23	5.31	5.39	5.46	5.53	5.59	5.65	5.70	5.75	95	19
6.34	6.43	6.51	6.58	6.65	6.72	6.78	6.84	6.89	99	
5.20	5.28	5.36	5.43	5.49	5.55	5.61	5.66	5.71	95	20
6.28	6.37	6.45	6.52	6.59	6.65	6.71	6.77	6.82	99	
5.10	5.18	5.25	5.32	5.38	5.44	5.49	5.55	5.59	95	24
6.11	6.19	6.26	6.33	6.39	6.45	6.51	6.56	6.61	99	
5.00	5.08	5.15	5.21	5.27	5.33	5.38	5.43	5.47	95	30
5.93	6.01	6.08	6.14	6.20	6.26	6.31	6.36	6.41	99	
4.90	4.98	5.04	5.11	5.16	5.22	5.27	5.31	5.36	95	40
5.76	5.83	5.90	5.96	6.02	6.07	6.12	6.16	6.21	99	
4.81	4.88	4.94	5.00	5.06	5.11	5.15	5.20	5.24	95	60
5.60	5.67	5.73	5.78	5.84	5.89	5.93	5.97	6.01	99	
4.71	4.78	4.84	4.90	4.95	5.00	5.04	5.09	5.13	95	120
5.44	5.50	5.56	5.61	5.66	5.71	5.75	5.79	5.83	99	
4.62	4.68	4.74	4.80	4.85	4.89	4.93	4.97	5.01	95	∞
5.29	5.35	5.40	5.45	5.49	5.54	5.57	5.61	5.65	99	

SOURCE: This table is abridged from Table 29 in *Biometrika Tables for Statisticians,* 3rd ed., vol. 1 (New York: Cambridge University Press, 1966), edited by E. S. Pearson and H. O. Hartley, by permission of the *Biometrika* Trustees.

Table A.8 Percentiles of Dunnett's t'-distribution

		Number of means (including control)								
df_2	P (%)	2	3	4	5	6	7	8	9	10
5	95	2.02	2.44	2.68	2.85	2.98	3.08	3.16	3.24	3.03
	97.5	2.57	3.03	3.29	3.48	3.62	3.73	3.82	3.90	3.97
	99	3.36	3.90	4.21	4.43	4.60	4.73	4.85	4.94	5.03
	99.5	4.03	4.63	4.98	5.22	5.41	5.56	5.69	5.80	5.89
6	95	1.94	2.34	2.56	2.71	2.83	2.92	3.00	3.07	3.12
	97.5	2.45	2.86	3.10	3.26	3.39	3.49	3.57	3.64	3.71
	99	3.14	3.61	3.88	4.07	4.21	4.33	4.43	4.51	4.59
	99.5	3.71	4.21	4.51	4.71	4.87	5.00	5.10	5.20	5.28
7	95	1.89	2.27	2.48	2.62	2.73	2.82	2.89	2.95	3.01
	97.5	2.36	2.75	2.97	3.12	3.24	3.33	3.41	3.47	3.53
	99	3.00	3.42	3.66	3.83	3.96	4.07	4.15	4.23	4.30
	99.5	3.50	3.95	4.21	4.39	4.53	4.64	4.74	4.82	4.89
8	95	1.86	2.22	2.42	2.55	2.66	2.74	2.81	2.87	2.92
	97.5	2.31	2.67	2.88	3.02	3.13	3.22	3.29	3.35	3.41
	99	2.90	3.29	3.51	3.67	3.79	3.88	3.96	4.03	4.09
	99.5	3.36	3.77	4.00	4.17	4.29	4.40	4.48	4.56	4.62
9	95	1.83	2.18	2.37	2.50	2.60	2.68	2.75	2.81	2.86
	97.5	2.26	2.61	2.81	2.95	3.05	3.14	3.20	3.26	3.32
	99	2.28	3.19	3.40	3.55	3.66	3.75	3.82	3.89	3.94
	99.5	3.25	3.63	3.85	4.01	4.12	4.22	4.30	4.37	4.43
10	95	1.81	2.15	2.34	2.47	2.56	2.64	2.70	2.76	2.81
	97.5	2.23	2.57	2.76	2.89	2.99	3.07	3.14	3.19	3.24
	99	2.76	3.11	3.31	3.45	3.56	3.64	3.71	3.78	3.83
	99.5	3.17	3.53	3.74	3.88	3.99	4.08	4.16	4.22	4.28
11	95	1.80	2.13	2.31	2.44	2.53	2.60	2.67	2.72	2.77
	97.5	2.20	2.53	2.72	2.84	2.94	3.02	3.08	3.14	3.19
	99	2.72	3.06	3.25	3.38	3.48	3.56	3.63	3.69	3.74
	99.5	3.11	3.45	3.65	3.79	3.89	3.98	4.05	4.11	4.16
12	95	1.78	2.11	2.29	2.41	2.50	2.58	2.64	2.69	2.74
	97.5	2.18	2.50	2.68	2.81	2.90	2.98	3.04	3.09	3.14
	99	2.68	3.01	3.19	3.32	3.42	3.50	3.56	3.62	3.67
	99.5	3.05	3.39	3.58	3.71	3.81	3.89	3.96	4.02	4.07
13	95	1.77	2.09	2.27	2.39	2.48	2.55	2.61	2.66	2.71
	97.5	2.16	2.48	2.65	2.78	2.87	2.94	3.00	3.06	3.10
	99	2.65	2.97	3.15	3.27	3.37	3.44	3.51	3.56	3.61
	99.5	3.01	3.33	3.52	3.65	3.74	3.82	3.89	3.94	3.99
14	95	1.76	2.08	2.25	2.37	2.46	2.53	2.59	2.64	2.69
	97.5	2.14	2.46	2.63	2.75	2.84	2.91	2.97	3.02	3.07
	99	2.62	2.94	3.11	3.23	3.32	3.40	3.46	3.51	3.56
	99.5	2.98	3.29	3.47	3.59	3.69	3.76	3.83	3.88	3.93

df_2	P (%)	Number of means (including control)								
		2	3	4	5	6	7	8	9	10
16	95	1.75	2.06	2.23	2.34	2.43	2.50	2.56	2.61	2.65
	97.5	2.12	2.42	2.59	2.71	2.80	2.87	2.92	2.97	3.02
	99	2.58	2.88	3.05	3.17	3.26	3.33	3.39	3.44	3.48
	99.5	2.92	3.22	3.39	3.51	3.60	3.67	3.73	3.78	3.83
18	95	1.73	2.04	2.21	2.32	2.41	2.48	2.53	2.58	2.62
	97.5	2.10	2.40	2.56	2.68	2.76	2.83	2.89	2.94	2.98
	99	2.55	2.84	3.01	3.12	3.21	3.27	3.33	3.38	3.42
	99.5	2.88	3.17	3.33	3.44	3.53	3.60	3.66	3.71	3.75
20	95	1.72	2.03	2.19	2.30	2.39	2.46	2.51	2.56	2.60
	97.5	2.09	2.38	2.54	2.65	2.73	2.80	2.86	2.90	2.95
	99	2.53	2.81	2.97	3.08	3.17	3.23	3.29	3.34	3.38
	99.5	2.85	3.13	3.29	3.40	3.48	3.55	3.60	3.65	3.69
24	95	1.71	2.01	2.17	2.28	2.36	2.43	2.48	2.53	2.57
	97.5	2.06	2.35	2.51	2.61	2.70	2.76	2.81	2.86	2.90
	99	2.49	2.77	2.92	3.03	3.11	3.17	3.22	3.27	3.31
	99.5	2.80	3.07	3.22	3.32	3.40	3.47	3.52	3.57	3.61
30	95	1.70	1.99	2.15	2.25	2.33	2.40	2.45	2.50	2.54
	97.5	2.04	2.32	2.47	2.58	2.66	2.72	2.77	2.82	2.86
	99	2.46	2.72	2.87	2.97	3.05	3.11	3.16	3.21	3.24
	99.5	2.75	3.01	3.15	3.25	3.33	3.39	3.44	3.49	3.52
40	95	1.68	1.97	2.13	2.23	2.31	2.37	2.42	2.47	2.51
	97.5	2.02	2.29	2.44	2.54	2.62	2.68	2.73	2.77	2.81
	99	2.42	2.68	2.82	2.92	2.99	3.05	3.10	3.14	3.18
	99.5	2.70	2.95	3.09	3.19	3.26	3.32	3.37	3.41	3.44
60	95	1.67	1.95	2.10	2.21	2.28	2.35	2.39	2.44	2.48
	97.5	2.00	2.27	2.41	2.51	2.58	2.64	2.69	2.73	2.77
	99	2.39	2.64	2.78	2.87	2.94	3.00	3.04	3.08	3.12
	99.5	2.66	2.90	3.03	3.12	3.19	3.25	3.29	3.33	3.37
120	95	1.66	1.93	2.08	2.18	2.26	2.32	2.37	2.41	2.45
	97.5	1.98	2.24	2.38	2.47	2.55	2.60	2.65	2.69	2.73
	99	2.36	2.60	2.73	2.82	2.89	2.94	2.99	3.03	3.06
	99.5	2.62	2.85	2.97	3.06	3.12	3.18	3.22	3.26	3.29
∞	95	1.64	1.92	2.06	2.16	2.23	2.29	2.34	2.38	2.42
	97.5	1.96	2.21	2.35	2.44	2.51	2.57	2.61	2.65	2.69
	99	2.33	2.56	2.68	2.77	2.84	2.89	2.93	2.97	3.00
	99.5	2.58	2.79	2.92	3.00	3.06	3.11	3.15	3.19	3.22

SOURCE: This table is reproduced from C. W. Dunnett, "A multiple comparison procedure for comparing several treatments with a control," *Journal of the American Statistical Association*, 1955, 50, 1096–1121; and C. W. Dunnett, "New tables for multiple comparisons with a control," *Biometrics*, 1964, 20, 482–491.

Table A.9 Coefficients of orthogonal polynomials

		a											
A	Polynomial	1	2	3	4	5	6	7	8	9	10	$\sum w_a^2$	λ
3	Linear	−1	0	1								2	1
	Quadratic	1	−2	1								6	3
4	Linear	−3	−1	1	3							20	2
	Quadratic	1	−1	−1	1							4	1
	Cubic	−1	3	−3	1							20	10/3
5	Linear	−2	−1	0	1	2						10	1
	Quadratic	2	−1	−2	−1	2						14	1
	Cubic	−1	2	0	−2	1						10	5/6
	Quartic	1	−4	6	−4	1						70	
6	Linear	−5	−3	−1	1	3	5					70	2
	Quadratic	5	−1	−4	−4	−1	5					84	3/2
	Cubic	−5	7	4	−4	−7	5					180	5/3
	Quartic	1	−3	2	2	−3	1					28	
7	Linear	−3	−2	−1	0	1	2	3				28	1
	Quadratic	5	0	−3	−4	−3	0	5				84	1
	Cubic	−1	1	1	0	−1	−1	1				6	1/6
	Quartic	3	−7	1	6	1	−7	3				154	
8	Linear	−7	−5	−3	−1	1	3	5	7			168	2
	Quadratic	7	1	−3	−5	−5	−3	1	7			168	1
	Cubic	−7	5	7	3	−3	−7	−5	7			264	2/3
	Quartic	7	−13	−3	9	9	−3	−13	7			616	
	Quintic	−7	23	−17	−15	15	17	−23	7			2184	
9	Linear	−4	−3	−2	−1	0	1	2	3	4		60	1
	Quadratic	28	7	−8	−17	−20	−17	−8	7	28		2772	3
	Cubic	−14	7	13	9	0	−9	−13	−7	14		990	5/6
	Quartic	14	−21	−11	9	18	9	−11	−21	14		2002	
	Quintic	−4	11	−4	−9	0	9	4	−11	4		468	
10	Linear	−9	−7	−5	−3	−1	1	3	5	7	9	330	2
	Quadratic	6	2	−1	−3	−4	−4	−3	−1	2	6	132	1/2
	Cubic	−42	14	35	31	12	−12	−31	−35	−14	42	8580	5/3
	Quartic	18	−22	−17	3	18	18	3	−17	−22	18	2860	
	Quintic	−6	14	−1	−11	−6	6	11	1	−14	6	780	

SOURCE: From B. J. Winer, *Statistical Principles in Experimental Design*, 2nd ed., 1971 (New York: McGraw-Hill Book Company).

APPENDIX B

Power Charts

These charts are from L. S. Feldt and M. W. Mahmoud, "Power function charts for specification of sample size in analysis of variance," in *Psychometrika* 23: 201–210 (1958).

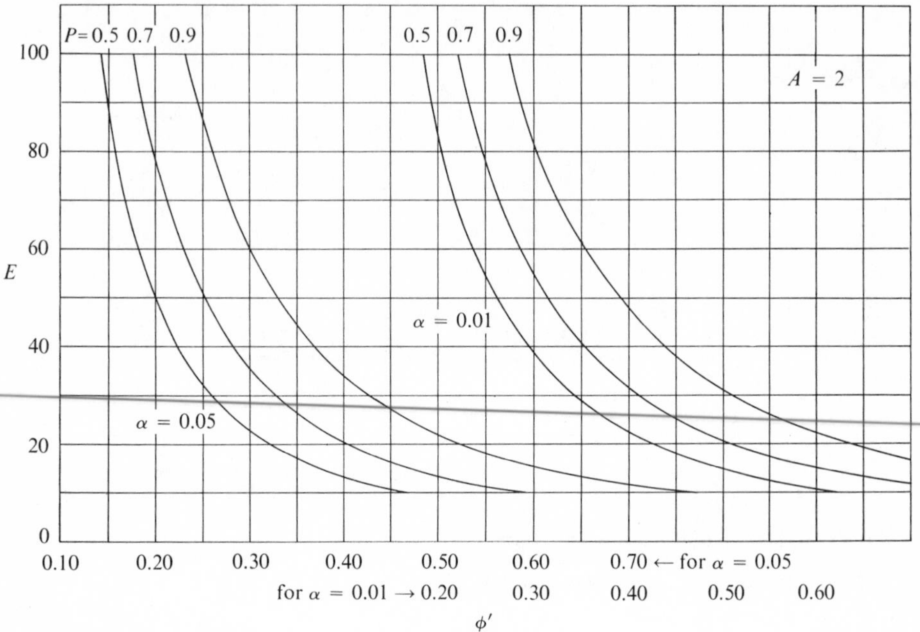

Figure B.1-1

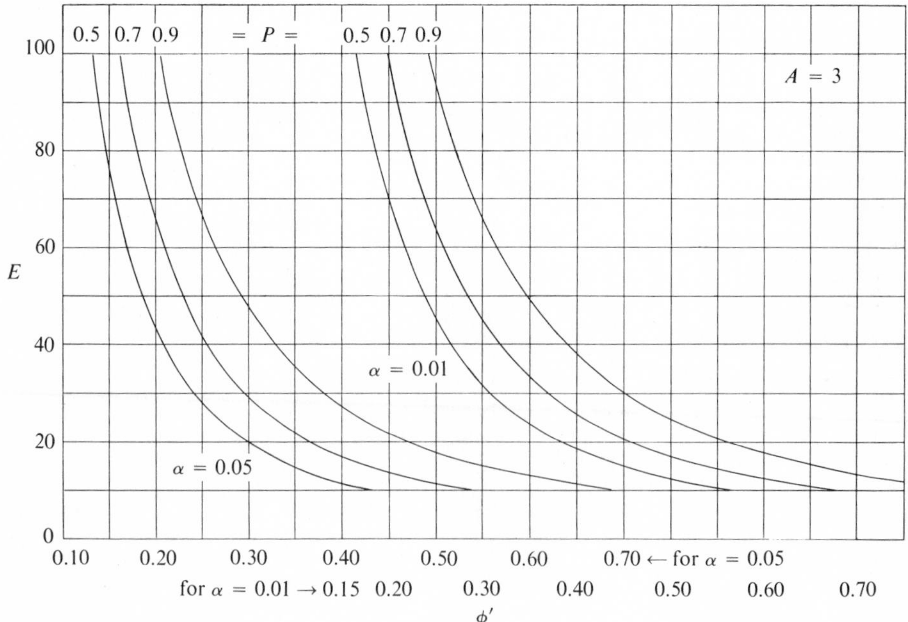

Figure B.1-2

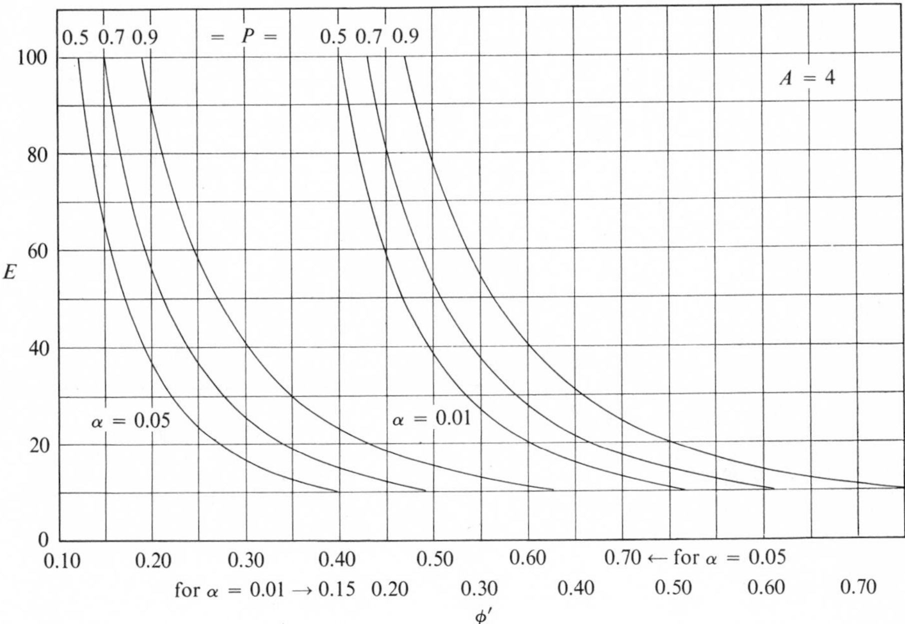

Figure B.1-3

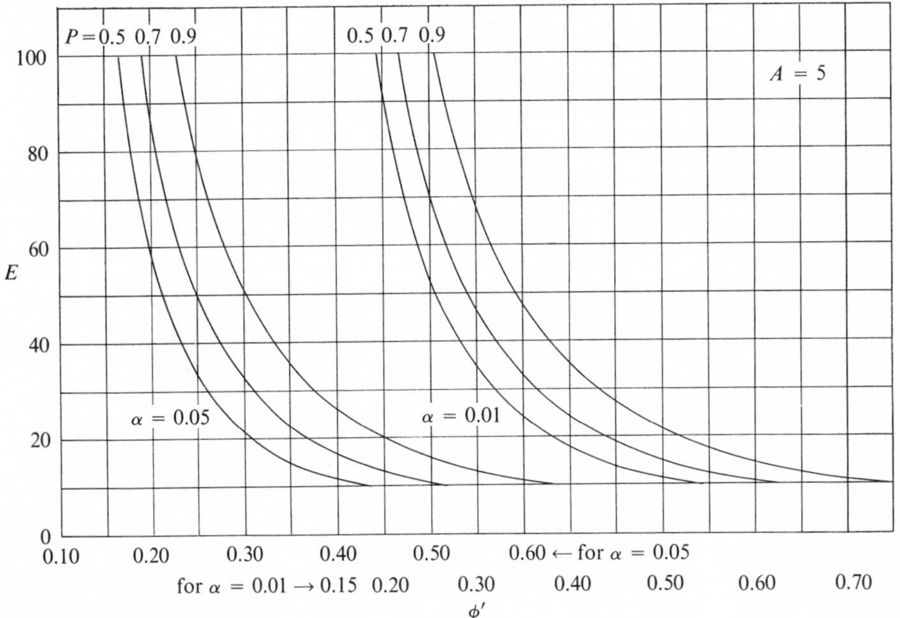

Figure B.1-4

These charts appear as Table 30 in *Biometrika Tables for Statisticians*, vol. 2 (New York: Cambridge University Press, 1972), edited by E. S. Pearson and H. O. Hartley, reprinted by permission of the *Biometrika* Trustees.

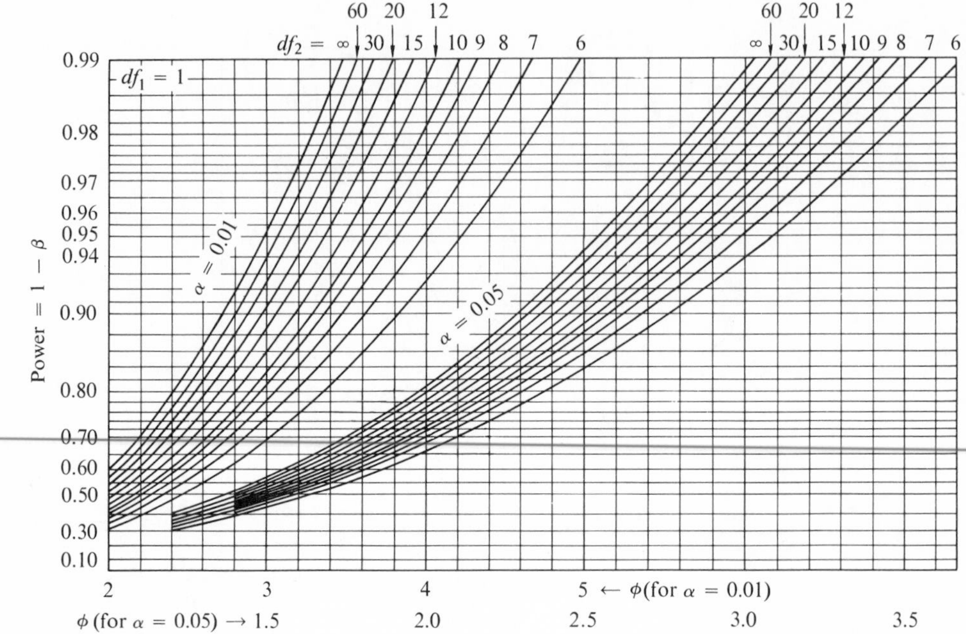

Figure B.2-1

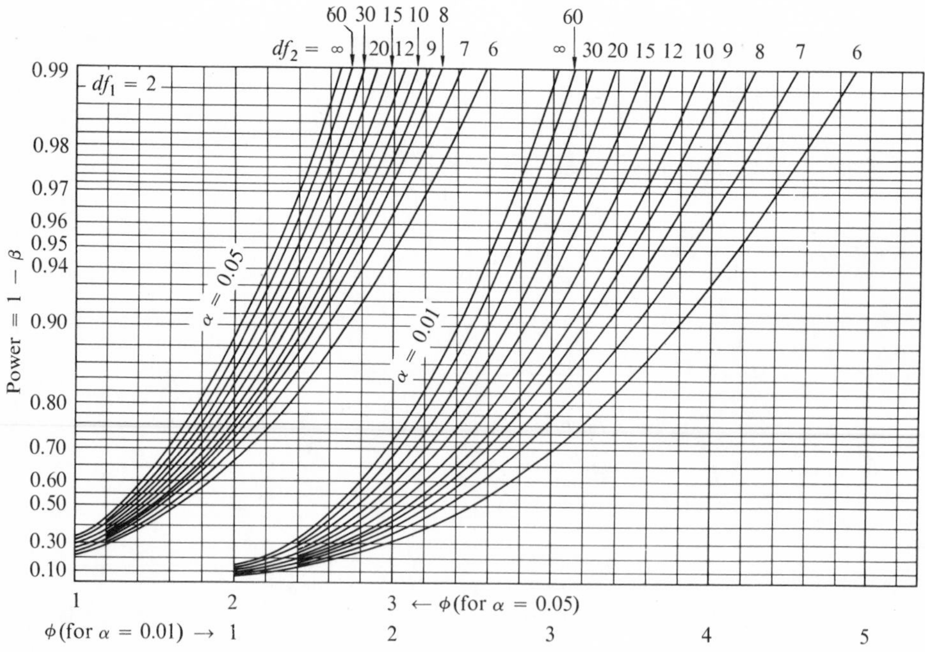

Figure B.2-2

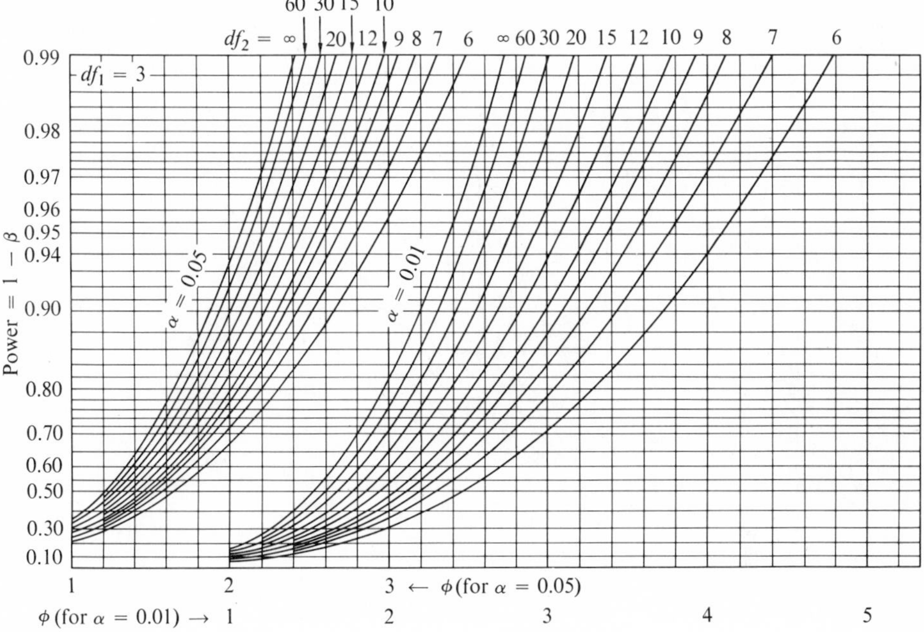

Figure B.2-3

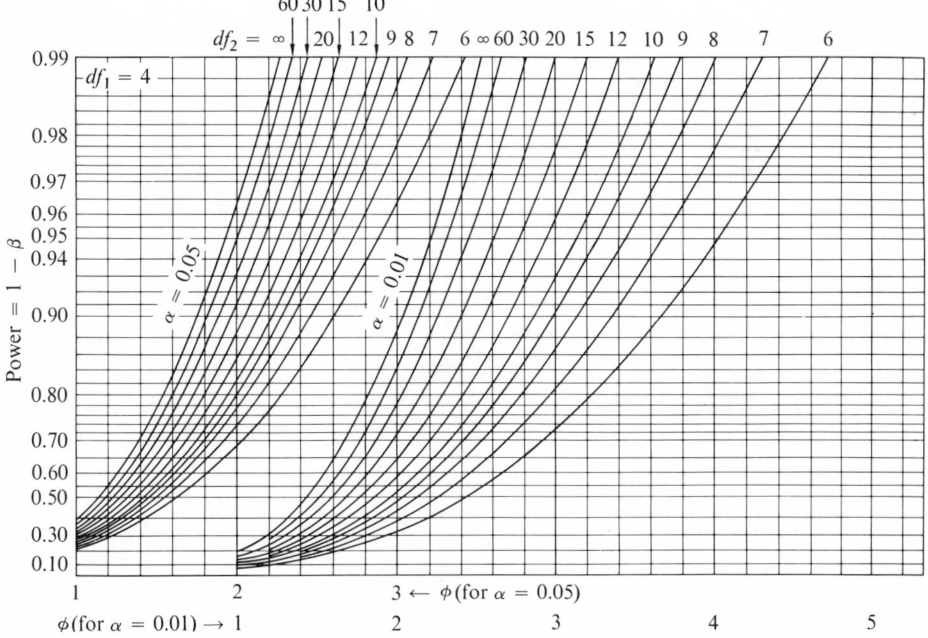

Figure B.2-4

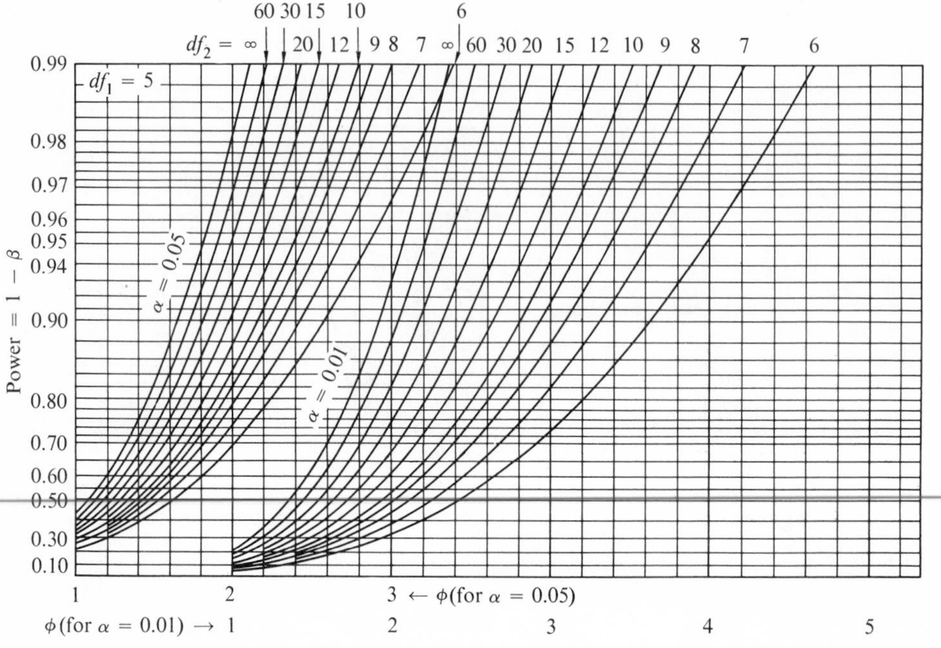

Figure B.2-5

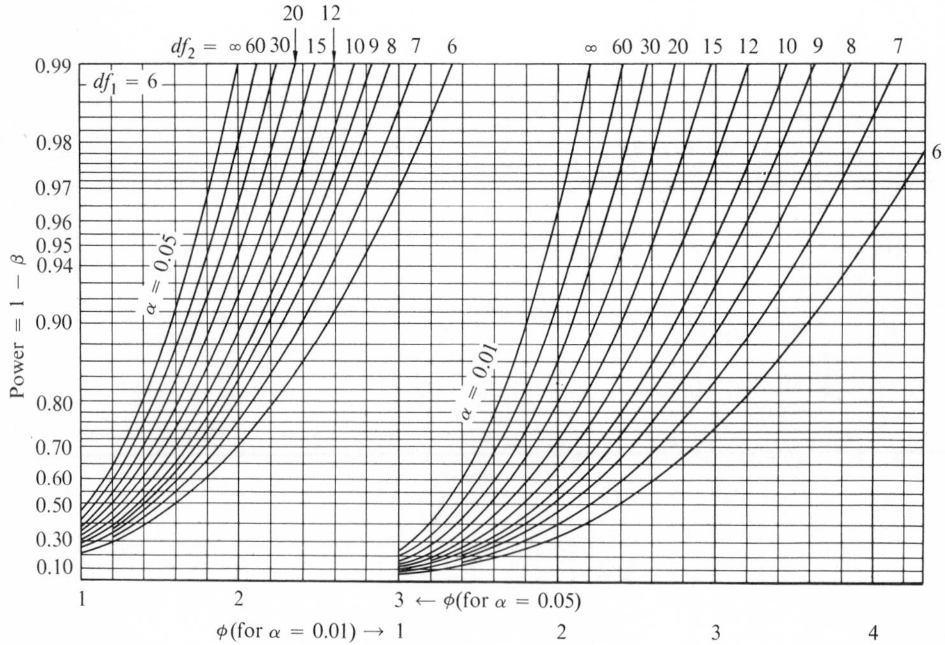

Figure B.2-6

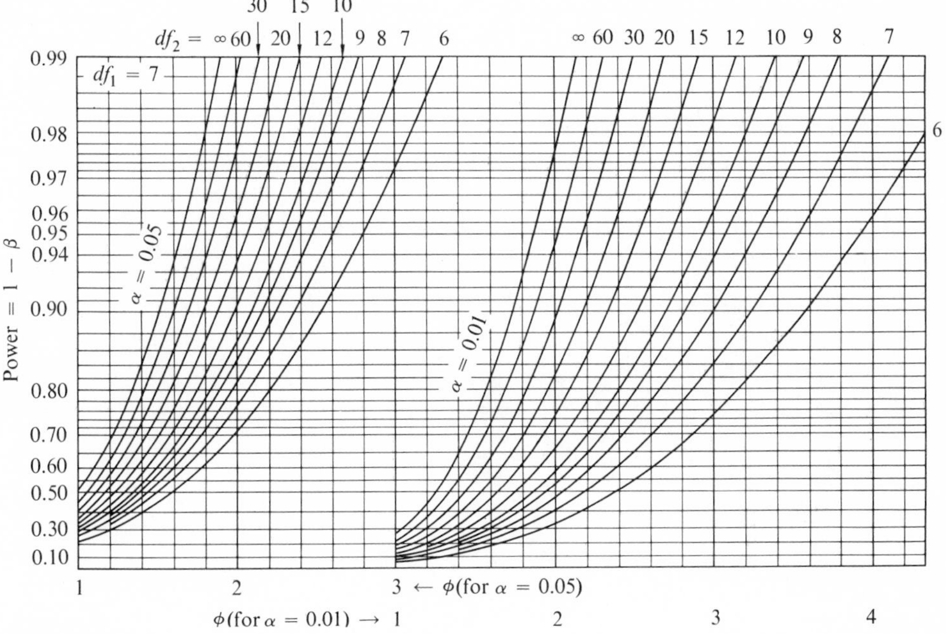

Figure B.2-7

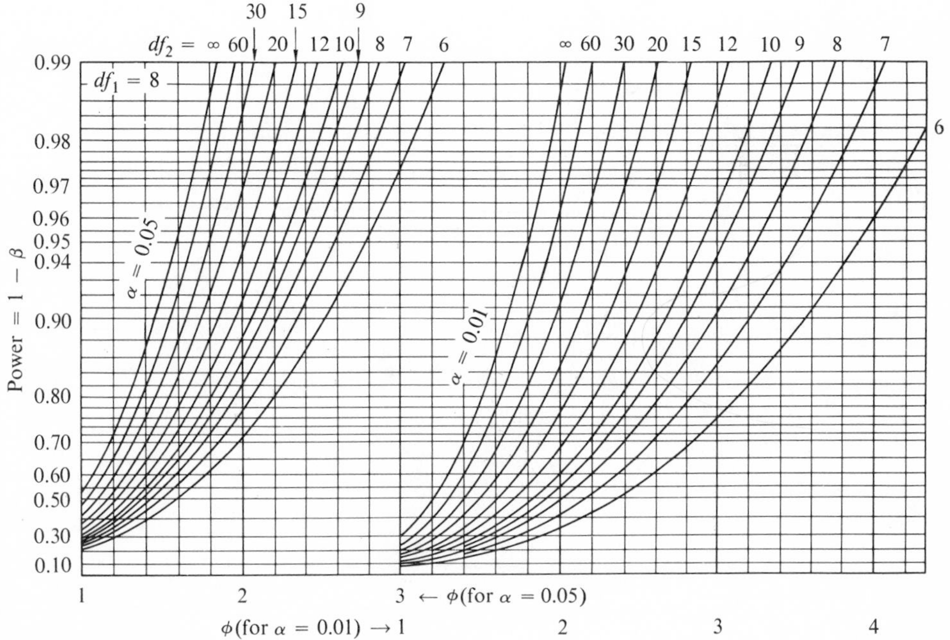

Figure B.2-8

SUMMARY OF NOTATION

This summary omits many symbols or special uses of symbols defined within a particular section and limited in use to that section. Also, no attempt is made here to summarize the material in Chapter 3 on the symbolization of experimental designs. Page references for most of the entries under "Abbreviation and Roman Letter Symbols" may be found in the Index.

FACTORS AND LEVELS

Italicized capital letters symbolize (1) factors, and (2) the number of levels of a factor. For nested factors, the letter represents the number of nominal levels of the factor, not the number of literal levels (p. 59).

Letters A, B, C, D, and H are used as general purpose factor symbols. If the design has a group factor, G is used to symbolize it. Otherwise, G is used as a general purpose factor symbol.

S is reserved to symbolize the subject factor (and the number of nominal levels of the subject factor).

E is reserved to symbolize the error factor (and the number of nominal levels of the error factor).

F is *not* used as a factor symbol; it is reserved to represent the F-distribution and F-ratio.

LEVELS AND COMBINATION OF LEVELS

Subscripted italicized lower-case letters symbolize the levels of a factor. For example, a_2 symbolizes the second level of factor A; a_a symbolizes a general, unspecified level of factor A.

If a factor is nested, the nest level or levels are shown in a subscript nest. For example, if the design is $S(A) \times B$, a general level of factor S is symbolized $s_{s(a)}$. A specific level is symbolized $s_{2(1)}$, the second subject within level a_1. If the design is $S(A(B)) \times C$, a general level of factor S is symbolized $s_{s(ab)}$; that is, although the design symbolization indicates levels of nesting, the level symbolization has only one subscript nest (one set of parentheses).

Combinations of levels are symbolized by combining the symbols for the separate levels in alphabetical order, collecting all subscripts at the end. For example, ab_{ab} symbolizes the combination of levels a_a and b_b. To symbolize a combination of the levels $s_{s(a)}$ and $b_{b(c)}$ we write $bs_{bs(ac)}$; that is, all nest subscripts are gathered together within one nest.

MODEL TERMS

Symbolization of model terms is analogous to symbolization of levels and combinations of levels, except that the lower-case letters other than the subscripts are printed in boldface. The main term (or effect) corresponding to level a_a is $\mathbf{a}_a$. The nested main term corresponding to level $s_{s(a)}$ is $\mathbf{s}_{s(a)}$. Interaction terms such as $\mathbf{ab}_{ab}$ or $\mathbf{bs}_{bs(a)}$ correspond to the combinations ab_{ab} and $bs_{bs(a)}$.

ABBREVIATIONS AND ROMAN LETTER SYMBOLS

ANOVA	analysis of variance
df	degrees of freedom
df_1	numerator df in F-ratio
df_2	denominator df in F-ratio
EMS	expected mean square
ESP	expanded symbolic product
ESS	expected sum of squares
H_A	alternative hypothesis
H_N	null nypothesis
k_τ	meaning k_A, k_B, . . . coefficients of σ_τ^2's in EMS formulas
L	the common number of levels for the three factors of a Latin square
$\mathbf{m}$	mean of normal distribution, or mean of a population of scores
MS	mean square
MS_1	numerator mean square in F-ratio
MS_2	denominator mean square in F-ratio
n.s.	not significant
P	percentile score
$\mathbf{res}_{abc}$	residual term in Latin square
r.v.	random variable
SP	symbolic product

SS	sum of squares
SS_a, SS_b, SS_{ab}, . . .	sum of squares elements for terms $\mathbf{a}_a$, $\mathbf{b}_b$, $\mathbf{ab}_{ab}$, . . . (subscripts on SS include both nest and non-nest subscripts of the key term)
SS_A, SS_B, SS_{AB}, . . .	sums of squares for terms $\mathbf{a}_a$, $\mathbf{b}_b$, $\mathbf{ab}_{ab}$, and so on (subscripts are included only for bold-faced letters of model terms, not for letters appearing only in a subscript nest)
SS_{tot}	total sum of squares
X_a, X_{ab}, . . .	scores

SUMMING AND AVERAGING

$\sum$ summation sign, p. 17

dot replacing subscript (for example, $X_.$, $X_{a.}$, $\mathbf{ab}_{.b}$) summation over the subscripts replaced by a dot, pp. 20–21

dotted subscripts and overbar (for example, $\bar{X}_.$, $\bar{X}_{a.}$, $\mathbf{ab}_{.b}$) mean over the subscripts replaced by dots, p. 21

OTHER SPECIAL SYMBOLS

α	significance level, pp. 32–33
β	probability of a type 2 error ($1 - \beta =$ power), pp. 32–33; or, slope of regression line, p. 257
σ	standard deviation, p. 26
σ^2	variance, p. 25
σ^2_τ	where τ is a capital letter or letters corresponding to bold-faced letters of a model term; variability of key term τ, p. 172
$\times$	cross, p. 57
( )	nest, pp. 57–58
^	circumflex, meaning estimate of, p. 29
∞	infinity symbol

DESIGN INDEX

Because of the organization of this textbook, material pertaining to a particular design is dispersed throughout various chapters. The Design Index will help the student and experimenter gain ready access to certain critical information pertaining to designs discussed in the book. Since material pertaining to each Latin square design discussed in this book is confined, for the most part, to one section of Chapter 7, such designs are not included in the Design Index. Refer, instead, to the Contents for the topics of Chapter 7 and to the general Index.

Where a page reference is lacking, general rules provided in the textbook enable the user of this book to derive the formulas desired. The rules likewise allow the user to construct and analyze designs more complicated than those listed here.

The Design Index is not meant to reference all information contained in the book about a design; see also the Contents and the general Index.

Design	Layout	Score model	Parameter estimation	SS elements formulas	SS elements table	SS formulas based on parameter estimates	df's (and/or SP's)[a]	EMS's[b]
A	79	80	87–90	144	145	141	144	180
$A \times B$	147	91, 101[c]	98	146	147	141–142	146, 171[d]	180, 183[e]
$A \times B \times C$	105	102, 108[c]	103–107					191
$S \times A$	109	110	110–112	151	152			184
$S \times A \times B$	58[h]	113	114	154	154		153	185
$S(A) \times B$	58[i], 118	119	121–126, 135	152	153	142	171[d]	186
$B(A)$		115						
$S \times A \times B \times C$	60[j]	129[g]						
$S(A) \times B \times C$	126	127						
$S(A \times B) \times C$	127	128						
$S(A(B)) \times C$	60, 127[f]	128						

[a] The SP (symbolic product formula) becomes the formula for the df when the lower-case letters are replaced with the corresponding capital letters. If the design has a factor of proportionality, the df formulas must be revised as explained in pp. 171–172. For revisions required when there are missing scores, see Section 9.1.

[b] EMS formulas given assume that all factors are fixed unless otherwise specified, except that factor S is always assumed to be a random factor.

[c] One score per condition.

[d] With factor of proportionality A.

[e] Factor B random.

[f] The layout for $S(A \times B) \times C$ also serves for $S(A(B)) \times C$, if $A(B)$ is specified instead of $A \times B$.

[g] The terms of the score model for $S \times A \times B \times C$ are displayed vertically in Table 4.9-1, rather than as an equation.

[h] The bottom part of Table 3.4-1 represents $S \times A \times B$ if the c_c's are replaced with b_b's, and if $S \times A$ is specified rather than $S(A)$. See p. 59 for additional explanation.

[i] To make Table 3.4-1 represent $S(A) \times B$, simply replace the c_c's with b_b's.

[j] Table 3.4-2 represents $S \times A \times B \times C$ if the labeling is interpreted as literal for factors S and A instead of initialized.

INDEX